IEC 61850 Demystified

For a listing of recent titles in the
Artech House Power Engineering Series,
turn to the back of this book.

IEC 61850 Demystified

Herbert Falk

ARTECH HOUSE

BOSTON | LONDON
artechhouse.com

Library of Congress Cataloging-in-Publication Data
A catalog record for this book is available from the U.S. Library of Congress.

British Library Cataloguing in Publication Data
A catalog record for this book is available from the British Library.

ISBN-13: 978-1-63081-329-1

Cover design by John Gomes

685 Canton Street
Norwood, MA 02062

10 9 8 7 6 5 4 3 2 1

I would like to dedicate this book to all of the engineers in the world that hear the phrase "it can't be done" as a challenge and prove that it can be done; Bill Blair (retired from the Electric Power Research Institute), who had the vision, foresight, and fortitude to sponsor the founding technologies and research for IEC 61850; all the dedicated people that have worked on various aspects of IEC 61850; to SISCO that invested my time in the standard; and to my wife Lydia who provides me balance in life.

Contents

Preface

Being involved with the standards that led to IEC 61850, IEC 61850 and security-related standards has been an intellectual stimulating and challenging experience. The overall participation required listening to other perspectives and a thirst for learning. Two personal experiences taught me the importance of these traits prior to involvement with IEC 61850.

My thirst for knowledge started at Deerfield Academy where Albert Einstein, David Howell, Charles Danielski, Wayne Turner, Martin the Martian, and Daniel Hodermarsky further refined my thirst for knowledge, learning, innovation, and taking risks. Dr. Einstein's quote, "Education is what remains after one has forgotten what one has learned in school." spoke to me. I interpreted it to mean that what you have left is the ability to learn. Knowing that one needs to continue to learn and change is at the core of the IEC 61850 effort.

During academics at Northwestern University, tests were not regurgitations; they were teaching moments. My master's thesis defense was another teaching moment. As an electrical engineering master's candidate, I was asked to solve a mechanical pivot, spring, and weight problem. I had done this before and could have provided the solution. However, the pivot was not pinioned to the frictionless wall. This was a Kobayashi Maru[1] scenario.

After trying to solve the problem for five minutes, I admitted that I could not solve the problem. When I asked my master's advisor what the answer was, he responded that the purpose of the problem was not to be solved. The purpose was to prove that there are always people smarter than you and we all know that there are people dumber than us. He continued the teaching moment in that asking questions is not a sign of weakness, nor is admitting that one does not know the answer. Additionally, if one has the knowledge and is asked a question, there are no dumb questions, and it is the responsibility of the knowledgeable person to share that knowledge. The lessons learned at that moment have made the work on IEC 61850 so much easier.

My involvement with standards was based on luck and not who or what you knew. As a design engineer at Westinghouse Numa-Logic, a programmable logic controller (PLC) manufacturer, I had just finished a project circa 1982. At the same time, General Motors announced an initiative to develop a new nonproprietary protocol for communicating with PLCs. I was assigned to work on the GM project.

1. The Kobayashi Maru scenario has appeared in two different Star Trek movies. It represents a no-win scenario.

However, my instructions were to not allow the other participants a competitive edge, attempt to gain Numa-Logic a competitive advantage, and to not disclose technologies that provided Numa-Logic a perceived competitive advantage. As it turns out, the other participants of the other vendors had similar instructions. It made the first six months of the activity frustrating. The engineers decided that the atmosphere had to change and that the GM problem had to be solved. Every vendor participant went back to their management and received the backing of management to work together to solve the problems. It proved that competitors could work together to solve a common problem. It also provided the perspective that a good standard is one in which everyone is equally dissatisfied. That experience and knowledge prepared me for the IEC 61850 working environment where competitors work toward the common good.

The story of IEC 61850 starts in 1982. It is not only a story of technology, but one of politics, personalities, failures, accomplishments, education, research, knowledge sharing, and a bunch of engineers who don't like being told that they can't solve a problem.

Introduction

This book is intended to provide information regarding IEC 61850 so that an overall knowledge of the business drivers, design methodologies, technology, functionality, testing, and maintenance of IEC 61850 systems is imparted to the reader. It is not intended to replace the standards, but rather to give a practical use perspective to the standard while having a little irreverent fun.

The book is intended to be used from many different perspectives, as are the IEC 61850 standards. The following provides information regarding where certain information can be found and what type of reader may be interested in that information. As with any engineering project, the actors' (e.g., the intended readership) needs should be identified along with the information that will be provided:

- *Management*: Impart enough information so that managers are willing to evaluate breaking the "if it ain't broke, don't fix it" mentality. Additionally, the book should provide enough guidance so that managers can put together the cost justification for such a change.
- *Financial*: Provide a basis for life-cycle cost analysis as opposed to hard asset costs. There is no doubt that the first IEC 61850 project will be costly. Information will be provided on how to minimize that initial cost and leverage the education that will be needed for the future along with the benefits of that future.
- *SCADA engineering*: Provide concepts for optimizing the supervisory control and data acquisition (SCADA) engineering process.
- *Substation and design engineering* (S&D): Provide enough knowledge so that there is comfort in the theory of IEC 61850, distributed functionality, and automation concepts.
- *Protection engineers*: Provide enough knowledge and proof so that the paranoia over performing high-speed protection functions over a network is minimized. It is recognized that no amount of education will remove the need for this paranoia as these engineers are the ones responsible for keeping the lights on.
- *Testing and field crews*: Provide enough knowledge so that there is some degree of comfort in how to test an IEC 61850 system. IEC 61850, due to its network connectivity, requires different methodologies and thought processes regarding testing of a distributed system.

Table I.1 Chapters of Interest versus Type of Reader

	Actor						
Chapter	*Management*	*Finance*	*SCADA*	*Substation Engineering*	*Protection*	*Test and Field Crews*	*Academia and Students*
1	x	x	x	x	x	x	x
2	x						x
3				x	x	x	x
4	x						x
5	x		x	x	x	x	x
6	x		x	x	x		x
7	x		x	x	x		x
8			x	x	x	x	x
9			x	x	x	x	x
10			x	x	x	x	x
11			x	x			x
12	x		x	x	x	x	x
13					x	x	x
14	As needed						
15	As needed						

- *Academia and students* (ACA): Provide enough urban knowledge so that the topic may be of interest as part of a class, or even an entire class.

Table I.1 shows which chapters may be of interest to the various actors.

CHAPTER 1

What Makes IEC 61850 Different?

The history of IEC 61850 will be covered in a different section. However, to understand what makes IEC 61850 different, a little history lesson is needed. The current IEC 61850 standards are not even close to the initial work that had started within the International Electrotechnical Commission (IEC). In truth, the initial work on IEC 61850 was intended to provide minor improvements to the IEC 60870-5 telecontrol (e.g., SCADA) protocol. This would have been the direction if there had not been an invasion of different ideas, philosophies, and technologies from the United States. Initially, it was not clear if the new approaches would have been adopted by IEC and this could have produced two competing standards. However, several organizations (Electric Power Institute (EPRI), Institute of Electrical and Electronics Engineers (IEEE), and IEC) were able to harmonize the "new" technologies and research from (EPRI and IEEE) and the rigorous engineering practices of the Europeans. The international cooperation laid the foundation of the current IEC 61850.

1.1 Designed for the Future

IEC 61850 was designed for the future. It was designed with the expectation that CPU processing power would increase, memory prices would decrease, and to concentrate on local area network/wide area network (LAN/WAN) technologies as opposed to the serial (e.g., RS-232) communication mechanisms that were prevalent. If one were to validate the design assumptions from 1995, there are some interesting factoids:

- In 1995, the top-of-the-line personal computer (PC) was based on an Intel 486 CPU clocked at 66 MHz. The 486 was one of the first 32-bit central processing units (CPUs). It is so old that our current CPU benchmark performance tests yield no results. Today we have Intel I5 and I7 CPUs that are dual or quad core (e.g., multiple processors) that are clocked at 3 GHz and beyond.
- In 1995, PC memory was approximately 8 MB. Now, you can't buy random-access memory (RAM) in less than 1 GB.
- If you are reading this book, you may never have seen or used a dial-up modem unless you watched the movie *War Games*. A dial-up modem is a device that allowed information to be exchanged over telephone lines (e.g., serial

technology). In 1995, Martin the Martian and Earth-based utilities were using 1,200-baud modems. A top-of-the-line 9,600-baud modem for a PC was approximately $500. If one was to buy a modem today, a 56-Kbps modem can be purchased for less than $20.

Also consider that today's laptops don't even include a serial port[1]; they all have Ethernet ports. In 1995, most Ethernet systems were limited to 10 Mbps. Today we have systems that are over 1,000 Mbps. Homes now have cable internet connections that are typically well over 10 Mbps. Our cell phones use 4G Long-Term Evolution (LTE) (or better) to exchange information at speeds beyond 5 Mbps.

The cost of bandwidth has certainly dropped radically.

Besides designing based on projected technology changes, there was a concerted effort to design to decrease the integration costs and not the automation hardware costs. This intent is accomplished through standardized semantics, functions, file formats, and engineering workflow.

1.2 Functions and Semantics Instead of Numbers

In 1999, the prevalent protocols of Modbus, Allen Bradley Data Highway, IEC 60870-5, and Distributed Network Protocol (DNP) had the ability to read or write indexes or registers. The actual meaning of the values for the registers (e.g., semantics) were needed to be assigned by the users and through programming logic. To assign and verify the semantics, users developed complicated processes typically involving spreadsheets, databases, and others. The workflow and processes required to maintain and coordinate changes is costly.

There is an inherent problem with indexes and the coordination of configurations based on field changes. As an example, consider a set of Modbus registers. Register 40007 contains a value. Without some type of external documentation, it is impossible to know that register 40007 represents the magnitude of Phase A Voltage. Now consider the integration issues if the voltage measurement is moved to register 40008. Such a move would cause reconfiguration and testing of all applications that use or relay the value.

This would not be tolerated in normal society. We use names instead of numbers without even thinking about it. As an example, think about phones and phone numbers. How many phone numbers do you have memorized and have immediately available to dial via number? Most people memorize the numbers of their family circle, work, and a couple of close friends. In the noninternet era, if you didn't know the phone number you used a telephone book.[2] In the internet era, we google restaurants and people by name. These mechanisms represent a binding of a name to an address (e.g., phone number). If the phone number changes, the binding changes, but not the name. It is still possible to find the new phone number by name. However, these mechanisms are centralized into a website or book.

With the advent of personal data assistants (PDAs), digital phones, and smartphones, the binding information is distributed into the contact list of your

1. Universal Serial Bus (USB) is a serial port but is not relevant for this discussion.
2. When was the last time you actually used the white pages to look up a person's phone number?

particular device. We maintain and create the bindings as needed. If we apply the same concept to protocols, a contact entry might represent the change in register number as shown in Figure 1.1.

The example shows how easy it is to update the local binding. The user can still access the register (e.g., phone number) using the same name. In terms of telecontrol and automation, it means if the name is used, there are no changes needed to the consumers of Phase A Voltage Magnitude. This decreases the amount of testing dramatically. If there are three consumers and one source of the register, only the source maintaining the binding needs to be tested, not it and the three other consumers (e.g., applications or devices).

The binding of a name provides the first step in improving integration and testing issues. However, if you put 10 people in a room and ask them to provide a name for something, you might get 30 names. William Shakespeare indicates the importance of a name: "What is in a name? A rose by any other name would smell as sweet." Although we know, through the context of the sentence, that Shakespeare is referring to a flower, "rose" could also describe a color. Well-defined semantics are the key to information exchange and ease of integration. Consider the fact that some utilities have the practice of naming what would typically be known as Phase A as Phase X. If the IEC 61850 standard allowed this type of flexibility, the cost savings in using names would be lost because the names—and their meanings—could vary based on device vendor and utility.

The Utility Communication Architecture (UCA) User Group identified the integration, testing, and maintenance issues caused by indexes to be approximately 80% of the life-cycle integration costs of a substation. The initiative set forth to solve this problem. In today's terminology, the solution was to use object orientation

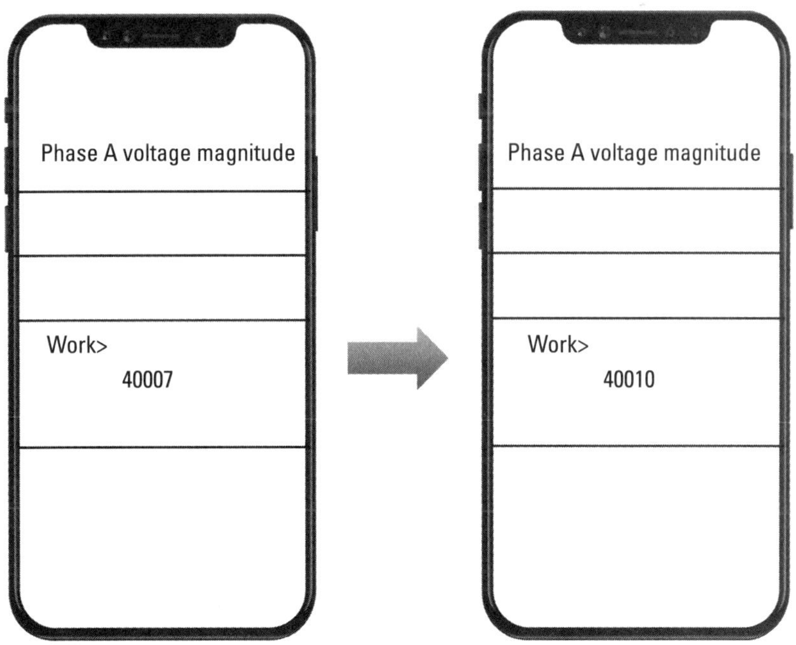

Figure 1.1 Using names to represent addresses. (Adapted image used under license from Shutterstock.com.)

and well-defined attributes and methods. This required standardization of object semantics as well as attribute semantics. This was an example of engineers being told "it can't be done." but it was. This object orientation and standardized semantic approach was proposed and adopted by the IEC. This is the founding principle of logical nodes: standardized object and standardized semantics that aggressively reduce the life-cycle integration costs. Sometimes no good deed or technology goes unpunished. It is probably this concept that causes the most resistance to adopting IEC 61850; people have to learn the standardized semantics instead of assigning registers.

Therefore, IEC 61850 standardizes a broad range of semantic names. When the name is used, its meaning is always the same.[3] This further reduces the cost of information integration. However, it is one of the attributes of IEC 61850 that is difficult for some utilities and people to embrace. Consider the utility that has been using the designation of Phase X for 20 years. This is the local standard for that utility. Embracing the change to Phase A is often difficult, distasteful, and requires education and patience.

The smartphone revolution also has introduced us to applications (apps). A smartphone is delivered with a set of standardized applications (email, web browser, camera, etc.). Then users add more applications to fulfill their particular needs. All applications fulfill a given need, perform a given function, and provide a given set of information. IEC 61850 also provides an equivalent to smartphone apps, called logical nodes in IEC 61850 vernacular. In 2016, there are over 2.2 million iPhone applications available. The standardized set of IEC 61850 logical node types is approximately 300. If an iPhone were an IEC 61850 device, the apps might look like a mixture of electric utility functions. The IEC 61850 iPhone shows that the user is using two measurement units (MMXU), two circuit breakers (XCBR), differential protection (PDIF), interlocking (CILO), and other functions. iPhone apps have names, and so does IEC 61850 (see Figure 1.2).

When IEC 61850 was first introduced, the concepts were foreign to most people. However, we now use these concepts in everyday life without even thinking.

1.3 Automation Instead of SCADA Focus

The initial scope of IEC 61850 was limited to internal substation functionality due to the structure and responsibilities of different technical committees (TCs) within IEC. A skeptical view of the restriction might be that IEC did not want IEC 61850 to compete with IEC 60870-5. The original scope can be seen in the titles of all the Edition 1 standards: "Communication networks and systems in substations." This initial scope constraint may partially account for why IEC 61850 is typically not used for communication between the substation and control center SCADA/EMS systems.

The restriction did the standard a huge favor. It allowed the focus on the functions and communication exchanges needed within a substation required for au-

3. There have been a few exceptions to this statement. But as of Edition 2.1 of IEC 61850, the statement should be true.

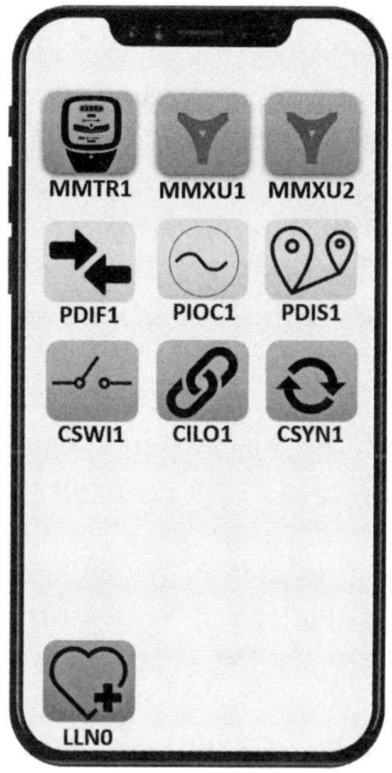

Figure 1.2 If logical nodes were smartphone applications. (Adapted image used under license from Shutterstock.com.)

tomation and protection but not SCADA. It provided the Petri dish for something new that can provide functionality into the future.

Protection, especially differential protection, requires high-speed information exchanges that are well beyond the capabilities of master/slave protocols such as Modbus, DNP, or IEC 60870-5. Rapidly, it became obvious that an architecture would need to be decided on. Basically, the simple choices were to centralize or distribute functions. Back in 1999, it would have been easy to choose centralized. However, it is clear that a centralized architecture can be accommodated by a distributed architecture, but the inverse is not true of a centralized architecture. The distributed architecture adopted an advanced conceptual model that functions could exchange information with other functions. A substation application was a grouping of cooperating functions and their information exchanges. Even though the conceptual model is that functions communicate with each other, the concrete implementation obviously involves devices exchanging information.

There was an additional objective that a multivendor automation solution to be developed. At the time of the start of IEC 61850, the major European vendors had all adopted different field bus technologies. These technologies all had different physical media and were not compatible with each other at any OSI Reference Model layer. Thrown into the mix was the research from UCA to use ISO/IEC 802.3 Ethernet for automation and control. Imagine if Profibus[4] had been selected; where would IEC 61850 be today?

There are several different types of control that needed to be addressed. Point-to-point control where a function, or user, can command an action (e.g., open or

4. Profibus is a field bus technology that provided network access through token rotation and used specialized wiring. It was a German national standard.

close a breaker). IEC 61850 supports this type of control as well as other SCADA functions using communication via TCP/IP and Ethernet. SCADA functionality follows the basic principles established with IEC 60870-5, of course with a different protocol that allows the use of names and functions.

Under emergency situations, there is a need to have high-speed coordination and control of potentially a hundred devices. Prior to IEC 61850, this coordination was done through hardware I/O and wiring of device outputs to device inputs. This creates substation trenches that are full of various combinations of wires and connections. An excerpt from a case study in Ghana provides a clear example:

> Case Study: IEC 61850 Application for a Transmission Substation in Ghana
>
> Substation wiring practices vary depending on the voltage level, equipment age, and associated apparatus technology. Traditionally, copper is the primary interface between components in the yard and a relay that is centrally located within a control house. Evaluation of traditional in-service installations finds that there are typically 44 conductors between the field and a relay in a control house. Normally, several multiconductor cables are used; separate cables are typically installed for breaker status (trip/close) and current transformer (CT) and potential transformer (PT) secondaries. Wiring runs are fairly long, spanning between 200 and 500 meters, as shown in Fig. 3 [1].

The trench wiring shown in Figure 1.3 is a combination of the device I/O interconnections and wiring to support sharing of the high voltage CT and PT signals.

The decision to use Ethernet opened up opportunities to use the networking technology to reduce the need for device to device I/O wiring and to provide a network-based infrastructure for protection coordination. The difference between this type of coordination and the SCADA type of control was the need for speed. Typical I/O wiring solutions operate in the realm of 8 to 20 milliseconds. The use of Ethernet to support signal exchange with 100 devices using point-to-point protocols could not meet the performance requirements. The protection engineers and consultants said this couldn't be done over Ethernet. This was one of those "can't" moments that pushed a dedicated set of communication engineers to develop publish and subscribe technology. The initial technology developed is the foundation

Figure 1.3 Example of wiring in a substation trench. (From [1].)

of today's IEC 61850 Generic Object-Oriented Substation Event (GOOSE) and Generic Substation Status Event (GSSE).

If you tweet today you are using publish and subscribe technology. As a subscriber, you declare your interest to follow the tweets of somebody or something. When somebody or something posts a tweet, it is delivered to you. This is analogous to what GOOSE provides. Substation devices that need the outputs of a device or function from another device register their interest in the information. When the publisher publishes the information, the network delivers the information. As in Twitter, the publisher does not need to know how many or who the subscribers are. This provides a scalable infrastructure where a publisher can support an unlimited number of subscribers. It also allows the replacement of I/O related wiring. The only constraint is network bandwidth.

The proposal to use Ethernet for protection was radical and met with skepticism within IEC. UCA and the United States provided research, results, and mathematical simulations that convinced the IEC committee to adopt Ethernet and the GOOSE concept. It is the GOOSE capability that differentiates IEC 61850 from other substation protocols. It can, and is, used on the same substation networks in parallel with DNP, IEC 60870-5, and other legacy protocols. If you use GOOSE, you are using IEC 61850.

The other culprit in trench wiring is the wiring for CTs and PTs. Analog CT and PT information need to be delivered continuously and to almost all the devices in a substation. The use of analog technology required point-to-point wiring and creating wiring complexity. Consider the wiring improvements if analog point-to-point signaling was replaced with digital signals used over the network; this would allow the distribution of the CT/PT digital information to be used in a point-to-multipoint manner. The other benefit of such a conversion is that signal calibration could be made easier. Sampled Values (SV) was developed provide this functionality. It is like GOOSE using publish and subscribe; however, the publications are stream-based and not event-driven.

Although the substation automation focus allowed engineering practices, communication exchanges, and other technologies to be the focus, the current IEC 61850 has moved well beyond the restrictions of the substation. Literally, the barn door is wide open for applicability now.

1.4 Engineering Workflow and File Exchange Standardization

There are typically three mechanisms used to exchange the functions and semantic definitions: paper or HTML; communication services; and files.

1. *Paper or HTML representation of the information.* Humans are visual driven beings. We need to see information in order to process it. As an example, we need to see a menu in order to know what food to order. That visual representation doesn't work so well for computers and computerized applications. These need some form of electronic information exchange.[5]

5. This statement ignores HTML screen scraping. Screen scraping concepts should be evaluated in a similar vein as communication services.

2. *Communication services provide the ability to dynamically exchange information between computers and devices.* The best examples of the pervasive nature of these services are called web services. Web service technology allows almost anybody to provide a definition that can be consumed by an application. That definition can then be used by the application to request the provider of the web service to provide information through that service. Since it is so easy to do a web service, everybody has their own definitions. Without common service definitions, we create a spiderweb of integration points and needs for translation (e.g., from one service to another). Examples of this issue can be found easily: consider environment information.

   Even within the National Oceanographic and Atmospheric Association (NOAA), there are different interfaces and technologies even for the same information. The integration problem is furthered if one looks at weather web service interfaces. Consider the weather interfaces from CDYNE (http://cdyne.com/downloads/SPECS_Weather:PDF—10/27/18) and NOAA Webservice (http://www.ncdc.noaa.gov/cdo-web/webservices/v2—10/27/18).Even though both are services that return weather information, the service interface takes different inputs. Additionally, similar information is returned but not in the same structure or format. Therefore, integration of the two weather sites requires different interfaces and logic to be written so that an application can use either source.

   Figure 1.4 has two service end points (e.g., CDYNE and NOAA) that provide weather information. Two different interfaces are required so that information can be acquired from each end point. Then specialized conversion, commonly called transformation, must occur to transform the information into a neutral format. The combination of interface and transformation functions is typically referred to as an adapter. It is this neutral representation that allows the application to utilize information from either end point. As the number of end points increase, the number of adapters increases. Each adapter requires additional design and maintenance activities.

   To minimize the integration effort, IEC 61850 defines a standardized set of services that minimizes the need for adaptation. One of the capabilities of the IEC 61850 services could be referred to as autodiscovery. The services allow one application to ask another application to return its semantics

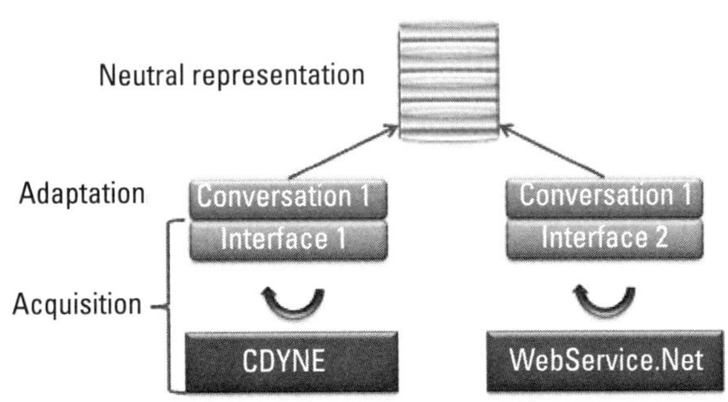

Figure 1.4 Example of required adaptation.

and functions in a standardized format. Grant County Public Utility District (PUD) used this capability to decrease substation integration time by over 20-fold, which translated into cost savings of over $100,000. However impressive these savings were, Grant County still had to do manual configuration of communication addresses and had no standardized methodology to record the configuration of the communication or exchange patterns for their system.

3. *File exchange.* Since the advent of the electronic age, files have been used to exchange programs, settings, and other information. As with web services, everybody can develop a file format that contains similar information with different layouts. Also as with web services, each different file format creates the potential need for an adapter to be created in order obtain the semantic definitions in order to account for the various formats. IEC 61850 standardizes how to represent the semantic and functional definitions in Extensible Markup Language (XML) format.

   However, standardized formats do not optimize the cost savings that such files can provide. To maximize benefits there is a need to not only standardize file formats but also their intended use, workflow, and exchange patterns. Substations and automation systems are designed. Most utilities and consultants have their own design processes, documentation, and potentially custom developed tools.

   The System Configuration Language (SCL) standard (IEC 61850-6) standardizes file formats, engineering workflows, and file exchanges and usage. The workflow is based on typical engineering design practices: specification; design; and deployment. Although SCL can be used to assist in testing, it does not dictate the test methodologies. The technologies and tooling standards allow for the specification and design phases to produce a system design and configuration without requiring actual devices.

The use of SCL is a major advance in the technology for implementation of a substation or automation system. It offers major cost savings to autodiscovery. Savings can be three- or fourfold that of autodiscovery if the life-cycle automation system costs are considered. This is one of the reasons that IEC 61850 is widely adopted outside of North America.

1.5 Adoption and Barriers

Figure 1.5 shows the regions where IEC 61850 is widely adopted or trending for adoption. It also shows that North America has little or no installed base of IEC 61850. Five countries (France, Germany, Italy, Russia, and Japan) of the eight G8 countries have widely adopted IEC 61850. It also shows that other major countries have wide adoption (e.g., India, Brazil, China, South Korea, and Mexico) also have adopted the standard. Of the major industrial countries only the United States and Canada have a limited installed base of IEC 61850.

There are some influences that can account for the differences in adoption: governmental substation implementation methodologies and regulations. In Europe, as

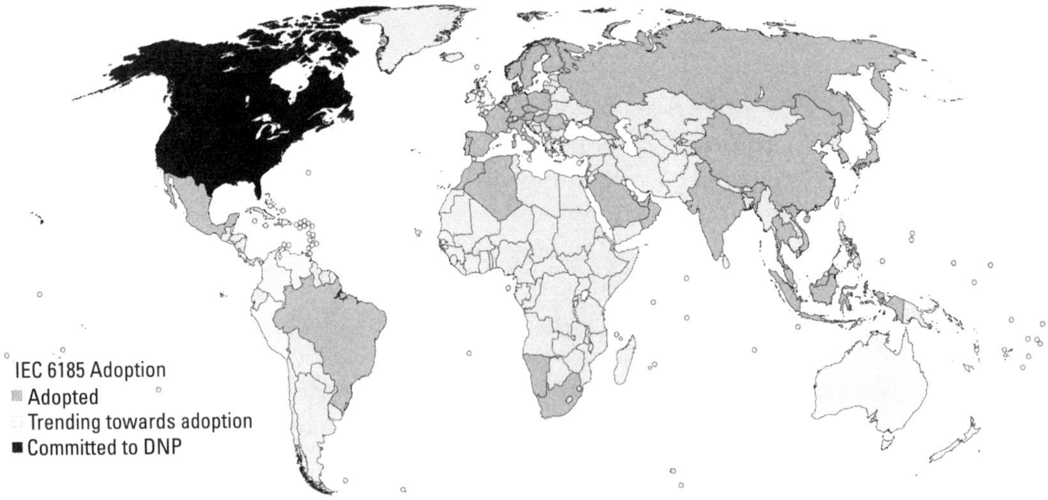

Figure 1.5 Worldwide adoption of IEC 61850.

an example, substation design and deployment is typically contracted to a vendor such as Siemens, ABB, Schneider, or Alstom-GE Power (formerly Alstom). Prior to IEC 61850, each of these vendors had their own proprietary field bus technology that was used for substation integration. When IEC 61850 and the engineering workflow was introduced, these companies embraced the cost and time savings to the point that IEC 61850 is now the technology that they prefer to utilize. In North America, utilities typically do their own design and use technologies that they are already familiar with.

Regarding the impact of regulatory authorities, the European Network of Transmission System Operators for Electricity (ENTSO-e) provides regulations pertaining to the transmission level grid of the entire European Union. It has a set of proposed regulations that will require the European Union (EU) transmission companies to begin standardization on IEC 61850. Additionally, the EU has smart grid regulations that also mandate the use of IEC 61850. IEC 61850 was either already in use by the transmission system operators (TSOs) or is being embraced due to regulation. This reaction is in stark contrast to what happened in North America. When the U.S. Federal Energy Regulatory Committee (FERC) proposed mandates for the use of IEC 61850, there was a major negative response.

Inquiring minds would want to know why the North American utilities are not rushing for the cost benefits that come with the adoption and implementation of IEC 61850. There are three major areas within North America where IEC 61850 could be readily adopted, but has not been: distributed energy resources (DER), substation to control center communication, and substation automation.

There have been two decisions made within the United States that will further complicate adoption of IEC 61850 for DER. The California Public Utility Commission (CPUC) is a regulatory body that controls regulations for California. When faced with a protocol choice for smart inverters due to the need for an immediate solution, the Smart Energy Profile (SEP) was adopted (http://www.energy.ca.gov/electricity_analysis/rule21/). SEP is no better or worse than the current IEC 61850 standard, but the Web Technology profile for IEC 61850 (e.g., IEC 61850-8-2) was still under development when this decision was made. The adoption of SEP will prevent the adoption of IEC 61850 in the future since its services and semantics do not align easily with IEC 61850. As with hotels, occupancy is nine-tenths of the law. Who would upgrade a previously installed smart inverter to support IEC 61850? Financially it would probably not be justified. Luckily, the CPUC decision is being met with resistance by at least one utility in California.

The other decision was regarding solar integration. The SunSpec Alliance, an industry consortium (www.suspec.org), decided to use Modbus to integrate to photovoltaic generations. Unlike the CPUC choice of SEP, SunSpec did adopt the semantics of IEC 61850 and defined specific registers (e.g., binding of standardized semantics to specific registers). This will make integration with IEC 61850 systems a bit easier in the future. However, this binding is typically performed by a gateway function. The actual vendor devices do not natively support this mapping. The selection of Modbus was driven by the U.S. vendors. The predelection of U.S. vendors to stay with legacy protocols was also demonstrated in the U.S. comments regarding IEC 61400-25-2 for Windpower. The majority of the U.S. Technical Advisory Group (TAG) wanted only Modbus, not even DNP, and definitely not IEC 61850. Until utilities or regulatory authorities require IEC 61850, the prevalence of registered-based protocols will exist in North America. There are some other issues that prevent utilities from mandating IEC 61850: education; fear of change; and again, "if it ain't broke, don't fix it."

To know that education is required, there must be an inherent knowledge that something different exists. Figure 1.6 is an extract from a Newton-Evans report that shows 33% of the respondents had little or no awareness of IEC 61850. Additionally, the report indicates that North American utilities don't understand the potential benefits to adoption of IEC 61850.

Of the common barriers, education is a true barrier. This barrier can only be overcome by hands-on experience. This means there must be an advocate for change that is willing to provide the training and experimentation facilities so that experience and confidence can be gained. This is little different from the change

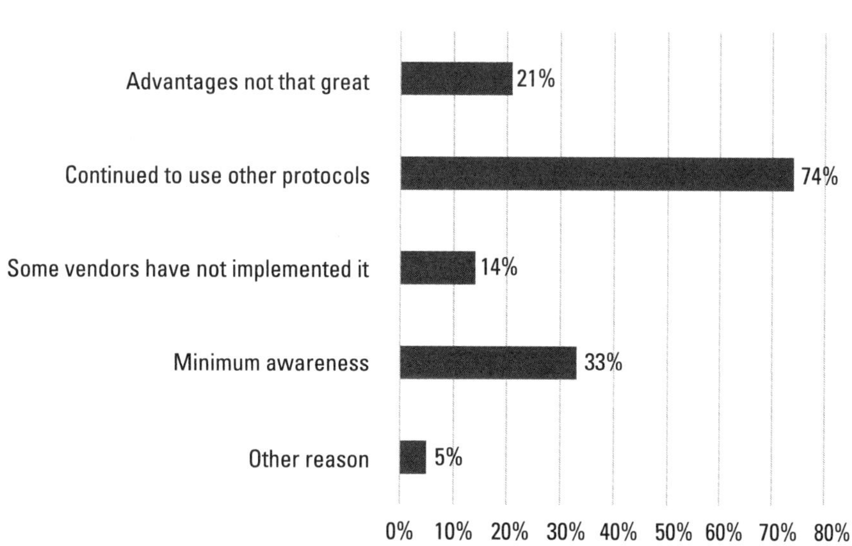

Figure 1.6 Newton-Evans respondent reasons for not implementing IEC 61850.

from mechanical protection relays to digital protection relays. It takes time, persistence, and a strategy to succeed. The cost benefits of IEC 61850 justify this investment. However, utilities are shrinking in manpower and the remaining personnel are always trying to accomplish more with less. Therefore, there is a true balancing act needed to adopt, embrace, and change to a new technology.

Figure 1.7 is an extract from a Newton-Evans report reflects the lack of adoption of IEC 61850 for substation to control center communications. The figure clearly shows that serial protocols (e.g., DNP 3.0 and Modbus Serial) are the predominant installed base of communication technology. However, DNP over LAN/WAN technology is becoming a preferred technology. A rational person would ask why IEC 61850 doesn't fare better.

The first hint is there is prevalent usage of serial links between the control center and substation. Remember that IEC 61850 was designed to be used over LAN/WAN technology. Therefore, if the SCADA link is not LAN/WAN technology, IEC 61850 is not a viable alternative. The Newton-Evans report does not determine the baud rates of the serial links but they are probably lower bandwidth links than WAN links. An additional impediment, at least in the United States, has been the North American Electric Reliability Corporation (NERC) Critical Infrastructure Protection (CIP) regulations. NERC CIP regulations are cybersecurity policies that allowed an exclusion from regulation if serial protocols were used. However, the latest NERC CIP (version 5) has removed that exclusion. Now, the use of LAN/WAN versus serial links is a matter of life-cycle costs.

Figure 1.7 also shows that the adoption of DNP 3.0 LAN is increasing, which also means the use of TCP/IP. A rational reader would ask if TCP/IP usage is increasing, why not use IEC 61850 instead of DNP, since it also uses TCP/IP? Some SCADA vendors, when asked in the past, responded that the scope of IEC 61850 was intrasubstation. This argument may have been politically correct, but not technically correct. This flawed argument has been removed with the publication of

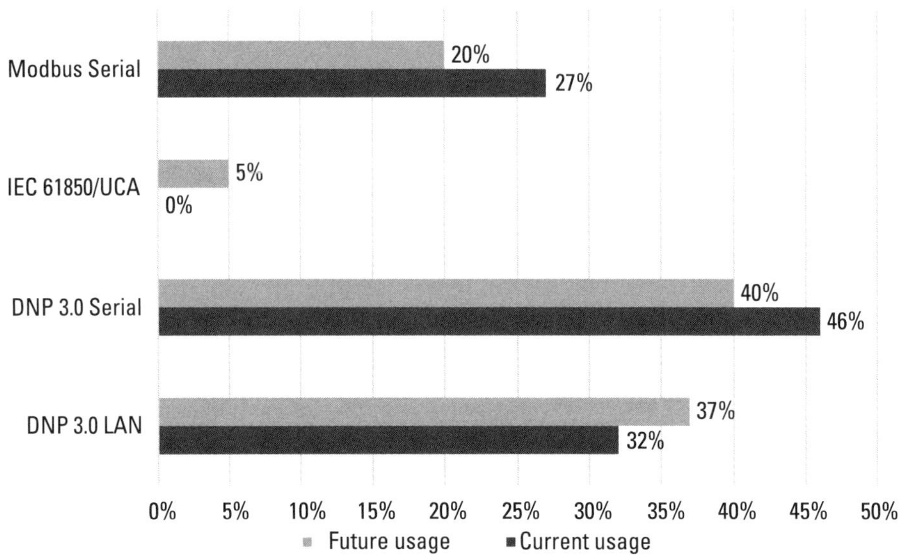

Figure 1.7 Technologies in use for SCADA in North America. (Courtesy of Newton-Evans.)

IEC TR 61850-90-2, *Using IEC 61850 for Communication between Substations and Control Centres.*"

Even within the TR, one of the prevalent architectures discussed is how to convert IEC 61850 information into DNP or IEC 60870-5 for exchange with the SCADA system. The non-IEC 61850 protocols use less bandwidth. Thus, if the substation to SCADA communication link is relatively low bandwidth, this architecture is justified. However, we all know that bandwidth cost is decreasing and there are advantages to using the increased bandwidth (e.g., decreasing the number of communication links required to support the substation). An inquiring mind would ask what about IEC 61850 has made SCADA vendors averse to providing an interface using the protocol.

EMS/SCADA systems are currently designed to manipulate flat, or register, oriented information. The systems provide the ability to bind a display name, which is in its own right a semantic, to an actual register. These SCADA display names are much smaller than the standardized semantics in IEC 61850. It could be hypothesized that the SCADA databases are not sized to support the semantics of IEC 61850. Additionally, the ability to receive and process objects, as opposed to flat registers, is often a foreign concept to the current generation of SCADA systems.

The use of IEC 61850 within North American substations has similar adoption issues to its use for SCADA. A Newton-Evans report, Figure 1.8, shows trends that are like the projected adoption of the various protocols. The report concentrates on the client/server communication that is like DNP, but does not include GOOSE where major hardware, performance, and reliability improvements can be achieved. The adoption of GOOSE requires even more education and experimentation since its benefits are for critical automation and protection function.

Although there are barriers for adoption within North America, there are several utilities beginning, or have completed, successful pilot projects. Southern California Edison (SCE) has committed to utilize IEC 61850 for its new substations as

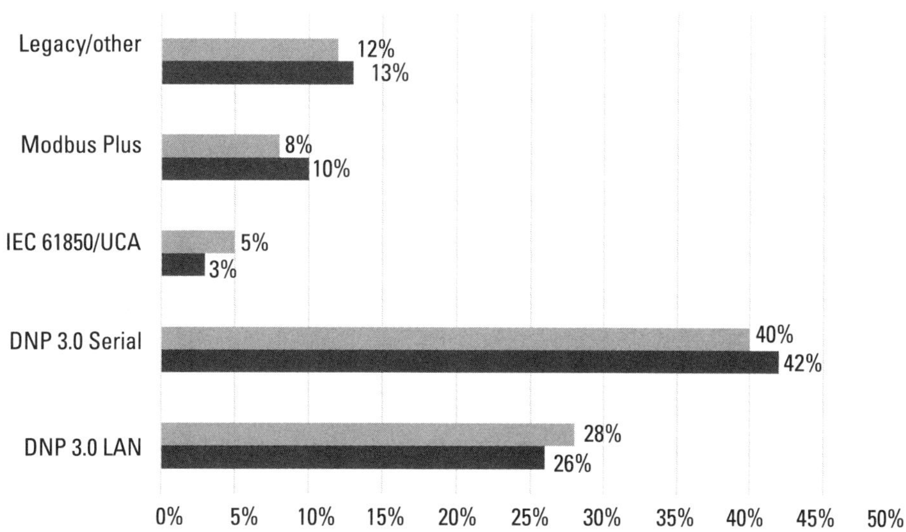

Figure 1.8 Technologies in used for intrasubstation client/server communications in North America. (Courtesy of Newton-Evans.)

well as its Centralized Remedial Action Schemes. Other utilities have successfully implemented it and are continuing to embrace the technology. Con Edison of New York continues to deploy new substations using IEC 61850 as does AEP. With education, publication of use cases, and publication of cost savings the North American trend toward IEC 61850 can be encouraged.

Reference

[1] https://cdn.selinc.com//assets/Literature/Publications/Technical%20Papers/6595_CS_IEC61850_SZ_20130213_Web.pdf?v=20151125-100101.

CHAPTER 2

History of IEC 61850

The history of IEC 61850 spans multiple decades. If one were to include the development of complex math and the initial protection and SCADA concepts, the history spans from the 1800s until today. IEC 61850 has been built on the back of projects that came before it, as simplistically shown in Figure 2.1.

Without the development of microprocessors automation and control would not be what it is today. These developments and that of LAN technologies allowed for the Manufacturing Automation Protocol/Technical Office Protocol (MAP/TOP) project to leverage these technologies and develop technologies that were developed to solve integration and cost issues in the automotive and business environment. The EPRI analyzed the MAP/TOP technologies and began adopting, refining, and extending those technologies for use within electric utility substations known as the UCA project. The concepts and technologies develop within UCA were proposed to the IEC 61850 project. IEC 61850 embraced the technologies and refined and extended them to meet the global need. Today, IEC 61850's usage has extended beyond the substation and continues its evolution.

There are at least two quotations that are relevant to the analysis of the history of IEC 61850. The first is, "Those who do not learn from history are doomed to repeat it" (George Santayana).

The quote has applicability to IEC 61850 in that without knowledge of the past, the present and the future will probably repeat the same mistakes. This has proven true in the history of IEC 61850. As an example, UCA and IEC 61850 were evaluating the use of a token bus technology (e.g., Profibus) instead of Ethernet. The GM MAP initiative faced the same choice and adopted the token bus technology. This proved to be the wrong decision from a market acceptance, price, and performance perspective. When faced with the same decision, the EPRI funded an effort to research and compare both technologies. Ethernet proved to be the better technological decision.

As you read the history of IEC 61850, there are several decision points that eventually proved to be the correct choice. In some instances, the initial choice was not the long-term solution and the standards have made mistakes that needed to be corrected and user inputs needed to be accommodated. This is a normal evolution of technology. Consider the evolution of the DNP or Microsoft Windows. DNP version 3 was released in 1995 but had two prior noninteroperable versions. Microsoft released Windows 1 in November 1985. It wasn't until Windows 3.1 was

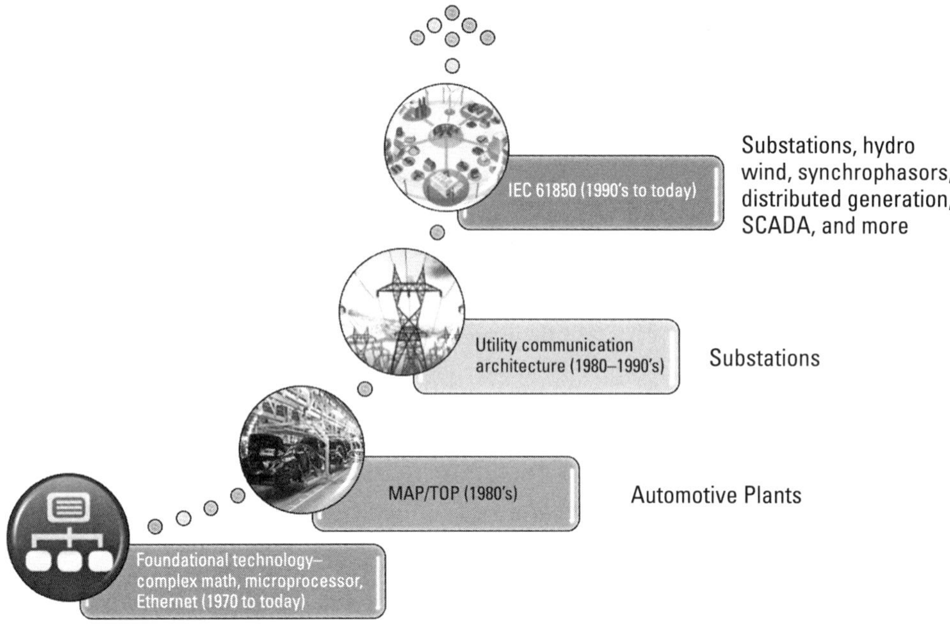

Figure 2.1 Overview of IEC 61850 history.

released in April 1995 where there was a stable and market accepted technology platform.

The second relevant quotation is, "History doesn't repeat itself, but it often rhymes" (attributed to Mark Twain).

The business drivers for IEC 61850 and the GM MAP initiatives were very similar. Similarly, both initiatives faced similar technological and North American market acceptance. Prior to the GM MAP imitative, register- or indexed-based protocols such as Modbus were used. Today, several U.S. power initiatives still desire to use this type of protocol. This attitude has permeated the Sunspec Alliance and U.S. comments on wind turbine controller standards.

(Author's note: It is understandable that a vendor desires to use or promote the technology that is currently being provided. This approach minimizes development and research costs. However, it limits the cost benefits that users could achieve and innovation in the industries.)

The motivations for General Motors and Boeing to start the MAP/TOP initiatives are still relevant in the world of IEC 61850: the need to integrate and exchange design and real-time information among heterogeneous tools and devices.

If you are reading this book, you will remember solving this type of problem when there was no internet, in the era of thermal[1] and dot matrix printers, and using carbon for copying[2] (don't worry about climate change) or mimeographed[3] documents. It was typical to exchange documents using paper using either facsimiles or postal mail. Imagine the problem of designing a widget in a computer-aided design (CAD) mainframe program, having to print out the design, send it to the

1. See https://en.wikipedia.org/wiki/Thermal_printing.
2. See https://en.wikipedia.org/wiki/Carbon_copy.
3. See https://en.wikipedia.org/wiki/Mimeograph. Mimeographing was the forerunner to photocopying.

manufacturer or manufacturing engineering to have the design converted into computer numerical control (CNC), robot, and production planning systems. The use of paper and human conversion was slow and prone to errors. The iterative design and implementation process was cumbersome at best. It was these types of integration issues that the MAP/TOP initiatives were targeted to solve through the definition of standardized electronic file formats and semantics (e.g., for CAD files), file transfer mechanism, and real-time automation control and automation information exchange. To solve the CAD issue, there was a standard file format defined that was exchanged between tools, and then the tools convert the file information into vendor specific information. If changes are required, the vendor-specific tooling outputs the standard file format in order to allow the changes to be conveyed to other tooling.

The scope of GM's integration problem, circa 1980, was large. When the MAP initiative started there were approximately 40,000 intelligent devices consisting of 2,000 robots and 20,000 programmable logic controllers (PLCs). Of the devices, only 15% had the ability to communicate with other devices or computers from other vendors. This created a large integration cost where 30% to 50% of the system cost was related to communication and information integration. This cost did not include the point-to-point wiring costs, which were substantial [3].

This provided the motivation to solve the same integration issue that GM and Boeing experienced in 1982, but 17 years later. The same issue existed for substation design. As time progresses technology offers different mechanisms to solve the problem, but good architectural solutions repeat. The engineering process and standardization on a file format to exchange of IEC 61850 is no different than the solution set forth by MAP/TOP.

The history of IEC 61850 is a cornucopia of dates and complex interrelationships of various technological areas: computer, programming, network infrastructure, communication protocol evolution, government mandates, power systems, automation, cybersecurity, and others. Thus, the history of IEC 61850 is more complex than many might believe. Additionally, history is more than dates; it is about the context in which decisions are made, motivations for the decisions, the people involved, and the stories of those people that make history engaging.

2.1 Prior to 1980

These decades laid the foundation for the technological future and IEC 61850, from the development and deployment of the first protective relay and SCADA system, the development of the mathematical concepts that are the foundation of synchrophasor technology and utility applications today, to the start of communication and computer technology laying the foundation for the future.

2.1.1 Foundational Principles

IEC 61850 and the utility industry still use concepts and math that were developed in the late 1800s and early 1900s. The concepts used for protective relaying, SCADA, and synchrophasor and angular measurements were all developed in this era.

IEC 61850 was initially developed to provide distributed automation and protection functions. The first protection relay concept (nondigital) was developed in the early 1900s. A protection relays was deployed in 1905. Although this relay is probably not in service at this time, protective relays are commissioned and intended to be in service 30 to 40 years. Digital protective relays have a lesser expected life span. However, it is true that there is a large expense in replacing and updating these devices once they are deployed in the field. They are typically geographically dispersed, may be on a pole top, and have a need to be upgraded or replaced without impacting the grid operational integrity.

As more automated controls and protections were deployed into the field, a centralized intelligence was needed to supervise the assets in the field. The first Supervisory Control and Data Acquisition (SCADA) system was installed in Chicago circa 1912. Obviously, it was not digital in nature and relied on human beings for providing the centralized intelligence. These human beings are now known as operators and provide grid reliability through analyzing information from the field and taking the appropriate actions.

These initial systems provided analog magnitude and status indications. However, for the more advanced applications of IEC 61850, vector quantity measurement, analysis, and math are required. The first article regarding the use of complex/vector quantities was published in 1893. This publication laid the foundation for phase and frequency calculations that is the foundation of synchrophasor measurements today [10].

2.1.2 The 1970s: Moore's Law and Computational Power

In the 1970s, Moore's law began having a significant impact. In 1971 there was a 4-bit microprocessor called the TI 4004. This CPU had approximately 2,300 transistors. The 8-bit 8008, with 3,500 transistors, was introduced in mid to late 1972. One of the first power-conscious microprocessors, the 1802, with 5,000 transistors, was introduced in 1976. Today, the Intel i7 quad-core and AMD K10 quad-core 6M L3 have approximately 731 million and 758 million transistors, respectively. The 4004 could perform 60 instructions/second whereas the i7 is capable of approximately 65,770 million instructions per second (MIPS).[4]

One of the most pervasive microprocessors developed, the Motorola 6502, was introduced in 1975. This microprocessor was the foundation for the first Apple products including the first programmable television remote control. Motorola also introduced one of the first 16-bit microprocessors, the 6809, in 1978.

In 1974, the first microcontroller was based on a 4-bit processor architecture that combined a CPU, read-only memory (ROM) for holding programs, and RAM was introduced. The TMS 1000 had a program memory of 8k (e.g., 1024 * 8 bits) and 64 4-bit nibbles of RAM. It could be purchased, in bulk, for $2. It also had 23 pins of I/O that could be utilized for various purposes. Microcontrollers, due to their low cost, became one of the drivers for digital automation and control.

Consider the issue of utilizing and programming communication support for RS-232. This would require allocating at least nine I/O pins. Then consider the

4. If you are curious about processing power over the years, see https://en.wikipedia.org/wiki/Instructions_per_second.

limited resources for programming the toggling of the I/O. To perform this task alone, it would require consumption of many of the TMS thousands of resources.

Therefore, the advent of the universal asynchronous receiver/transmitter (UART) was important. Western Digital provided the first generally available UART in 1971. This provided a semiconductor chip that could be commanded to perform RS-232 functions as a coprocessor to the main microprocessor. Technology continued to evolve and UART technology began to be embedded in microcontroller chips such as the 8051.

As the 1970s ended, the prevalent computer communication technology was either RS-232, RS-485, or bit-synchronous. There was no Ethernet, no Bluetooth, and no internet. However, there was intense competition between Intel and Motorola for establishing dominance for microprocessors. This competition pushed both companies to revolutionize the industry, much in the same way Intel and AMD compete today.

2.1.3 Computer Communication and Internet

There was no standardized digital exchange mechanism until 1962 when Recommended Standard (RS) 232 was published by the Electronic Industries Association (EIA). This was the specification for asynchronous communications between data terminal equipment (DTE) and data communication equipment (DCE). It was the milestone that allowed transitions from bit-synchronous protocols to asynchronous protocols and it is widely deployed in the power industry today. It is used for radio and direct communications today. Its derivatives are RS-422, RS-485, and RS-449. When RS-232 was published, computers had to be programmed to toggle and sense the bits specified in the specification; there was no semiconductor chip available to assist in its implementation. Until recently, personal computers were provided with serial ports (e.g., RS-232). However, today the USB has replaced the serial ports of old. If you need to interface to DCEs through RS-232 today, you will probably need a USB to RS-232 converter. As semiconductor technology has progressed, the voltage levels on the RS-232 pins have decreased from +/– 12V to +/– 3V. In some cases, the connections today are 0V to 3V. This may create communication problems with conformant RS-232 older devices. EIA ceased operation in February 2011.

The 1970s saw the precursor of the internet take its infant steps. In 1973 to 1974, the Advanced Research Projects Agency Network (ARPANET) was established and began to be realized as an internet service provider, where two entities were able to communicate over a packet-switched network instead of modems. This communication was done without the use of what we know as IP or TCP today. Therefore, ARPANET demonstrated ability and was utilized by the defense industry but was not widely available to the public.

Computers, like human beings, need to exchange information with peers. Telephone systems allowed humans to communicate in the late 1870s and a computer modem was one of the first long-distance mechanisms for computer communications. The first modem was developed by AT&T in 1958. In 1962, the first commercial modem (also known as an acoustic coupler modem) was produced. It was able to exchange information at rates up to 300 bits per second. This is a far cry from the 40 Mbps of 4G that our cell phones use today.

2.2 1980 to 1989

The advances of the 1980s included the creation and deployment of Ethernet, standards developments that led to the internet as we know it today, a completion between Open Systems Interconnection (OSI) protocols and Internet Engineering Task Force (IETF) protocols, the introduction of the first personal computers, and a major initiative to exchange information between the office/design and manufacturing environments using a single set of international standards.

2.2.1 Foundational Principles

There were two activities that had a direct impact on IEC 61850 design:

- In 1980, Westinghouse, as part of EPRI project RP-1359-1, prepared a requirements specification for substation and protection control systems. These requirements were used as the basis of the Utility Communication Architecture (see Section 2.3.3).
- The mathematical concepts of complex math from 1893 were refined in 1983 to provide guidance for voltage phasors. This is the basis of current synchrophasor measurement techniques today [11].

These requirements and math are still relevant today.

2.2.2 Computers: Personal Computers

The 1980s were the start of the ongoing competition we see today between Apple and PCs. Apple was the first computer company to introduce personal and business computers on a wide scale. Apple introduced the Apple II in 1977 and the Apple III in 1980. Both were based on the 6502 microprocessor. The Apple III+, introduced in 1983, could have a maximum of 256K of RAM.

The first IBM PC, shown in Figure 2.2, was introduced in 1981. It had an Intel 8088 CPU operating at 4.77 Mhz. The maximum memory was 256K. There was no hard disk and storage was done via two 320K 5-1/4-inch floppy drives. The first 100% IBM-PC-compatible portable PC was introduced by COMPAQ Corporation in 1983. It weighed 28 pounds, which is luggable, not portable. There was no hard disk in the computer. IBM released its first laptop computer in 1986

Figure 2.2 First IBM PC. (Image used under license from Shutterstock.com.)

(see Figure 2.3). It weighed 13 pounds, which was a significant improvement from the COMPAQ portable. The maximum RAM available was 640K. This is a poor cousin to what we have as laptops,notebooks, and tablets today, but it was state of the art in 1986.

2.2.3 Communications

In the 1980s there was progress regarding computer communications in the areas of networking technology, the beginning of the internet becoming pervasive, and protocol developments that had a direct impact on the utility industry.

2.2.3.1 802 Networking

During the late 1980s and early 1990s, several technologies emerged and competed for commercial success. These can be categorized as media access and control (MAaC) and protocol selection.

Within IEEE there were two categories of standards emerging and standardized for MAaC: token passing and Ethernet. There were two competing token passing standards—IEEE 802.5, which was used by IBM (also known as token ring), and IEEE 802.4 (also known as token bus). Token bus was chosen by the General Motors for use in the MAP due to token passing determinism and vendor neutrality. This choice by MAP eventually lead to the MAP movement's demise (see Section 2.2.4.4). The most prevalent networking technology in the world today is Ethernet.

Bob Metcalf is the recognized father of Ethernet. In 1973, he published a memo within Xerox that promoted the idea of network-based communications. He left Xerox and patented the concept of network collision detection, which became the basis of IEEE 802.3 Ethernet. In 1979, he founded 3com, which became one of the primary Ethernet vendors in the 1980s and 1990s. Until 1983, there was no standard for Ethernet. At that time, IEEE published a standard IEEE 802.3 regarding carrier sense multiple access and collision detection (CSMA/CD). It is the IEEE 802 set of standards that provide Ethernet standard connectivity today, including Wi-Fi, WiMAX, and others.

Figure 2.3 Portables. (Image used under license from Shutterstock.com.)

The specified media in 1983 was known as Thicknet. The connection to the RG/8 cable was done using what was called a vampire. The vampire tap consisted of two teeth that penetrated the cable and connected to the shield and core conductor. The impedance of the cable was 50 ohms and the maximum distance for the cable was 500m. The cable needed to be terminated with a 50-ohm resistor and the shield needed to be grounded. This was the equivalent of a high-speed RS-485 multidrop network/flat bus structure.

In 1985, IEEE published IEEE 802.3a (also known as 10Base2). This standard provided Bayonet Neill-Concelman (BNC) connection for Ethernet networks. They still needed to be terminated in a similar fashion to 10Base5 and the length of the network was limited to 200m instead of 500m. This specification still was flat-/bus-like and was limited in the distance that a single segment of a LAN could service.

The flat nature of Ethernet LANs began to change in 1986 when AT&T published the specification for StarLAN. Another company, Synoptics Communications, published a specification in 1987 known as LattisNet, which the precursor of was 10-Mbps Ethernet as we know it today, which uses twisted pairs. The actual standard of IEEE 802.3e was published in early 1990. The transition to twisted pairs and the desire for nonhalf-duplex connections (e.g., repeating Ethernet hubs) gave rise to the development of Ethernet switches and the LAN infrastructure that is pervasive today.

2.2.3.2 Networking Protocols, and the Industry Makes a Decision

During the 1980s there were primarily two different protocol stacks being utilized. One was based on the IETF RFCs that constitute the internet today. The other was a suite of ISO protocol standards. It is interesting that the historical order of the internet standards started with applications, which is shown in Table 2.1.

Although MAP/TOP and governments around the world adopted the full ISO protocol stack, at the end of the 1980s it became clear that the internet RFCs were what the world was adopting. However, Manufacturing Message Specification (MMS) and X.500 are still in use today. MMS is one of the application protocols used by IEC 61850.

One of the major differences between the IETF and the ISO protocol stacks is that the IETF application protocols tend to be bound directly to TCP (e.g., no intervening protocols). The ISO protocols have protocols that are used at the Presentation and Session OSI reference model layers. The OSI reference model (also

Table 2.1 Equivalence of Internet and ISO Protocols

Purpose	*Standard Protocol*	*Date*	*Standard Protocol*	*Date*
File transfer	FTP (RFC 354)	1972	FTAM (ISO 8571)	1988
Network protocol	IP (RFC 791)	1981	CLNP (ISO/IEC 8473)	1988
Transport protocol	TCP (RFC 793)	1981	ISO TP4 ((ISO/IEC 8073)	1986
Manufacturing	Not Available		MMS (ISO 9506)	1990
Mail	SMTP (RFC 821)	1982	MHS (ISO/IEC 10021)	1990
Terminal communications	Telnet (RFC 854)	1983	Not available	
Directory services	LDAP (RFC 2251)	1997	X.500 (ISO/IEC 9594)	1990

known as the seven-layer model) was published [5] in 1984 as ISO 7498. This reference model is still used today to differentiate functions that are needed to create a reliable communication exchange. The referencc model consists of seven layers with the layers being divided into Application (layers 5–7) and Transport profiles (layers 1–4).

Marshal Rose had a stroke of brilliance in the recognition that it was possible to allow the execution of packet-oriented ISO applications over the stream transport protocol TCP. This was the final nail in the coffin (the start of the demise) of the ISO T-Profile. RFC 1006 was published in 1987 and it provided this capability. Today, the ISO T-Profile is long forgotten.

2.2.3.3 Utility-Specific

The foundations for DNP and IEC 61850 GOOSE were developed during this decade:

- The IEC began work on a telecontrol protocol IEC 60870-5-101 in 1989. This was a serial-based protocol intended to be used over noisy networks with a Hamming distance [13] of three or better. This standard did not progress quickly enough for certain North American companies or projects. In 1990, Westronics[5] decided to use the current state of the 60870-5 technology to develop their own protocol. This eventually became the DNP. DNP is not interoperable, even at a Layer 2 (e.g., OSI Datalink) basis. Thus IEC 870-5 and DNP serial protocols are not able to share the same serial or multidrop channel.
- Documentation of connectionless OSI profiles began to emerge [5] in 1987. This work became the foundation of the UCA GOOSE (now IEC 61850 GSSE) profile. The development of UCA GOOSE eventually provided the guidance for the IEC 61850 GOOSE and a change of name of the UCA GOOSE within the context of IEC 61850. When the IEC 61850 working groups (e.g., IEC TC57 WG10, WG11, and WG12) embraced the UCA GOOSE multicast construct, it was decided that UCA GOOSE was not object-oriented by the concepts of IEC 61850. Therefore, the UCA GOOSE became the IEC 61850 Generic Substation Status Event (GSSE).

2.2.4 The Rise and Demise of MAP/TOP

The MAP/TOP initiative was innovative in an era where the concepts we take for granted today had to be develop. The development was based on emerging standards at the time, selection of implementation standards based on technical superiority,[6] and much sweat equity. General Motors assigned a substantial staff to manage the MAP project. The spokesperson and driving force for MAP was Mike Kaminski. The GM staff is shown in Figure 2.4.

5. Eventually became GE Harris.
6. "Technical superiority" doesn't guarantee market acceptance.

Figure 2.4 GM MAP Project Staff. Back row: David Weisskopf, Kester Fong, Mike Buckowski, Vish Narayanan, Rich Gerhardt, Gary Workman, Ron Floyd, Praveen Dwivedi, Ali Baharaloomi, Mark Adler, and Howard Fingeroot. Front row: John Tomlinson, Chuck Groff, unknown. Missing: Katheleen Sturgis, Prasad Mantripragada, Mike Kaminski, David Greenstein, Hsin Way Chin, Pat Gorski, Bill Riker, Atul Kapoor, Kiumi Akingbehin, Pat Amaranth.

Most of the technical work was performed GM's Technical Center located in Warren, Michigan. Manufacturing Building B's basement was where most of the meetings that developed General Motors Manufacturing Format Specification (GMMFS), Manufacturing Message Format Specification (MMFS), and MMS were held. In terms of working conditions, the basement was not the best environment. There were no windows and the main walkway was dimly lit and had a musky or mildew smell at times. Although it wasn't a dungeon, those of us who worked consistently on MAP typically lost track of time. In 1982, the office and meeting space needed to be remodeled. No real changes occurred through the MMS work, which terminated in 1990. It is unclear how many updates occurred until recently when the basement flooded and much of the documentation on MAP was lost. The silver lining was that the office space did get remodeled.

The MAP/TOP imitative had three major demonstrations and two major deployments. The demonstrations occurred in 1984, 1985, and 1988. The deployments were known as Factory of the Future [4] and GM Truck and Bus's GMT400 projects.

At Autofact 1985, a small integration demonstration was staged. The object/widget that was designed and produced in real time was a toy known as the "Towers of Hanoi,"[7] as shown in Figure 2.5.

The culmination of the vision came in a demonstration known as Enterprise Networking Event (ENE). ENE occurred in June 1988 in Baltimore, Maryland. It

7. See https://en.wikipedia.org/wiki/Tower_of_Hanoi.

Figure 2.5 Towers of Hanoi game. (Image used under license from Shutterstock.com.)

was a simulation of the design-to-manufacturing process integration vision [3]. The object/widget that was designed and produced was a desktop set. The demonstration was a success through the efforts of hundreds of committed engineers, over 50 companies, bailing wire, and chewing gum. MAP/TOP was addressing the "can't talk, have to talk" syndrome for design and manufacturing, which is exactly what IEC 61850 is attempting to do for the utility industry.

GM Truck and Bus, as well as Opal, embraced and deployed the MAP solution. However, in 1998, GM decided to move away from standards-based integration and toward a different solution. This was almost at the same time where the Inter-Control Center Protocol (ICCP) was becoming prevalent in the electric utility industry. ICCP, or IEC 60870-6 TASE.2, is based on the Manufacturing Message Protocol (ISO 9506) that was developed as part of the MAP initiative, and UCA Version 2.0 and IEC 61850 were beginning to progress. The parent walked away from the child, but the child matured and grew into something much grander than the parent envisioned.

The concepts and architectures embraced by MAP laid the foundation of IEC 61850. Learning and adaptation typically comes from failures or mistakes. Some of the most relevant MAP/TOP mistakes, which IEC 61850 learned from, were

- Technical superiority does not guarantee market acceptance. MAP/TOP embraced the ISO suite of protocols known as Open Systems Interconnection protocols. Today, the world uses TCP/IP instead of ISO Transport Class 4 (ISO/IEC 8072, ITU-T Rec. X,214[8]) and Connectionless Network Protocol (ISO/IEC 8348, ITU-T Rec. X,233[9]) Protocol. Many of you may have never heard of the latter. The ISO transport was packet-, not stream-based, which makes it better suited to packet-oriented guaranteed delivery systems as opposed to the stream approach of TCP. The ISO network layer was well ahead of its time. It had/has 20 octets of address space. IPv6 only has 16. Although technically superior, neither of the ISO protocols is typically taught today. This grand technology was usurped by the easier-to-implement and market technologies of the IETF.

8. Available from http://www.itu.int/rec/T-REC-X.214/en/.
9. Available from http://www.itu.int/rec/T-REC-X.233/en/.

- Expense of network media drives vendor acceptance. MAP chose the use of broadband IEEE 802.4 Token Bus.[10] The use of broadband was difficult to maintain and almost immediately carrier-band technology was developed. However, it is rare to hear mention of Token Bus or Token Ring (e.g., IEEE 802.5) today. We are an Ethernet-based society today. However, the concepts from both of these technologies, especially in regard to redundancy, are still prevalent today.
- Conformance test tool certification. When MAP/TOP was being deployed, there was a single MMS conformance test tool provided by Fraunhofer Institute.[11] There was no oversite/certification process of the tool. The tool was incorrect in one aspect that required all vendors to change to pass the test, become nonconformant, but still be interoperable.
- Conformance test certificate equivalency. Although there was a single tool, the regional test centers (e.g., North America, Europe, and Japan) did not recognize the certification of the others. This meant that vendors typically needed to get certified by more than one test center depending on the region that the products were being sold into.

Besides the lessons learned, the historical developments prior to 1995 also laid the technological foundation for what IEC 61850 is today:

- ISO/IEC 9506. The Manufacturing Message Specification standards were developed as part of the MAP initiative. This application-level protocol is the foundation of the IEC 61850 client/server communication profile.
- Connectionless OSI profiles. During the MAP initiative, research was being performed regarding three-layer and connectionless profiles to improve data exchange performance. This research provided the foundation for what eventually became UCA GOOSE and IEC GOOSE.
- Function chart and sequential chart programming of PLCs. This research and deployment provided the technological precursor to IEC 61131.
- Methodologies for documenting implementations and conformance testing were developed. This included but was not limited to, the development, of implementation agreements, protocol implementation conformance statements (PICs), and ISO 9646 conformance testing.

2.2.4.1 Ecosystem

An entire ecosystem evolved to support the massive initiative. Governments developed specifications referred to as GOSSIP that mandated the MAP/TOP technologies as part of that effort the National Institute of Standards and Technology (NIST) OSI/OSE Implementers' Workshop (OIW) was formed in 1983. The OIW was at the request of industry with its purpose being to bring together future users and

10. See http://www.rwth-aachen.de/cms/root/Forschung/Einrichtungen/Institute-und-An-Institute/~qgw/Fraunhofer-Institute/lidx/1/.
11. See: https://en.wikipedia.org/wiki/Token_bus_network.

potential suppliers of OSI protocols and to develop agreements that allowed deployment and definition of Federal Information Processing Standard 146: Government OSI Profile (GOSIP) [7, 8]. In a similar time frame, the European Workshop for Open Systems (EWOS) and Asia-Oceania OSI Workshop (AOW) were formed. The coordination of the workshop's implementation agreements resulted in ISO/IEC ISP 11188. Additionally, the precept of three regions continued for conformance testing.

In 1986, the Corporation for Open Standards (COS) was founded with the purpose to advance the interoperability of OSI/ISO protocols. COS was heavily involved in the promotion of MAP/TOP and GOSIP, including an attempt to coordinate conformance testing across multiple regions (North America, Europe, and Asia) and a single conformance mark that was globally recognized. Unfortunately, the regional test centers were allowed/had different test cases, methodologies, and results that prohibited regional acceptance of other regions tests results. Thus, products delivered into different regions typically had to be tested by a recognized test center of that region.[12] This substantially increased the cost of the products and lowered overall acceptance.

Governments also joined in the support for the MAP/TOP initiatives through different published procurement governance (GOSIP). The government of the United Kingdom published its requirements for OSI profile-based communication in 1988 after ENE '88, This standard excluded TCP/IP-based communication profiles. Later in 1988, a European Procurement Handbook (EPHOS) was created. The government of the United States published its GOSIP, FIPS 146-1, in 1991.

Vendors of office and manufacturing technology, motivated by the potential market and ease of integration, flocked to support the initiated, as is shown in the various demonstrations discussed in the next section.

2.2.4.2 Demonstrations

There were multiple demonstrations of MAP/TOP technology during this decade. The intent, as with most demonstrations, was to promote the technology, show progress, and to convert disbelievers into supporting the technology.

National Computer Conference Demonstration of MAP/TOP prototypes occurred in 1984. Participating companies were ACDS, Allen Bradley (now Rockwell Automation), ASEA (now ABB), AT&T, Charles River Data Systems, Concord Data Systems, Digital Equipment Corporation (now Hewlett-Packard), Gould Electronics (became Modicon but now part of Schneider Electric), Hewlett-Packard, Honeywell, IBM, Intergraph, Industrial Networking Incorporated (no longer in existence), Motorola, NCR, Northern Telecom, Siemens, and Sun Microsystems (now Oracle). Table 2.2 shows the protocols selected for the MAP specification in 1984 and which layers were implemented for the demonstration within the General Motors portion of the demonstration. It is worthwhile to note the commitment to ISO protocols known as OSI [12] (e.g. non-TCP).

The General Motors demonstration was connected to a National Bureau of Standards (NBS) and Boeing demonstration. This connection was used to

12. This complexity and cost is why UCA accredits all test centers for IEC 61850 and also provides the conformance test certificate approval in order to ensure global equivalence.

Table 2.2 MAP Specification Protocols

ISO Layer	*MAP Protocol Selection*	*Implementation at Demo*
7. Application	ISO 8571 File Transfer (FTAM)	Subsets implemented
	GM Programmable Device Messaging (GMMFS)	
6. Presentation	ISO/IEC 8822 and ISO/IEC 8823	Not implemented
	GM Presentation Layer	
5. Session	ISO/IEC 8326 and ISO/IEC 8327	Not implemented
	GM Session Layer	
4. Transport	ISO/IEC 8072 and ISO/IEC 8073	Implemented
3. Network	ISO/IEC 8348 and ISO/IEC 8473	Not Implemented
2. Data Link	IEEE 802.2	Implemented
	IEEE 802.4 Token Media Access	
1. Physical	IEEE 802.4 Broadband	Implemented

demonstrate file transfer capability from an office Ethernet-based environment to a manufacturing environment using File Transfer Access and Management (FTAM). The file caused action and production in the General Motors portion of the demo using IEEE 802.4 networking.

There are several observations and comments during the preparation of the specification and implementation for NCC. The first, and most notable, was that the application layer protocol for manufacturing was General Motors specific and not based on an international standard. It was obvious, to gain international acceptance, that any global acceptance would require international standardization. Thus, the MAP effort focused on developing standards at the application layer for manufacturing and adopting international standards for the other layers.

The Autofact MAP/Top Demonstration occurred in 1985. This demonstration was typically referred to as a demonstration of partial MAP 2.1 capability. The protocols in use for the demonstration are shown in Table 2.3.

Participating companies were Allen Bradley, ASEA Robotics, AT&T, Charles River Data Systems, Computervision, Concord Data Systems, Digital Equipment Corporation, Gould, Hewlett-Packard, Honeywell, IBM, Industrial Networking

Table 2.3 MAP 2.1 Capability

ISO Layer	*MAP Protocol Selection*	*TOP Demo*
7. Application	FTAM	FTAM
	Manufacturing Message Format Specification Network Management	
6. Presentation	ISO Presentation	Not implemented
	GM Presentation Layer	
5. Session	ISO Session	Not implemented
	GM Session Layer	
4. Transport	ISO Transport Class 4	Implemented
3. Network	ISO Network Layer	Not implemented
2. Data Link	IEEE 802.2	Implemented
	IEEE 802.4 Token Media Access	
1. Physical	IEEE 802.4 Broadband	Implemented

Incorporated, Intel, Intergraph, Machine Vision International, Motorola, NCR, Northern Telecom, Siemens, and Sun Microsystems. This was the first demonstration of manufacturing operations being integrated with design processes and applications involving CAD, robotics, and vision systems.

Baltimore, Maryland was the location of another large demonstration of MAP/TOP technology in 1988. This was known as Enterprise Networking Event '88. This demonstration of MAP/TOP technology was global in scope. Not only were their five manufacturing and design areas staged in Baltimore, but there was also a manufacturing area in England connected to Baltimore. It took the collective resources of over 50 companies to stage the demonstration. Preparation took over a year of calendar time. This represented a major investment by many companies. The vendor participation page of the demonstration and conference brochure follows.

The biggest problem was that each of the discrete demonstration areas were staged at different sites prior to on the floor integration in Baltimore. The wide area networking and resource-sharing technologies available at this time made the coordination and preparation cumbersome at best. Since this was first time that all five areas were integrated in one location, it could have been a disaster. It turned out that with long hours, some bailing wire, and luck the demonstration was a success. In the evenings, during integration of the areas, work typically stopped at 1 or 2 a.m. Several of us wandered into downtown Baltimore in search of food, but there was nothing open at that time.

2.2.4.3 Factory of the Future

Given the state of MAP/TOP, microprocessor technology, and networking technology, one of the most forward-looking automation projects was started in 1984. General Motor's idea was ground-breaking for the era: design and implementation of a manufacturing facility where parts could be manufactured with no humans being required for the manufacturing process.

General Motors announced plans for the Factory of the Future [4]. The factory was staged on floor space allocated within Saginaw Steering Gear's facility located in Saginaw, Michigan. The plant is supposed to be the first General Motors facility that was paperless and 100% automated. In the description of the facility, General Motors intended to use 50 robots and 40 manufacturing cells to transport and product axles. The manufacturing floor was intended to be 100% automated, including the sweeping of the floors [6].

Achieving these objectives was difficult at best. Automated guided vehicle technology was not adequate to achieve the 100% automation objective and costs of other appropriate technologies proved very expensive, immature, and time-consuming to implement. However, many of the major objectives were achieved and the facility offered an incubator for even more advanced ideas.

As part of the Factory of the Future project, Maxitron was chosen as the programmable logic controller for the project. In 1985, Maxitron, a French programmable logic controller Manufacturer, filed U.S. Patent US4742443 A, Function Chart Programming. This was the precursor for IEC 61131 programming language. Figure 2.6 is an extract from the U.S. patent.

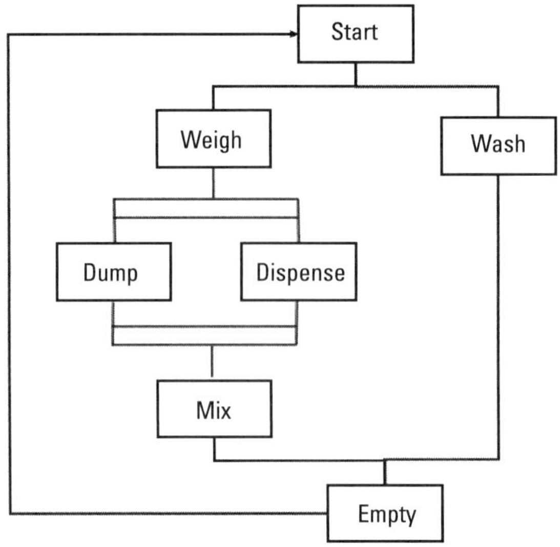

Figure 2.6 IEC 61131.

With the demise of MAP and other technological problems, the parts produced by the Factory of the Future were too expensive. The experiment had failed and the Factory of the Future closed in 1992.

2.2.4.4 The Beginning of the End

Not admitting a mistake is a bigger mistake. —*Robert Half*

During the preparation for the ENE '88 demonstration, the MAP/TOP initiative had an ecosystem of over 100 different companies. MAP technology was being installed in many of the General Motor's Truck and Bus Plants as well as Factory of the Future. Most of the automation and control vendors had committed to support MAP. This represented a meteoric support rate and an overall success.

There had been an ongoing dispute from the onset of the MAP initiative with Digital Equipment Corporation (DEC) [1, 2]. As with most disputes, it came down to money and market. Manufacturing, by its nature, is a smaller global market when compared with other markets. DEC was faced with needing to support Ethernet and a low-volume Token Bus for manufacturing. Since Token Bus was low volume, there were few manufacturers of the integrated circuits required for implementation. The promise of sales did not offset the relentless global movement toward Ethernet. In 1987, DEC and General Motors engaged in a public dispute over the use of IEEE 802.4 [3]. The article represents the first public dispute between MAP and DEC. It was not a surprise since there had been ongoing disputes about the selection of Token Bus from the onset of the MAP initiative. The discontent continued to grow and is probably one of the primary issues why MAP eventually became extinct.

The death knell of Token Bus was assured after ENE '88 when Ungerman-Bass announced it was dropping support for IEEE 802.4 [2]. Ungerman-Bass and

General Motors were partners in the founding of Industrial Networking Incorporated (INI) and the Ungerman-Bass retraction put the support of IEEE 802.4 on the shoulders of Concord Communication and Computrol. Instead of adoption of Token Bus, support was decreasing.

The MAP protocol stack could easily run over Ethernet and there were multiple vendors of Ethernet to Token Bus bridges. General Motors could have saved its initiative by relenting on the choice of Token Bus and adopt Ethernet. The ability to use Ethernet, in addition to Token Bus, did not occur until 1991, and this change was too late. MAP was doomed because of this and the market adoption of TCP/IP.

IEC 61850 faced a decision in choosing a token passing technology (e.g., Profibus) or Ethernet. After intensive testing and network performance modeling, Ethernet was chosen. IEC 61850 learned from MAP's failure.

2.2.4.5 Manufacturing Message Protocol

The MAP project started working on converting the Manufacturing Message Format Specification into an international standard that eventually became ISO/IEC 9506 MMS in 1985. The initial intent was to take the MMFS syntax and convert it into Abstract Syntax Notation (ASN) Backus-Naur Format (BNF) in order to achieve an acceptable grammar to the dictionary of semantics that had been created in MMFS. The core technical contributors were Allen Bradley (David Sweeton), General Motors (John Tomlinson), Industrial Technology Institute (George Schimmel, Michael Schumacher), Kodak (Chris Williams), Siemens (Karlheinz Schwarz), Texas Instruments (Dan Moon), and Westinghouse (Herbert Falk). Today, only two of the contributors (Herbert Falk and Karlheinz Schwarz) are involved in IEC 61850.

The initial estimate for the conversion was 3 months. The actual effort required almost 3 years. There were several reasons for the extra effort:

- Creating a concrete grammar from the MMFS dictionary of verbs and nouns was difficult, as many of the verbs and nouns were not well defined. Humans could infer what the semantics were intended to be, but computers required more concrete definitions. Therefore, the effort included not only syntactical transformation, but the creation of rigid definitions. As an example, MMFS had a verb "read" that could obtain data from a "variable." The definition of "variable" was not concrete enough to differentiate between an "address," "symbolic" (e.g., index), or a named variable. The effort to define the meaning of "variable" caused the team to think about the future and the complexity of data that could need to be supported in the future such as objects with complex structures (e.g., arrays, structures, structures of arrays of structures, or multidimensional arrays). As with most reworks, design activities sometimes lead in unanticipated directions.
- The concept of object orientation was becoming more prevalent in computer programming. Although the C++ programming language was nowhere ready for use, there was plenty of academic and journal information available that indicated this would be the way forward. The shift toward object orienta-

tion was not trivial. Learning newtechnology and applying it appropriately requires education, diligence, and a lot of effort.

Many people believe that standards development is stodgy and contentious. In many cases this is true. Every so often, consensus can't be reached and either a vote or some other mechanism is used to decide. In 1988, the development of MMS was at such a point. There was a technical change proposed between the Draft International Standard (DIS) and International Standard (IS) version of MMS/ISO 9506. General Motors already had an installed base of DIS-based equipment and was against the technical change. The committee could not reach consensus. To decide, a shilling was tossed. The changes were adopted based on the toss and this coin flip cost GM millions of dollars.

This "decision" allowed MMS to be published as IS ISO/IEC 9506 in 1990. ISO 9506 is like most ISO protocols; there is a services document, ISO/IEC 9506-1, and a protocol document, ISO/IEC 9506-2. Most implementers believe conformance is to part 2. This is an oversight as MMS has conformance requirements in part 1 (e.g., 9506-1).

2.2.5 The Rise of the Utility Communication Architecture

In 1986, Bill Blair of EPRI started workshops to develop integration and information exchange requirements for the electric utility industry. The initiative was based on similar needs that motivated MAP/TOP. There were so many vendor-specific protocols that were not compatible, and this made exchange of information and automation complex. Since MAP and GOSSIP attempted to address the similar issues on the manufacturing side, these technologies were the underpinnings of what became the UCA version 1.0. The version 1.0 specifications represented adoption of the MAP/TOP/GOSSIP profiles while extending them to address specific utility needs. In 1988, EPRI began sponsoring of projects based on the draft UCA 1.0 documents. The issues found through implementation,with the draft specification were corrected and UCA 1.0 was published.

GM decided to move away from standards-based integration and toward a different solution in 1988. This was almost at the same time where the ICCP was becoming prevalent in the electric utility industry. ICCP, or IEC 60870-6 TASE.2, is based on the Manufacturing Message Protocol (ISO 9506) that was developed as part of the MAP initiative, and UCA Version 2.0 and IEC 61850 were beginning to progress. Once again, the parent walked away from the child, but the child matured and grew into something much grander than the parent envisioned.

2.3 1990 to 1999

The 1990s provided several foundations for IEC 61850, saw the hand-off of UCA to IEEE, and harmonization of IEEE and IEC 61850.

2.3.1 Founcational Technologies

In the 1990s, several advances occurred that had a direct impact on IEC 61850 today. The concepts of object-oriented (OO) programming was starting to take hold as the C++ programming standard was published. MAP/TOP dealt with global testing issues and the complexity of their standards leading to the need to create subsets.

2.3.1.1 Conformance Testing

Even though the demise of MAP was impending, testing of the GOSIP protocol suite continued. In 1990, the United States' test plan for GOSIP is found to be lacking and having gaps [9]. The abstract preciseness of the test cases needed to be revisited as well as the testing architecture.

Sometimes good things come to those that wait, and ISO 9646 was published in 1991. ISO 9646 was a standard for conformance testing methodology and frameworks. This framework was used to improve the GOSIP testing and a draft GOSIP framework, based on ISO 9646, was published in 1993. The UCA initiative and the EPRI were involved in producing this framework. It is the foundation of the testing methodologies used for IEC 61850 today.

IEC 61850, in the auspices of the UCA International Users Group IEC 61850 Test Procedure Working Group (TPWG), utilizes ISO 9646 as its testing architecture today. The TPWG expends significant resources to minimize ambiguities in the test cases and to respond to those that are found.

2.3.1.2 The Concept of Profiling

MAP was complicated and different plants desired different functionality. In 1992, General Motors began working on the concept of defining profiles for interoperability. The concept was to allow different manufacturing plants a mechanism to define the functions, variables, and communication options that were to be supplied by vendors. The conceived process was called "Process to Support Interoperability" (PSI) and was defined as in Figure 2.7.

A product user profile is the concept that a user, or consortium, can specify the set of services and protocols to be utilized. Conformance testing is the methodology through which implementations can be tested as to adhering to the standard and the specified user profile. Interoperability evaluation represents the ability to

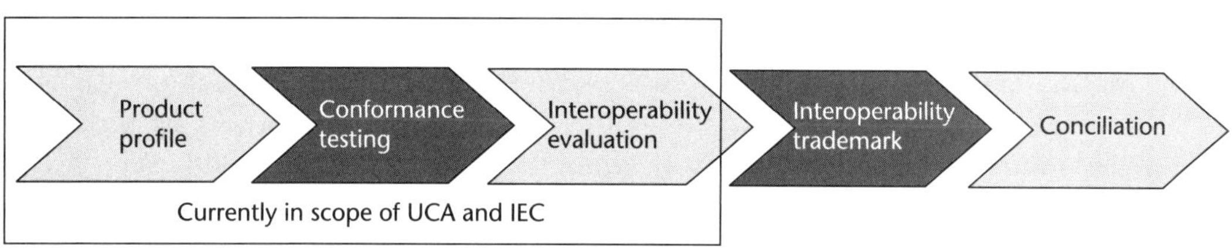

Figure 2.7 General Motor's process to support the interoperability process.

test that different implementations can exchange information. Conciliation is the process by which issues are resolved.

The concepts of product profile, conformance testing, and interoperability evaluation are currently in scope of UCA and IEC. Profiling is a difficult process. IEC is struggling for a consistent methodology of creating and documenting user profiles known as business application profiles (BAPs). Other aspects are addressed in IEC 61850 through the Substation Configuration Language, PICS, and Protocol Implementation Conformance Extra Information for Testing (PIXIT).

2.3.2 Communications

There were a couple of firsts in the 1990s:

- In 1990, Westronics developed and deployed the first DNP implementation.
- In 1995, IEC 60870-5-101, the first international serial telecontrol protocol standard, was published.
- In 1995, IEEE publishes IEEE 1344, a synchrophasor standard.
- In 1998, the first standard for digital wide area communication was released by the International Tele Union. It was known as the Integrated Services Digital Network (ITU Q.391). It allowed wide area network communications at a blazing speed of 64 Kbps.

The DNP developed by Westronics now has an associated users group, it is at version 3 (DNP3), and it is widely used in North America. ISDN has gone the way of the dinosaurs and been replaced with higher-speed technologies.

The adoption of TCP/IP based profiles became an unstoppable force. The IETF realized the need to support the OSI upper layers as specified by GOSIP and published an implementation strategy for minimal OSI upper layers (RFC 1698). This concept was developed prior to publication and was embraced by the NIST OIW. It is the minimal OSI upper-layer construct and RFC 102 (how to execute OSI upper layers over TCP/IP) that are the foundation of what IEC 61850 is today. Due to the impending change in scope of GOSIP with its adoption of IETF specifications (e.g., TCP/IP), the last NIST OIW Workshop was in 1994.

The U.S. federal government published FIPS 146-2 in 1995. This revised GOSIP specification was the first time that open voluntary standards (e.g., IETF standards) would be used instead of OSI protocols. The urban legend regarding this decision was that prior to 1995, both TCP/IP and OSI Transport and Network were paid software licenses. Prior to the FIPS 146-2, the IETF transport protocols (e.g., TCP/IP) began being embedded in operating systems basically free of charge. It was at this point that the use of OSI Network and Transport protocols began to decline.

IEC 61850 originally codified both OSI and IETF transports. This was short-lived and the current IEC 61850 Client/Server profiles utilize TCP/IP exclusively.

2.3.3 Utility Communication Architecture

EPRI released the UCA specification in 1991. The specification consisted of six volumes:

- Volume 1: Included use cases and functional descriptions that were used as part of the technological evaluation.
- Volume 2: Included communication requirements that were used as part of the technological evaluation.
- Volume 3: A standards assessment of the available protocols and technologies that might be used in the actual specification.
- Volume 4: The actual UCA Specification for Communication Profiles. The actual specification (Volume 4) was primarily a subset of the protocols from the MAP/TOP specifications.
- Volume 5: A user's guide that provided business justifications and case studies based on specific projects.
- Volume 6: A project summary that extracted important sections from the other volumes.

As with any new specification, there was a need to involve users and to access gaps in the publications. The MMS Forum was established to fulfill this need and to promote UCA 1.0. One of the participants of the forum decided to utilize parts of UCA 1.0 for a nonterrestrial application. In 1992, the Jet Propulsion Laboratory (JPL) had a project to utilize MMS for satellite communications [8].

2.3.3.1 Semantic Development Begins

During the forum meetings, semantic differences used for information exchange became obvious. These semantic differences could cause integration issues that increased cost. The concept of standardized semantic exchange was missing in UCA 1.0.

As an example, there was no standardization of how to retrieve Phase A Volts in a unique and standardized semantic. This is of no surprise since the MAP/TOP initiative also did not address semantics for the manufacturing environment. This meant that there was much work to be done between the release of UCA 1.0 and UCA 2.0 and IEC 61850. The semantic work started as part of the UCA Forums but came to a culmination in Las Vegas during 1996.

In 1996, Bill Blair of EPRI started a grueling sequence of meetings. The sequence of meetings lasted for 6 months with a meeting in Las Vegas every other week. The meetings consisted of a core team: Don Berkowitz (Energy Line Systems), Frances Cleveland (UCI), John Day (Bolt Beranek & Newman), Steve Dalyai (QEI), Herbert Falk (SISCO), Dennis Holstein (SDI), Dan Nordel (NSP), Jeff Robins (Cycle), and George Schimmel (Tamarack).

During these meetings there was a big "can't" challenge. The team was told that it was impossible to standardize semantic names within the power industry. The team took this as a challenge and the concepts developed in Las Vegas laid the foundation for UCA 2.0's Generic Object Model for Substations and Feeders (GOMSFE) and the semantics in today's IEC 61850. In this case, what happened in Vegas did not stay in Vegas. At the end of the meetings, Bill Blair created an "I Survived Vegas" t-shirt that had caricatures of all the participants. I still wear it proudly today, although only for special occasions.

2.3.3.2 Inter-Control Center Protocol

Within the United States there were two protocols used to exchange information between control centers: Inter-Utility Data Exchange Consortium (IDEC) and Western Coordinating Council (WSCC) specifications. The communication integration problems between U.S. control centers mimicked those that MAP attempted to address. There were similar issues in Europe. The resulting work became known as the ICCP specification.

EPRI published its ICCP specification in 1994. EPRI decided to leverage the contents of UCA 1.0 to create a neutral protocol that could meet the interchange requirements of both IDEC and WSCC. MMS and the profiles specified in UCA 1.0 were chosen. The EPRI specification was used as the initial draft input to IEC TC57 WG05. In 1997, IEC 60870-6 TASE.2 was published. This MMS-based standard is currently used for most of control center to control center real-time information exchanges globally.

2.3.3.3 The Inception of Multicast for Automation and Control

In 1995, the EPRI Report RP 3599 Substation Integrated Protection, Control, and Data Acquisition–Requirements Specification was published, which included a 4-msec requirement for some applications. There was no protocol available that could meet the 4-msec requirement of RP 3599.

Between 1995 and 1998 several attempts were made to specify a protocol that could meet the 4-msec requirement. The network infrastructure choice was known to be a key factor in being able to meet the 4-msec criteria. There was an ongoing discussion within UCA and IEEE about the selection of an appropriate substation medium for high-speed automation. The two contenders were Profibus and Ethernet. Profibus was a token bus technology and thus the issue of guaranteed performance of token bus and Ethernet was once again the caldron of trouble (e.g., GM's selection of 802.4 instead of Ethernet). The industry was unconvinced that Ethernet could meet the performance requirements. Unlike the MAP selection of 802.4, a critical and detailed network simulation was funded and the Fraunhofer simulations indicated that Ethernet could meet the requirements. The skeptical industry, along with the IEEE, indicated that simulations were not sufficient and that actual testing would be required. Therefore, EPRI sponsored a set of meetings and tests conducted by SISCO that was performed on both Ethernet and Profibus. The testing and resulting presentations occurred between October 1996 and May 1997. The results were the foundation of selecting Ethernet for the substation and will be discussed in Chapter 3.

Final EPRI test results on Ethernet were presented in 1997. It showed that Ethernet could meet the performance specification. Due to continued skepticism, another simulation was funded by EPRI and performed. These additional results, confirming that Ethernet could meet the performance requirements, were published in November 1997. These results were reviews at an IEEE Power System Relaying Committee (PSRC) meeting and Ethernet was adopted as a viable media moving forward within the United States.

In 1998, the Chicago 7+1 met in Wood Dale, Illinois. The companies represented were Alstom, UCA International Users Group (UCA IUG), SISCO, Tamarak,

ComEd, OPUS Publishing, and Ontario Hydro. This was the first meeting where a GOOSE message was initially defined that could meet the performance requirements using multicast. This concept and the choice of Ethernet became the foundation for the UCA GOOSE design and eventually IEC 61850 GOOSE and Sampled Values.

2.3.4 IEC 61850 Begins

Three new IEC TC57 work item proposals were submitted. These represented the start of the actual IEC 61850 initiative. The proposals were to

- 57/210/NP: Proposal for part 1 to establish functional architecture, communication structure, and general requirements.
- 57/212/NP: Proposal for part 3 to establish communication within and between process and unit levels.
- 57/213/NP: This document proposed to develop a companion standard to the IEC 60870-5 protocol and to use the standard for protection and control. The actual text of the proposal was, "This proposed companion standard will apply to protection equipment with coded bit serial transmission for exchanging data with control systems (informative interface of protection equipment). It defines a companion standard that enables interoperability between protection equipment and devices of a control system in a substation. This proposed companion standard utilizes documents of the International Standard (IS) IEC 870-5."

Note that the proposal included for the use of IEC 60870-5 was intended to use serial communications. IEC 61850 didn't end up with a serial profile nor did it use IEC 60870-5. One of the major reasons for the change was an American invasion, bringing with it the communications profiles from UCA 1.0, research regarding the use of Ethernet, and a functional (i.e., object-oriented) information model. Unfortunately, the IEEE effort to standardize UCA and the IEC 61850 activities had no knowledge of each other for some time. This created harmonization concerns when both organizations discovered each other's activities. During an IEC 61850 meeting in Edinburgh, the question was posed to the U.S. members about what would happen if the harmonization failed. The response was then IEEE would standardize UCA, IEC would standardize 61850, and there would be two competing global standards.[13] The U.S. members went further to state that they were there to insure that harmonization succeeded. There was success and the results were a much stronger international standard, the IEC 61850 we have today. The agreement to harmonize also resulted in IEEE documenting UCA 2.0 as a technical report (IEEE TR 1550). This Technical Report was published in 1999 as was the final progression of UCA 2.0.

The LAN congestion paper and EPRI results were presented to IEC TC57 WG10 and WG12. This is one of the first introductions to Ethernet within IEC as opposed to fieldbus technologies, where it was asserted that Ethernet could be

13. This would have resulted in a similar situation of DNP and IEC 60870-5.

utilized for high-speed peer-to-peer data exchanges and meet the performance requirements of substation protection. In 1999, the concept of Ethernet multicast began for the distribution of CT/VT and this work culminated in IEC 61850-9-2.

2.4 2000 to 2009

The period of 2000 to 2009 saw the start of concerns regarding security, the impetus for synchrophasor deployments, and the continued evolution of IEC 61850 to address requirements outside of the substation.

2.4.1 Security

Utilities have always been protective of their information. In 2001, EPRI published a cybersecurity assessment of the IEC 60870-6 TASE.2 (ICCP). This was part of the thought leadership of EPRI to recognize that cybersecurity would eventually become an issue for the power industry and this laid the foundation for further projects to develop secure ICCP.

Based on the EPRI recommendations, several vendors implemented the recommendations and EPRI hosted interoperability tests for secure ICCP (e.g., cybersecurity) in Colorado in 2003. The results in the testing lead to EPRI publishing its secure ICCP profile specification. The work done for EPRI secure ICCP resulted in the IEC standards of IEC 62351-3 and IEC 62351-4. These standards are also the foundation of securing IEC 61850 that started in 2004. The security standards for IEC 61850 and ICCP were published by IEC as IEC 62351-3, 62351-4, and 62351-6 in 2007.

There were other entities becoming concerned about the cybersecurity of utility information. In 2006, NERC published a set of regulatory requirements known as CIP version 1. This specified that communication to "critical assets" must be done in a secure manner. This represents the first of many requirements for cybersecurity.

The security standards for IEC 61850 and ICCP were published by IEC as IEC 62351-3, 62351-4, and 62351-6 in 2007.

2.4.2 Synchrophasor

There had been previous theoretical and utility practices developed for synchrophasors. However, there was no standard measurement technique or protocol defined for synchrophasors, nor was there a driving motivation to solve these issues. That changed in August 2003.

The Northeast blackout (see Figure 2.8) provided the impetus to develop the needed standards. During the postmortem analysis, it was not possible to correlate different utility measurements, due to different time synchronization and measurement techniques, to determine the actual sequence of events that lead to the blackout. This spawned an effort to define a new synchrophasor standard that started the development of IEEE C37.118.

Note that while the blackout was occurring, the secure ICCP testing was happening in Colorado. The test witnesses were from the NY ISO (Glenn Sheffer), Western Area Power Authority (Dave Ambrose), and Southwest Power Pool (Kevin

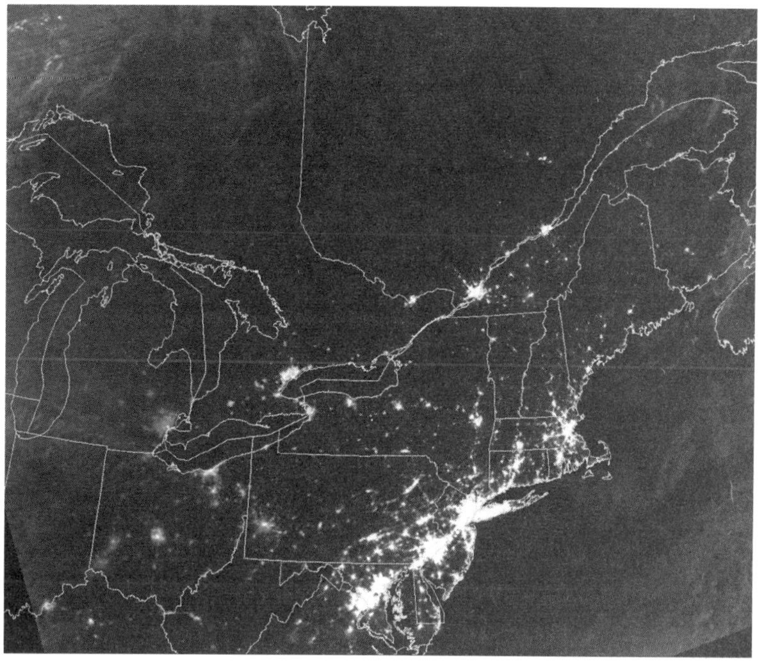

Figure 2.8 2003 Northeast Blackout. (Source: NASA Earth Observatory.)

Perry). The testing was proceeding well until 12:15 p.m. local time, when all of the pagers for all of the witnesses started beeping. This coincided with the Northeast blackout. Although it was initially thought that the blackout was caused due to a cyberattack, that was disproved.

IEEE C37.118 was published in 2005. It included measurement technique standardization as well as a packet format definition that could be used to convey the measured information. The packet format was not IEC 61850 based. In 2009, IEEE requested that IEC accept IEEE C37.118 as an IEC standard (e.g., dual logo request). IEC rejected the request since IEC 61850-9-2 could convey similar information. This rejection caused IEEE to split IEEE C37.118 into a measurement technique document (IEEE C37.118.1) and a protocol document (IEEE C37.118.2). IEEE C37.118.1 is the common measurement technique for synchrophasors worldwide. The split allowed a secure IEC TR 61850-90-5[14] protocol to convey synchrophasors.

2.4.3 IEC 61850

IEC 61850 development continued with the first IEC version of GOOSE, which was defined in 2001. The definitions were created by SISCO and Tamarack. However, the performance requirement was changed by IEC 61850-5 and the 4-msec requirement became 3 msec in 2003.

Edition 1 of IEC 61850 was formally completed with the publication of IEC 61850-8-1 (including GOOSE and GSSE) and IEC 61850-9-2 (for streaming digital

14. IEC TR 61850-90-5 no longer exists. Its contents have been split into several different IEC standards including IEC 61850-8-1 and IEC 62351-9.

CT/PT information) in 2004. CIGRÉ 2004 was the first IEC 61850-9-2 implementations and were demonstrated at the CIGRÉ Session in Paris. These implementations were based on implementation agreements produced by the UCAIug interest group. The UCAIug implementation agreements were known as 9-2LE (Light Edition).

There had been a political constraint placed on the applicability of IEC 61850 to be within substations. In 2005, work began to allow its use between substations and from substation to control center. These recommendations became IEC TR 61850-90-1 and IEC TR 61850-90-2. The TC responsible for Windpower (TC88) also adopted IEC 61850 in 2006 and began expanding the 61850 semantics/objects to support this domain in standard IEC 61400-25.

2.5 2010 to Today

IEC 61850 and synchrophasor technology continued to evolve. IEC 61850 continues to expand into other applications and domains besides operations and substations.

The joint logo publication between IEC and IEEE for the synchrophasor measurement technique (IEEE C37.118.1) occurred simultaneously to the publication of IEEE C37.118.2 (packet format). IEC TR 61850-90-5 was approved and sent to IEC for publication in the same time frame. It references IEEE C37.118.1 as the measurement standard. The communication protocol/profile is intended to replace IEEE C37.118.2.

The IEC standard process requires that any standard be revisited every 5 years. At this time, it may be reaffirmed, retracted, or published with changes. Six years after the publication of Edition 1, Edition 2 of IEC 61850 was published. In 2018, it is expected that an addendum (i.e., corrections) to Edition 2 will be finalized and published.

Different application and application domains have been adopting IEC 61850 and developing object extensions. Two such application domains are condition-based monitoring and distributed energy resources. Additionally, more IEC 61850 standards are under development to standardize human machine interfaces (HMIs), role-based access control (RBAC), redundancy, and many others. IEC 61850 is a living, breathing, and ever-evolving standard.

References

[1] Fagan, M., "Big Two Row over Factory Automation," New Scientist, February 5, 1987.

[2] Wallace, B., "MAP Fouls DEC," Network World, April 28, 1986.

[3] Buck Rogers: A.P.M.E.S. M.A.P. video, produced for General Motors (date unknown).

[4] Holusha, J., "GM 'Factory of Future' Will Run with Robots," New York Times Business Day, October 20, 1984.

[5] Ominicom Newsletter Service, "Open Systems Data Transfer: Transmission," October 30, 1987.

[6] Heuser, W. R., "An OSI Architecture for the Deep Space Network," TDA Progress Report 42-109, The Telecommunications and Data Acquisition Report, May 15, 1992.

[7] "Manufacturing in Deep Space," LAN Magazine, December 1992.

[8] Messmer, E., "NIST Falters in GOSIP Test Plan," Network World, September 3, 1990.

[9] Steinmetz, C. P., "Complex Quantities and Their Use in Electrical Engineering," In Proceedings of the International Electrical Congress, AIEE Proceedings, Chicago, IL, 1893, pp. 33–74.

[10] Phadke, A., J. Thorp, and M. Adamiak, "A New Measurement Technique for Tracking Voltage Phasor, Local System Frequency, and Rate of Change of Frequency," IEEE Transactions on Power Apparatus and Systems, Vol. PAS-102, No. 5, May 1983, pp. 1025–1038.

[11] Russell, A. L., "OSI: The Internet That Wasn't," http://spectrum.ieee.org/computing/networks/osi-the-internet-that-wasnt.

[12] Hamming Distance Calculation and Explanation, http://classroom.synonym.com/calculate-hamming-distance-2656.html.

CHAPTER 3

The Need for Speed: Networking versus Hardwire

What is meant by the term "real-time" or "fast" in terms of communications and control? Critical thinkers and engineers would respond that there is no single definition because "fast enough" or "real-time" are defined by the use case or application being considered. Even the term SCADA implies different performance criteria based on the speed and criticality of the processes being monitored or controlled. However, the performance requirement of typical SCADA is a 1- to10-second update rate. This is nowhere near the performance requirements achieved by hardwired I/O.

In the electric transmission environment (typically considered 230 kV and above), hardwired I/O is typically used to convey: currents for CTs, voltages from PTs, and digital signals used for monitoring and control. CT and PT information provides the information that is used to determine the loading (e.g., power in MW); it is this information in conjunction with switch and circuit breaker status that determine if protection functions for overloading need to be invoked.

Protection is about protecting high-value transmission assets such as power transformers and transmission lines. Transmission power transformers (see Figure 3.1) typically cost more than a million U.S. dollars.

The cost makes it prohibitive to stock enough local spares and in many instances, utilities share spares, making the repair of a burned out or damaged transformers difficult. The failure of either a power transformer or transmission line, due to overloading, impact users (e.g., blackouts and brown outs), are costly to repair or replace and are labor-intensive to replace or repair. It is the purpose of protection functions to prevent transformers and transmission lines from catastrophic failure due to power system conditions.

To protect these assets, physics are quite important. Power must be removed in less than 9 to 10 power cycles. Each cycle is approximately 16.7 msec for a 60-Hz nominal power system for an allotted 150 msecs. Of the total overall time budget, the circuit breaker operation is typically 80 msec. This leaves 70 to 80 msec for detection, processing, and control. From a hardwired perspective, a hardwired protection system might have time allocation similar to Figure 3.2.

Figure 3.2 shows a typical hardwired tripping circuit consisting of two protection relays. The monitoring relay is responsible for determining if the conditions on a specific power line require protection by shedding load or generation (e.g., over

Figure 3.1 Transmission level transformer. (Image used under license from Shutterstock.com.)

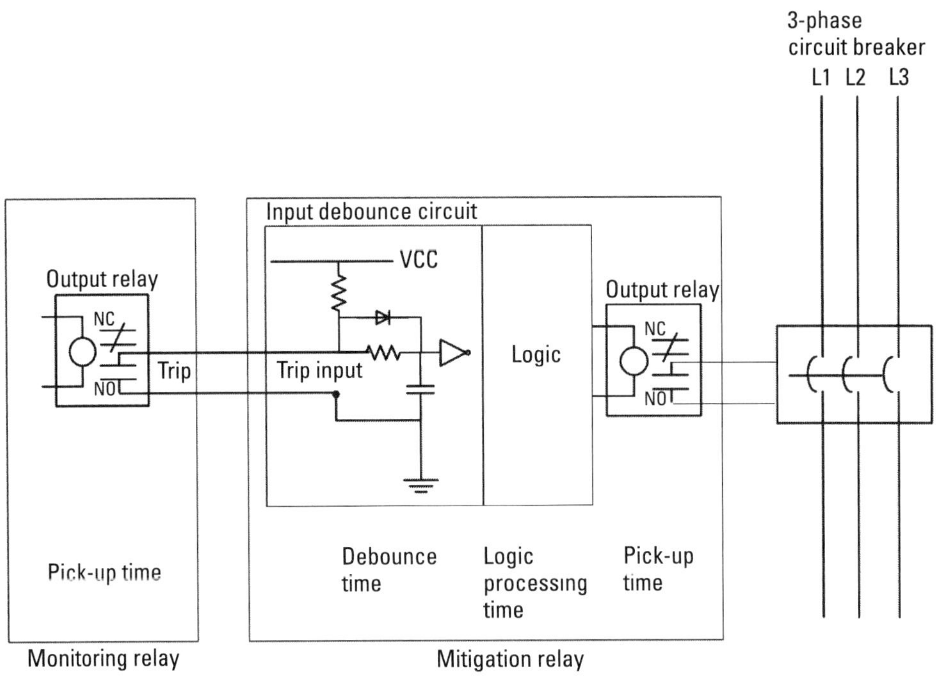

Figure 3.2 Schematic of hardwire tripping of circuit breaker.

current, under frequency, power overload). Once the monitoring relay determines that protection is required, it activates one of its output relays.

Typically, these relays are mechanical in nature. When the relay coil has power applied to it, the output contacts change state. There are two output contacts provided:

1. *Normally open* (NO): This contact is open when the coil does not have power applied to it. When power is applied to the coil, this contact closes and conducts current (i.e., closes the circuit).
2. *Normally closed* (NC): This contact is closed and allows power to flow through it when the coil does not have power supplied to it. When power is applied to the coil, this contact opens and does not allow current to flow (i.e., opens the circuit).

Since there is physical movement of the contacts, there is a time delay between the power being applied to the coil and the contacts achieving their intended state. This time is known as the pickup time.

The mitigation, or circuit breaker controller, has an input that the output of the monitoring relay is wired into. Typically, the input is called the trip input. If one analyzes the schematic, the input consists of a resistor/capacitor (RC) filter that debounces, or smoothes, the input. This is needed to prevent inadvertent trips while the physical contacts of the output coil are moving and to filter out noise. The time delay associated with the RC filter is known as the debounce time.

The total expected time from decision to trip to actual completion of circuit breaker movement can be expressed by the summation of the various component times as shown in (3.1):

$$\begin{aligned} ExpectedTime &= PickupTime_{Mon} + DebounceTime_{Mit} + Logic_{Mit} \\ &+ PickupTime_{Mit} + CircuitBreaker_{Time} \end{aligned} \tag{3.1}$$

The values for debounce and pickup times are provided in the actual specifications of the deployed devices. As an example, typical metrics are pickup times of 12 msec, debounce times of 6 to 8 msecs, logic processing times of 5 to 10 msec, and circuit breaker times of 80 msec. Based on these typical values, the *ExpectedTime* can be calculated as 122 $msec_{Max}$.

Of interest, during the UCA research, was the timing required to replace the monitoring output relay to mitigation tripping input hardwire with network communications. The maximum time allotted can be calculated to be 20 msec. It should be noted that in today's digital relays debounce times can be adjusted.

As with any good engineering design, it all starts with a use case to allow the development of the requirements and then the appropriate design.

3.1 Use Case for Digital Network-Based Protection

There are two aspects to the use case developed for digital networked based protection: signal distribution requirements and timing requirements.

3.1.1 Signal Distribution Requirements

To understand the signal distribution requirements, the use case was based on an actual event that had occurred within Illinois's Commonwealth Edison (ComEd) utility. ComEd had a substation that had provided connectivity and protection for three high-voltage three-phase transmission lines. Such substations are large and contain many devices that need to exchange signals. In this instance the substation had approximately 100 devices. The actual event was a tornado that tore through the substation. During this event, approximately 60% of the devices needed to exchange signals within the allotted protection interval. In reality the distribution of the signal generation occurred within approximately 20 msecs. It was also identified that multiple digital signals needed to be delivered from one device to different devices in parallel.

The replacements of hardwire with digital network technology needed to be analyzed to develop network event loading characteristics.

3.1.2 Timing

During the development of the use case, one of the consultants had a unique perspective on the timing performance requirements of the ComEd use case. His name was John Tengdin and his perspective follows:

> With few exceptions, electromechanical protective relays for transmission lines had only one tripping contact. A tripping contact, by definition, was one with the capability to make and hold closed a highly inductive tripping circuit up to 30 amps, 250V DC. These contacts could not interrupt that current. That function is accomplished by the 52a contact on the tripped HV circuit breaker's operating mechanism so that when the HV circuit breaker opened, the trip current is interrupted by the breaker.

But a substation with a ring bus, breaker and a half, or double breaker double bus configuration will have two circuit breakers serving each protected element (e.g. transmission line or transformer). When a fault occurs, the protective relay operates an auxiliary multicontact tripping whose isolated contacts close to trip both breakers without the need to parallel the two trip circuits. In other applications such as interlocking, these same auxiliary relays were applied when they were energized for long periods of time. Thus, their operating coils were designed for minimal battery drain (with thousands of turns of fine wire). The result was a relay with a very high inductance (65 Henries, in one case) and so a slow operating time (over 8 msec). This of course added to the total clearing time of a fault.

With the advent of solid state analog transmission line protection, those units were designed with at least two SCR trip outputs—eliminating the need for auxiliary tripping relays to multiply their trip output. However, in many cases, the backup relays were electromechanical, so the worst-case total clearing times used in system stability studies still had to include the operating time of these auxiliary relays. To improve the clearing times, some utilities spent many dollars replacing three cycle HV circuit breakers with those with a fault clearing time of two cycles, but the auxiliary tripping relay was still in the picture.

Then, in the mid-1970s, ASEA introduced a high-speed auxiliary tripping relay (the RXMS) available with an operating time of just 4 msec and with contacts suitable for trip circuit duty. If installed, this reduction of at least a quarter cycle in total clearing time was very attractive to the system planners and many relays were installed as a retrofit, replacing the slower relays. A gain of at least 4 msec was important.

Shifting now to the 1990s, the pressure was still on to improve overall power system performance. EPRI, as a part of the UCA Project, commissioned work on a RP 3599, titled “Substation Integrated Protection, Control, and Data Acquisition–Requirements Specification,” which was published on October 31, 1995, and defined the performance requirements (message delivery times) for a substation communications network. For protective relay applications, several tables (2-6 and 2-7) required message delivery times of—you guessed it—4 ms. This number was thoroughly vetted with protective relay engineers, who grudging accepted the concept if it could replace the need for auxiliary tripping relays. It may be folklore or an urban legend, but it is this writer’s opinion that the magic number of 4 msec came from substituting message delivery time for the operating time of the RXMS relay: 4 msec.

John Tengdin, in these paragraphs, provided a summary of over 8 months of meetings to determine the actual UCA performance requirements. UCA specified a 4-msec performance requirement. The actual meaning of the requirement can be seen when digital communications is overlaid with the hardwired diagram.

Figure 3.3 shows that the digital communication performance requirement (i.e, 4 msec) is being used to replace the pickup time and debounce time combination between the monitoring relay and mitigation relay. As with the pickup time, timing starts when the monitoring relay logic decides that an action is required (i.e, trip). In the hardwired case, it activates the trip coil. In the digital messaging case, the decision triggers the encoding and transmission of a message.

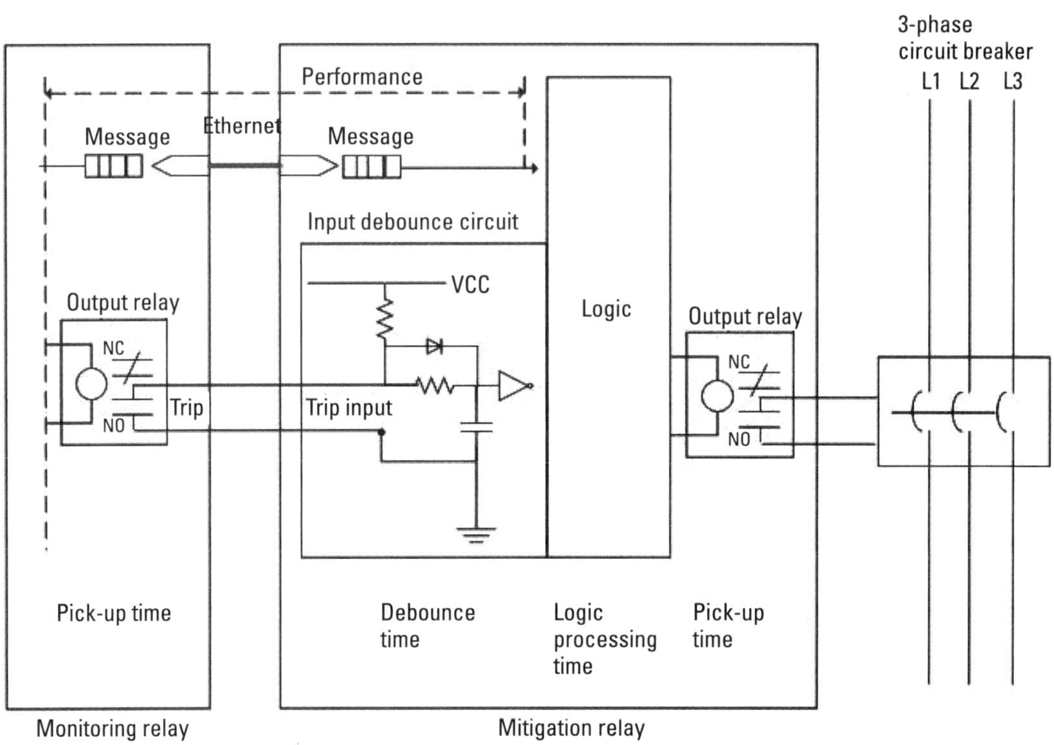

Figure 3.3 Hardware versus communication.

Performance timing ends when the decision (i.e., trip) is presented to the logic of the mitigation relay but does not include the execution of the logic. The timing of the reception of the information is determined by the debounce time or message parse time.

With this definition of performance, UCA specified a requirement that was four- to five (5) fold faster than the hardwired systems of the time (i.e., 20 msec vs 4 msec). IEC 61850-5 eventually developed a 3-msec requirement to replace the UCA requirement.

3.1.3 Quality of Message Delivery Service

During the development of the 4-msec message delivery requirement, there were a couple of very interesting questions that continued to be debated: What delivery means in the case that a message is not delivered and how to recover from the failure of delivery.

Communications engineers know that there is a principle of how to guarantee delivery and that is that messages are acknowledged by the receiving entity. This is the mechanism employed by the Transmission Control Protocol (TCP) that is pervasive in the internet. However, TCP is a connection-oriented (e.g., point-to-point) solution and given today's CPU and networking performances, it might have been able to be used for the point-to-multipoint message delivery requirement of the use case. However, there is another aspect of acknowledged message delivery that is not desirable from a protection environment perspective. Since the message delivery mechanism was developed to replace hardwire, a comparison of an acknowledged delivery mechanism and hardwired systems reveals the issue.

In a hardwired system, the monitoring relay logic can change the output state when a state change is needed. If the pickup time is ignored and the logic scan time is 2 msec, this means that the monitoring relay could change its output every 2 msec. In an acknowledged messaging system, messages are buffered and retransmitted until there is an acknowledgment or the fact that the remote is detected to be offline.

Figure 3.4 shows that the logic in the monitoring relay changes state at the 2-msec time (relative). At this time, the trip signal is activated and the digital message indicating TRUE is sent. If the message does not receive an acknowledgment, it needs to be retransmitted to elicit an acknowledgment in the case the initial message was not delivered. The diagram shows this retransmit coincident with another change of state of the monitoring relay. This state change generates another message: FALSE. FALSE is not allowed to be delivered until TRUE is acknowledged. This causes the initial transmission of FALSE to be delayed by approximately 7 msec in the figure and violates the 4-msec delivery requirement.

3.1.4 Requirements and Decisions Based on Use Cases

There are three major requirements that were derived from the use cases:

1. There is a need to deliver current information as rapidly as possible and the delivery of stale information is not needed for control purposes;
2. There is a need to deliver the current information to many other devices;

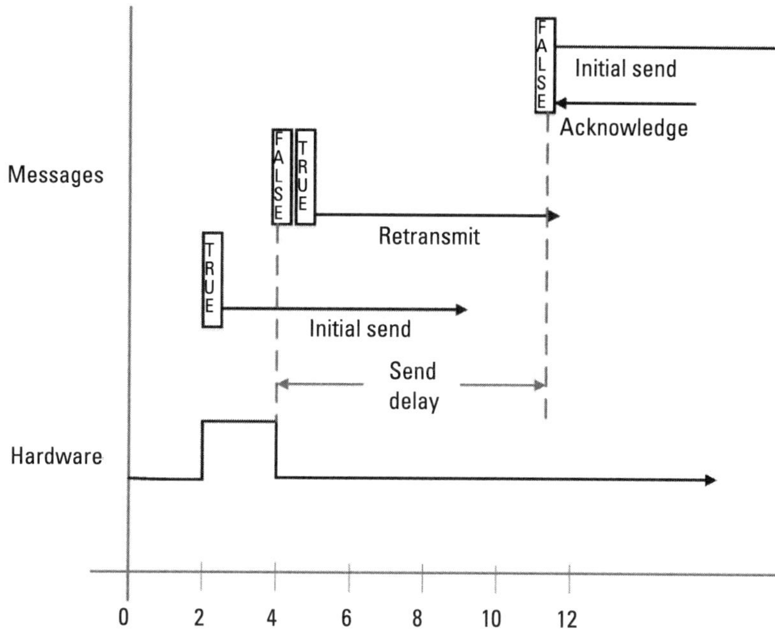

Figure 3.4 Hardwired versus TCP/IP retransmit performance.

3. The delivery of the information needs to occur in less than 4 msec (IEC revised this to 3 msec).

Based on these distilled requirements, several design decisions were made in 1998 at a meeting in Chicago. The first and most controversial design decision was to develop a guaranteed delivery multicast delivery technique. The premise of the design was to retransmit the current information in a manner that guaranteed delivery in less than 4 msec.

Inquisitive minds might wonder how a design of a multicast infrastructure (i.e., without acknowledgments) can guarantee delivery. The key to the design is that the use cases do not require packet guaranteed or noncurrent information delivery, but rather the delivery of the current information.

There was a simplicity to the design based on retransmission of the current information. Figure 3.5 shows the concept. If a value changes, there is an initial transmission of the value. The publishing device has an internal retransmit algorithm and retransmits the same value periodically. When the value changes, the new value are transmitted and the retransmission process starts anew. The state machine for this algorithm is depicted in Figure 3.5.

The figure shows a state machine that needs to be enabled and disabled. Once enabled the values are continuously published based on the publisher's retransmission timer or retransmission curve. This design means that the number of packets delivered to the other devices is based on the publisher's retransmission algorithm and the network latency. To meet the 4-msec delivery requirement, it was decided that there would be a need for at least one or two retransmits within the 4 msec.

The 4-msec delivery requirement spawned a discussion regarding the network technology to be utilized to achieve the requirement. There were two proposals: Profibus and Ethernet. Profibus was based on token bus technology. As with the

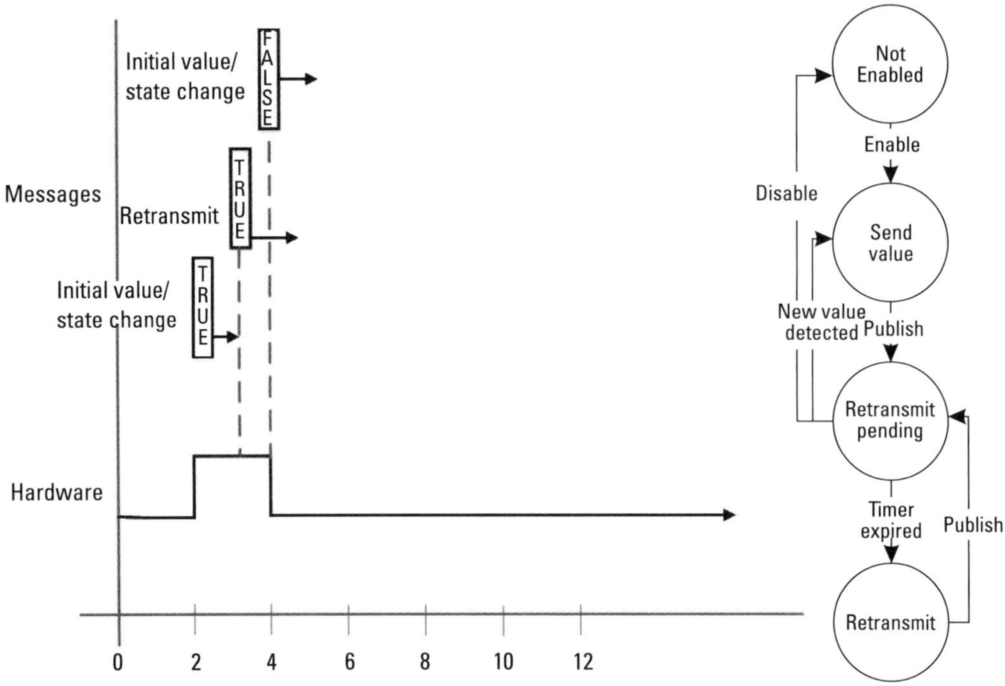

Figure 3.5 Unacknowledged retransmission and state machine.

GM MAP initiative, token bus was thought to be deterministic. Ethernet was not considered to be deterministic. The debate about choosing the appropriate network technology spanned a 2-year period from 1996 to 1998.

There are several analogies that can be used to highlight the issues with each technology and then mathematical formulas that can be used to calculate network performance.

3.2 Mathematical Analysis of the Technologies

The electric utility industry is typically conservative and requires extensive proof or experience to select and adopt a new technology. In 1996, the industry was divided into two groups supporting different technologies: the IEEE PSRC was advocating Profibus; and EPRI was advocating the use of Ethernet. This industry split became very evident at an IEEE PSRC meeting circa June 1996. During this meeting, EPRI presented a strong case to use Ethernet based on a mathematical analysis of Profibus showing that the technology could not meet the requirements.

The disclosure of that information started a sequence of testing and mathematical analysis that utilized engineering practices to determine the appropriate technology to select. The mathematical analysis of both technologies were reworked and verified several times. Additionally, actual benchmark testing was performed to validate the expected results.

The following sections detail the mathematical analysis of Profibus and a mathematical analysis and simulation of Ethernet.

3.2.1 Profibus

In 1996, Profibus was a master/slave network technology where the masters used token passing technology to gain access to the network to poll slaves. The media is based on RS-485[1] multidrop serial technology (see Figure 3.6).

As such, Profibus used multiple conductors to achieve full duplex operation. It was developed by Siemens in 1987 and standardized as a German National standard Deutsches Institut für Normung (DIN) 19245 in 1991. It has since become IEC 61158 and IEC 61784. The standard allows for signaling rates of 9.6 Kbps to 12 Mbps. Profibus comes in three different versions:

1. *Distributed Peripheral* (DP) was introduced in 1993 and is limited to only 32 devices per DP segment. There can be multiple masters on a DP network. Each of the master gains access to the network through the passing a token that allows the receiving master to transmit on the network. The master that holds the token then polls the slaves as needed and then passes the token to the next master. The amount of time that any master can hold the token is known as the token hold time (t_{hd}).
2. *Process Automation* (PA) is typically an all Profibus master network where the masters directly exchange information. Master access to the network is still determined through passing and holding the token as with Profibus DP. Since the same token passing protocol is used in Profibus DP and Profibus PA, the two technologies can coreside on the same physical cable.
3. *Field Bus Message Specification* (FMS) is used for peer-to-peer (i.e., master-to-master) information exchange.

Profibus does offer the multicast ability that is needed in order to meet the delivery requirements that were developed as part of the digital network based protection use case. The typical performance of Profibus can be approximated by calculating the token rotation time. In a perfect world, the maximum token rotation time (t_{ROT}) can be approximated through the simple formula shown in (3.2):

$$t_{\mathrm{ROT1}} = t_{hd} * num_{\mathrm{Masters}} \tag{3.2}$$

where $\mathrm{num}_{\mathrm{Masters}}$ is the number of masters on the network.

A more precise calculation needs to consider the time that is required to actually pass the token. This can be calculated based on the number of masters and the time required to transmit a token. In the case of Profibus, the size of the token pass PDU ($\mathrm{size}_{\mathrm{token}}$) and the size of the token acknowledge ($\mathrm{size}_{\mathrm{tokenACK}}$). There is also

Figure 3.6 Profibus cable. (Image courtesy of Pro Wire and Cable.)

1. See https://en.wikipedia.org/wiki/RS-485.

an acknowledge response delay that needs to be considered. The time required to pass a single is calculated by (3.3):

$$t_{tokenPass} = \left(size_{token} + size_{tokenACK}\right) / bps + delay_{Ack} \rightarrow 0 \tag{3.3}$$

The more precise token rotation time can be expresses as shown in (3.4):

$$t_{ROT} = \left(t_{hd} + t_{tokenPass}\right) * num_{Masters} \tag{3.4}$$

For a Profibus network of 32 masters, using signaling of 2 Mbps, and a token hold time sized to allow three telegrams to be sent by each master maximum, the maximum token rotation time can be calculated per (3.5):

$$t_{telegramTx} = max_{telegramSize} / bps$$

where $max_{telegramSize}$ is 2,440 bits.

$$t_{hd} = 3 * t_{telegramTx} = 3 * \left(max_{telegramSize} / bps\right)$$

$$t_{ROT} = \left(t_{hd} + t_{tokenPass}\right) * num_{Masters} = \begin{pmatrix} 3 * \left(max_{telegramSize} / bps\right) \\ + \left(size_{token} + size_{tokenACK} / bps\right) \end{pmatrix} * num_{Masters} \tag{3.5}$$

$$t_{ROT} = \left(3 * max_{telegramSize} + \left(size_{token} + size_{tokenACK}\right)\right) / bps * num_{Masters}$$

where

$size_{token}$ = a constant of 3 bytes or 30 bits;

$size_{tokenACK}$ = a constant of 1 byte or 10 bits.

For a Profibus network operating and 12 Mbps with 32 masters:

$$t_{ROT} = \left(3 * 2400 + (30 + 10)\right) / 12 * 10^6 * 32 = 19.6 \text{ msec} \tag{3.6}$$

The maximum rotation time, calculated in (3.6), is in excess of the 4 msec that exceeds the maximum performance requirement. If the t_{hd} is reduced to one maximum size telegram, the t_{ROT} is 6.6 msec, which is beyond the 4 msec.

The performance of Profibus is further brought into question if one accounts for 100 masters, 20 of whom need to transmit due to an event. In this case, the formula becomes

$$\begin{aligned} t_{ROT} &= \left(max_{telegramSize} * 20 + \left(size_{token} + size_{tokenACK}\right) * 100\right) / bps \\ t_{ROT} &= \left(2400 * 20 + (30 + 10) * 100\right) / 12 * 10^6 = 140 \text{ msec} \end{aligned} \tag{3.7}$$

which is even further beyond the 4 msec.

Although the token rotation calculation showed that Profibus would have difficulty in the best circumstances, error recovery and station (e.g., master) addition or removal cause even more degradation of performance. Token passing is like a game of hot potato, where each player (master) attempts to pass the ball (token) on to the next player as rapidly as possible. If a player drops the ball (a dropped token), there is an inordinate amount of time to recover from the dropped ball or token.

To know the list of available masters to which to pass a token, a master must listen to the communication occurring on the bus. This is known as the listen token state. As active communication occurs on the bus, masters who are not transmitting continue to build a list of available masters. A new station that is added to the bus also builds a list of active stations. It does not become an active participant on the bus until it receives a Request FDL Status message inviting the new node to join the bus. The station claims the token and becomes an active member of the bus. If a station does not receive the token within an expected period of time, it performs a process similar to being a new station so that it can claim the token in order to transmit.

If no transmissions are detected within a configured t_{Idle} (i.e., idle time), every master enters an algorithmically determined wait time before automatically claiming the token (t_{Claim}), which can be calculated per (3.8):

$$t_{Claim} = t_{Idle} = 6 * t_{MaxResponse} + 2 * MasterAddress * t_{MaxResponse} \quad (3.8)$$

There are three use cases that are addressed through this recovery technique and Table 3.1 shows the typical claim times for each at different transmission rates.

The mathematical analysis of Profibus shows that a Profibus segment of 100 nodes will exceed the 4-msec delivery requirement developed by the use case. Additionally, the maximum rotation time is not guaranteed due to the dropped token recovery time. The analysis clearly showed that Profibus was not a valid candidate. However, skepticism in the industry and IEEE persisted and there was a need to provide actual testing results. This need resulted in the benchmarking of Ethernet versus Profibus for the Digital Protection use case.

For more supporting information regarding Profibus, please see the following:

- https://en.wikipedia.org/wiki/Profibus
- http://instrumentationtools.com/profibus-communication-interview-questions-answers/
- http://www.rtaautomation.com/technologies/profibus/

3.2.2 Ethernet

The media access mechanism of Ethernet is vastly different from Profibus. Profibus controls access for transmission through the passing of a software token. Only master nodes that hold the token can transmit. Ethernet is like the old telephone party

Table 3.1 Typical Token Claim Time for Profibus

Recovery Use Case	*Typical Recovery Time at 1.5 Mbps*	*Typical Recovery Time at 12 Mbps*
One transmission failure	500 μsec	100 μsec
One station drops off the bus or fails	700 μsec	300 μsec
All stations drop off the bus (worst case)	33 msec	6.6 msec

lines.[2] In the case of party lines, the telephone line was shared, and multiple people could talk at the same time, and this was prior to the ability to conference call. If you have ever been on a conference call or party line, you will know the problem of understanding people when multiple people talk at the same time. In the instance where multiple people talk simultaneously, some of the message is lost because each person's conversation becomes garbled. The same is true with Ethernet; when multiple nodes transmit, messages are garbled.

The initial Ethernet media was 10Base5 (also known as thicknet Ethernet). Eventually, 10Base5 transitioned to 10Base2 (also known as thinnet). Both technologies were multidrop coaxial-based cable systems with a single conductor and ground. As such, the initial Ethernet could only operate in half-duplex mode. To minimize the probability of multiple nodes transmitting at the same time, Ethernet utilizes a technique, CSMA, to detect if a node on the network is transmitting and CD to determine if the transmitted message and how to recover from a collision. Thus, Ethernet is also known as CSMA/CD.

The introduction of IEEE 10BaseT (also known as twisted-pair Ethernet) offered multiple conducting pairs, but was still restricted to half-duplex operation. However, 10BaseT and its newer versions are not multidrop media solutions. 10BaseT is designed as a point-to-point solution. The initial 10BaseT cables contained two conducting pairs. The pairs are allocated for transmit and receive. However, since there are transmit and receive conductors, in order to provide a node-to-node connectivity the transmit pins of a node must be connected to the receive pins of the other node (e.g., a crossover cable). To achieve connectivity beyond two nodes, an Ethernet Hub is required. An Ethernet Hub is an Ethernet Repeater and typically does not have any intelligence embedded. Hubs accept straight-through 10BaseT cables and provide bit level repeating and do not provide store-and-forward capability. 10BaseT Hubs are therefore restricted to half-duplex operation. Thus, the CSMA/CD network includes all nodes and hubs that are interconnected.

Category 5 10BaseT cables offer four conducting pairs and therefore full duplex operation, so there is no probability for collisions as there is a point-to-point CSMA/CD link (between the node and the switch port). In the case of operating in full-duplex mode, that link does not operate in party-line mode as there are separate transmit channels for the node to the switch port and another channel for the switch port to transmit to the node. The use of an Ethernet Switch whose ports operate in full duplex mode alleviates the issue of collisions even though CSMA/CD is still utilized on the link from the node to the Switch.

To evaluate Ethernet against the communication requirements, Ethernet Hubs and Switches and various media speeds needed to be evaluated. In order to perform the mathematical analysis, EPRI hired Fraunhofer Institute IITB to perform the mathematical analysis and simulation.

The simulated scenario included background traffic to simulate SCADA TCP/IP communication and multicast tripping messages. The SCADA communication load was calculated based on the following assumptions:

2. See https://en.wikipedia.org/wiki/Party_line_(telephony) for more information.

- There may be one or two SCADA masters (S) that poll up to 98 nodes (N) and six PLCs.
- The six PLCs also perform polling functions of up to 15 nodes each.
- There is a monitoring/display station (HMI). The HMI polls information from the nodes in the substation and provides the information so that manual action can be taken.
- Phasor data was to be acquired by up to two IEDs 10 times a second.[3]
- Requests are assumed to be 65 bytes.
- Responses are assumed to be 105 bytes.
- Poll rate vary per polling type.
- Reports by Exception (RBE) messages are assumed to be 105 bytes.
- RBE messages are assumed to be sent based on 10% of the total nodes at 10 times/second.

The worst-case background traffic is shown in Table 3.2.
The worst-case RBE load can be calculated per (3.9):

$$\begin{aligned}&(\mathrm{N}+\mathrm{PLC}+\mathrm{IED})*0.1*PR*105\\&=(98+6+2)*.1*10*105=11{,}130\ \text{ bytes/second}\end{aligned} \tag{3.9}$$

The worst-case load of the combined polling and RBE traffic is 31% of 10 Mbit (i.e., Ethernet) and 3.1% of 100 Mbit Ethernet. This traffic is distributed due to the asynchronous polling nature and this aspect was accounted for in the Ethernet simulation.

However, the objective of the simulation was to determine the probability of achieving the 4-msec multicast delivery requirement as up to 60% of the 100+ nodes decide to transmit a trip message. The EPRI study denoted this type of traffic as Trip Load. The probability of a node sending simultaneous multicast trip message is based on the number of nodes in the system and can be expressed via Table 3.3, where B is the node and *n* represents the number of nodes in the system.

Table 3.2 Worst-Case Simulated Background Traffic Due to Polling

Polling Type	*Quantity*	*Poll Rate (per Second)*	*Nodes Polled*	*PLCs Polled*	*IEDs Polled*	*Number of Poles/Second*
SCADA	2	1	98	6	2	212
PLC	6	10	15			900
HMI	1	10	98	6	2	1,060
IED	2	10	2			40
Total poll messages (per second)						2,212
Bytes per poll						170
Total bytes polled traffic/second						376,030

3. Phasor data is now typically provided at a rate greater than 30 times a second.

Table 3.3 Probability of a Node Transmitting a Multicast Trip

Node	B_1	B_2	B_3	...	B_n
B_1		$1(n{-}1)$	$1(n{-}1)$	$1(n{-}1)$	$1(n{-}1)$
B_2	$1(n{-}1)$		$1(n{-}1)$	$1(n{-}1)$	$1(n{-}1)$
B_3	$1(n{-}1)$	$1(n{-}1)$		$1(n{-}1)$	$1(n{-}1)$
...	$1(n{-}1)$	$1(n{-}1)$	$1(n{-}1)$		$1(n{-}1)$
B_n	$1(n{-}1)$	$1(n{-}1)$	$1(n{-}1)$	$1(n{-}1)$	

The mathematical equations developed by Fraunhoffer show the relationship of arrival time to the probability of communication, shown in (3.10):

$$\text{Arrival rate } \overline{\lambda} = \begin{pmatrix} \lambda_1 \\ \lambda_2 \\ \cdots \\ \lambda_n \end{pmatrix} \tag{3.10}$$

Arrival rate for all stations S_1 to S_n: k
For each S_i, the arrival rates are calculated as follows:

$$k/n \, \frac{messages}{second}$$

Communication Probability:

$$p_{i,j} = (\text{p.rate, p.size})_{i,j}$$

Communication Matrix:

$$\overline{\overline{p}} = \begin{pmatrix} p_{1,1} & p_{1,2} & \cdots & p_{1,n} \\ p_{2,1} & \cdots & \cdots & \cdots \\ \cdots & \cdots & \cdots & \cdots \\ p_{n,1} & \cdots & \cdots & p_{n,n} \end{pmatrix}$$

The following condition must hold:

$$\sum_{j=1}^{n} (p.rate_{i,j}) = 1$$

Based on Table 3.3, mathematical equations, and the distribution of the background traffic, simulated results for various Ethernet technologies could be performed. The simulations included 10-Mbit shared Ethernet and are shown in Figure 3.7.

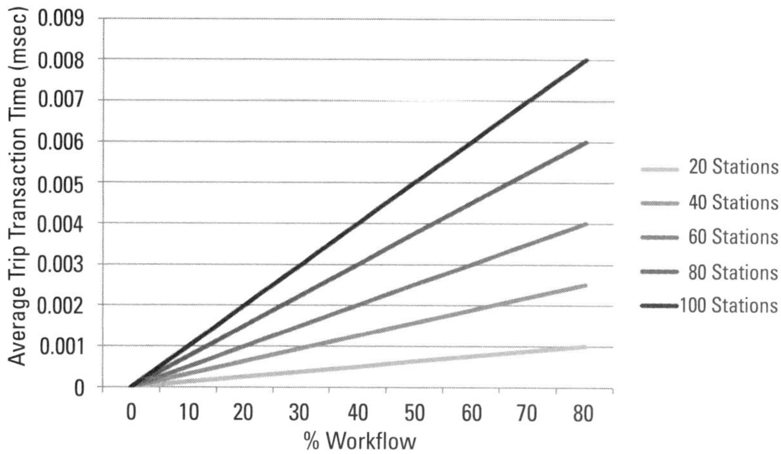

Figure 3.7 Estimated average trip arrival times versus workflow for 10Base5.

10Base5 is a shared half-duplex Ethernet. The simulation results shown in Figure 3.7 show that for this type of Ethernet, the 4 msec requirement could be met for substations with 60 nodes or less.

Since the arrival time is related to the speed of the Ethernet, the same graph for 100BaseT is similar to the one for 10Base5. However, Figure 3.8 shows that a 100BaseT network can achieve the 4-msec requirement.

Although the average arrival times indicated that 100BaseT, shown in Figure 3.8, could satisfy the requirement, there were substantial numbers of trip messages that the simulation indicated would be beyond the 4-msec requirement. For a simulation consisting of analyzing more than 10,000 trip messages, the exponential impact of shared Ethernet collisions can be seen.

Figure 3.9 shows the number of trip messages that exceeded the 4-msec requirement. The maximum percentage of expected trip messages beyond the 4-msec requirement can be calculated based on the number of samples simulated. The

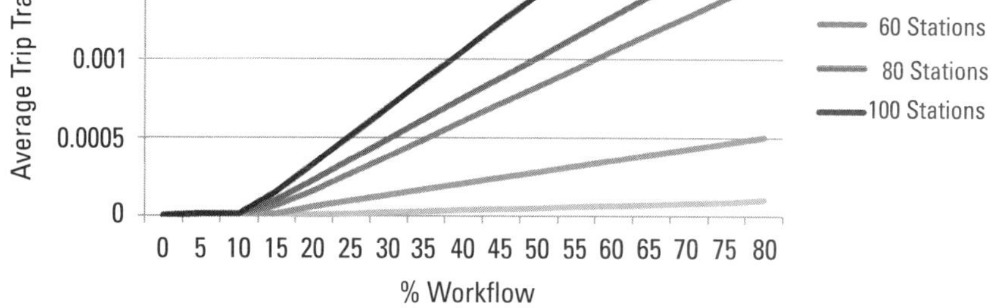

Figure 3.8 Estimated average trip arrival times verss workflow for 100BaseT.

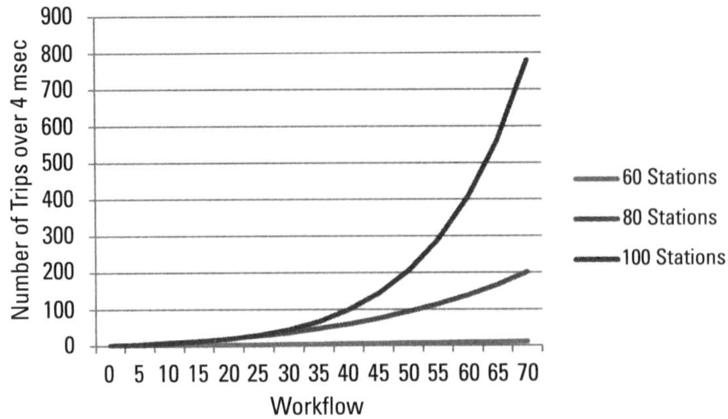

Figure 3.9 Number of trip messages beyond 4 msec versus workflow for 100BaseT.

worst-case percentage appears to be 8% for 100 stations. The distribution of the delta beyond the 4 msec can be seen in Figure 3.10.

The simulation indicated that the worst-case delta from the requirement was 1 msec and most of the messages beyond the 4-msec requirement was less than 100 μsec from the requirement. Thus, there was a conclusion that 100BaseT half-duplex shared media could meet the requirements with the appropriate retransmit methodology.

The simulation was also executed for an emerging technology known as an Ethernet switch. The formula for switch performance is vastly different from the half-duplex shared media calculations. This is due in large part to the fact that there are no collisions when using a switch in full-duplex mode. The performance is a simple ingress, egress, and switch transfer time calculation. Ethernet switches at the time of the simulation were assumed to have a 30 μsec transfer time/message. If the trip messages are 200 bytes in length and a maximum of 10 trip messages need to be delivered per node, the following can be calculated:

- The ingress time for a single message is approximately 160 μsec;

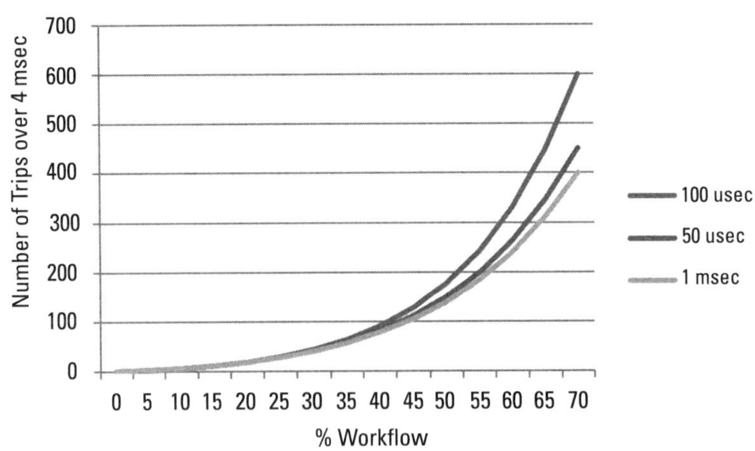

Figure 3.10 Distribution of trips beyond 4 msec based on time beyond the requirement for 100BaseT.

- The maximum egress time is 1.6 msec;
- The maximum transfer time is 30 μsec.[4]

Therefore, the gross approximation of trip message performance can be a straight addition and results in 1.79 msec. Not only is this metric the average, but it also represents an approximation of the maximum trip performance.

The mathematical simulation for Ethernet revealed the following conclusions:

- 10Base5 shared media (hub or cable) was adequate for substation networks of 20 nodes or less;
- 100BaseT shared media (hub) was adequate for substations of 60 nodes without any issue;
- 100BaseT shared media (hub) could be made to meet the 4-msec requirement with appropriate trip retransmission;
- 10BaseT full-duplex switched Ethernet could meet the 4-msec requirement with no issue for 100 nodes.

For more supporting information regarding the performance of Ethernet, please see the following: http://www.hpl.hp.com/techreports/Compaq-DEC/WRL-88-4.pdf.

3.3 Mathematical Truth but Numbers Can Lie

Prior to the development of the Ethernet simulation, the math regarding Profibus was presented to the IEEE and the industry. The skepticism expressed at the math was unexpected and therefore the industry requested that a set of benchmark testing be performed. EPRI funded Systems Integration Specialists Company (SISCO) to perform the benchmark. There was an initial benchmarking that tested four nodes of Profibus versus four nodes of 10BaseT Ethernet (shared media). The results showed that Ethernet performed better than Profibus. When these initial results were presented, the response was that four nodes did not represent the scale of a substation of 100 nodes. EPRI was asked to stage 100 nodes for Ethernet testing.

The cost of staging 100 nodes for Ethernet testing was prohibited. Therefore, there was an agreement reached to stage 20 nodes and perform simulations that scaled to 100 nodes. This agreement was the reason that the Fraunhofer Institute was funded to perform the simulation. The simulation results, see Section 3.3.2, showed that with constraints Ethernet could meet the performance requirements. The benchmark results were intended to verify the validity of the simulation results.

3.3.1 Profibus versus Ethernet Test Results

The initial testing was intended to determine the performance of Profibus versus Ethernet. The same set of computational resources was used to test both Ethernet

4. Ethernet switches today have a typical transfer time of 10-15 usec.

and Profibus. Table 3.4 contains the resource information of the two laptops and two desktops used.

The purpose of the testing was to compare the ability of the Profibus Link Layer versus that of Ethernet (i.e., CSMA/CD). Thus, the protocol stack used for testing needed to be the same in order to compare the link layer performance. ISO 9506 (i.e., MMS) was selected as the application protocol and the UCA Trim 7 profile, defined in Table 3.5, was used over both Profibus and Ethernet.

For each test profile, two computers acted as clients (generating MMS requests) and the other two computers were slaves. The request generation was synchronized through an external printer port connection.

The setup provided Profibus with two masters and two slaves on a 12-megabit link was tested. The 12M Profibus link was the target of the test to minimize the potential transactional difference due to bandwidth. The test setup provided an intentional Ethernet collision on the 10BaseT and 100BaseTx shared media (hub) technology that was tested.

The executed tests were based on the retrieval of certain types of information. The tests were the retrieval of

- A single analog value (e.g., a floating-point value);

Table 3.4 Computer Hardware Utilized for Benchmark Testing

	Laptop	*Desktop*
Manufacturer	Texas Instruments	Gateway
CPU	133-Mhz Pentium CPU	133-Mhz Pentium CPU
RAM	32 Megabytes	32 Megabytes
Harddrive	1.6 Gigabytes	2 Gigabytes
Operating System	DOS	DOS
Profibus Hardware	PCMCIA PROFIcard-KOMBI	ISA PROFI-IF-KOMBI
Ethernet Hardware	PCMCIA Xircom Adapter 10/100	PCI 3com PCI 10/100

Table 3.5 Definition of Trim 7 Communication Stack

ISO Layer	*Profibus*	*Ethernet*
7. Application	ISO 9506 Manufacturing Message Format	ISO 9506 Manufacturing Message Format
	ISO/IEC 8649 and ISO/IEC 8650: Association Control Service Element	ISO/IEC 8649 and ISO/IEC 8650: Association Control Service Element
6. Presentation	Fast-Byte presentation	Fast-Byte presentation
5. Session	Fast-Byte session	Fast-Byte session
4. Transport	ISO/IEC 8072 and ISO/IEC 8073	ISO/IEC 8072 and ISO/IEC 8073
3. Network	ISO/IEC 8348 and ISO/IEC 8473	ISO/IEC 8348 and ISO/IEC 8473
2. Data Link	IEEE 802.2 LLC 1	IEEE 802.2 LLC 1
	DIN 19 245 Part 1: Fieldbus Datalink Layer	IEEE 802.3 CSMA/CD
1. Physical	DIN 19 245 Part 1	IEEE 802.3

Note: The Fast-Byte Presentation and Session specifications were draft specifications and were never codified as international standards.

- A list of analog values (e.g., an array of floating point values);
- A list of SCADA status points (e.g., an array floating point data);
- A list of SCADA accumulators (e.g., integer data);
- A scattered list of analog values (e.g., not an array of data);
- A trip status (e.g., a Boolean).

There were 1,000 iterations of the tests performed so that appropriate minimum, maximum, and average transaction times were able to be calculated. Figure 3.11 correlates the average transactional performance of several of the different tests.

Figure 3.12 provides a high-level look at the minimum, maximum, and average distribution for the trip test is revealing. The results in Figure 3.12 show that Profibus does not meet the 4-msec requirement even for the minimum measured transactional value. The Ethernet results showed that even the maximum transactional performance for 10BaseT meets the use case performance requirements.

Figure 3.11 shows that the transactional performance of Profibus with two masters is much worse than that of a single collision on either Ethernet media. The test setup and implementation provided an environment where the performance of the MMS communication profile could be assumed to be the same since the code/implementation was the same except for the interface to the link layer. Thus, the delta of performance is created by the different technologies and bandwidth, as shown in Figure 3.13.

If one were to discount the trip outlier for the Profibus trip, the average transactional difference between Profibus and 10BaseT Ethernet is 12.24 msec. This delta can be attributed, in its entirety, due to the different media access methodologies (e.g., token rotation versus CSMA/CD). The comparison also shows that 100BaseTX Ethernet has better performance than 10BaseT Ethernet, which is an expected result.

This average delta clearly indicates that Profibus is a much worse transactional performer than Ethernet. When all the math and test results were presented to IEEE

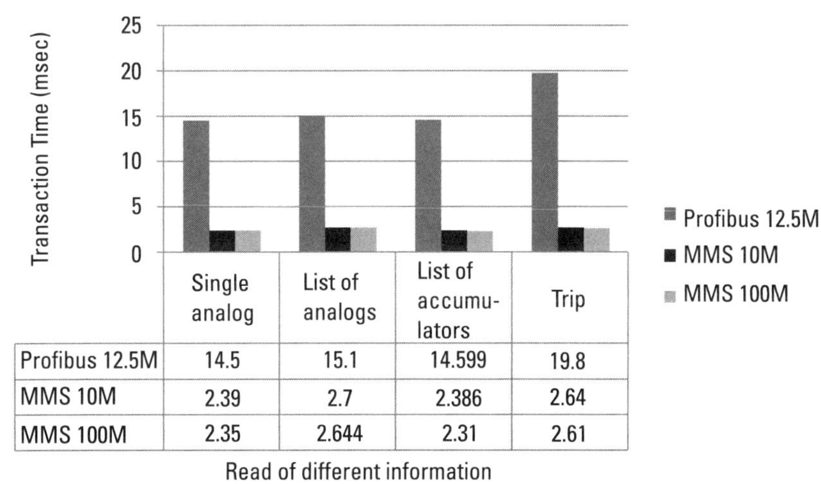

	Single analog	List of analogs	List of accumu-lators	Trip
Profibus 12.5M	14.5	15.1	14.599	19.8
MMS 10M	2.39	2.7	2.386	2.64
MMS 100M	2.35	2.644	2.31	2.61

Figure 3.11 Average performance of MMS over Profibus and Ethernet.

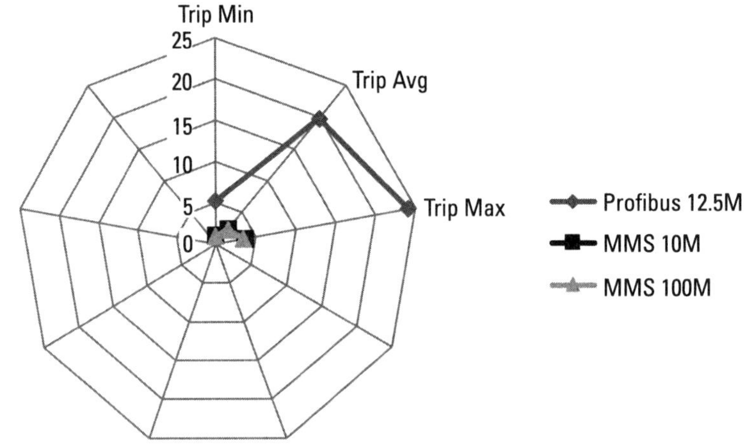

Figure 3.12 Distribution of performance for trips.

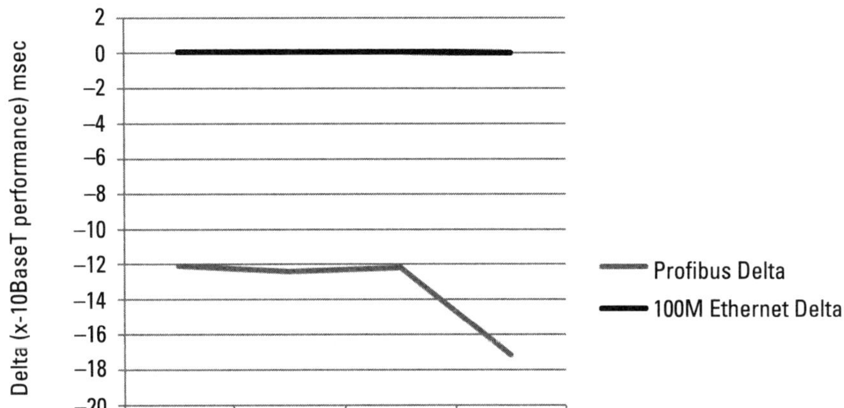

	Single analog	List of analogs	List of accumulators	Trip
Profibus Delta	−12.11	−12.4	−12.213	−17.16
100M Ethernet Delta	0.04	0.056	0.076	0.03

Figure 3.13 Normalized transactional performance of Profibus and 100M Ethernet.

PSRC, skepticism persisted. The skepticism was focused on the fact that only a single Ethernet collision was being generated during the tests.

3.3.2 Skepticism and Ethernet Scalability Test Results

The industry skepticism expressed of the initial benchmark resulted in the industry asking EPRI to perform a larger-scale test for Ethernet that intentionally caused more collisions. In response to this request, EPRI took the following actions:

- Contracted Karlheinz Schwartz, George Schimmel, and the Fraunhofer Institute to product a set of simulated Ethernet results for a 100-node substation. These results were already detailed in Section 3.2.2.

- Contracted SISCO to modify the test setup so that more Ethernet collisions could be created and to execute a set of tests and to produce a report on the results.
- Formed a technical review and advisor team consisting of several different companies.

Although names and affiliations have changed, the project members and affiliation are shown in Table 3.6. This team produced results that altered an industry.

The following is a spoiler alert and will be covered in more detail later. During most experimentation, sometimes generated results don't match with expected results. In some cases, it is just the fact that the expected results were incorrect. In the case of the scalability test, a problem with the operation of the Ethernet cards being used was detected. It took a team of dedicated individuals (see Table 3.7) from 3com to help determine and correct an anomaly in the 3com Ethernet cards. Without the assistance from 3com, the test results would not have been as widely accepted as they eventually were.

3.3.2.1 Test Setup

The test setup (see Figure 3.14) consisted of 22 computers that had the same resources, network cards, and 3com Ethernet Network Interface Cards (NICs).

Table 3.6 Team Members of Profibus and Ethernet Testing Initiative

Individual	*Affiliation in 1996*	*Responsibility*
Herbert Falk	SISCO	Test Setup, execution, and report
Dan Bingham	SISCO	Test Setup, execution, and report
Jack Robinson	KEMA-ECC/AEP	Reviewer
Karlheinz Schwarz	SCC	Simulation and mathematical analysis
George Schimmel	Tamarack	Simulation and mathematical analysis
Al Colcer	CISCO	Technical project support
Glenn Harmon	Basler Electric	Reviewer
David Wood	SEL	Reviewer
Jim Schnegg	AEP	Reviewer
James Whatley	Ontario Hydro	Reviewer
Bill Blair	EPRI	Funder and visionary
John Lytle	3com	Technical project support

Table 3.7 3com Anomaly Resolution Team

Individual	*Position at 3com*
Robert (Bob) Metcalfe*	Founder and owner of 3com
Jim Sanchez	Provided technical support for anomaly resolution
Sergio Arzate	Technical customer support contact for anomaly resolution

*If you don't recognize this name, Bob Metcalfe is commonly referred to as the creator of Ethernet. His contributions to the computer and communication industry resulted in him being awarded the National Medal of Technology and Innovation. He is also in the Internet Hall of Fame.

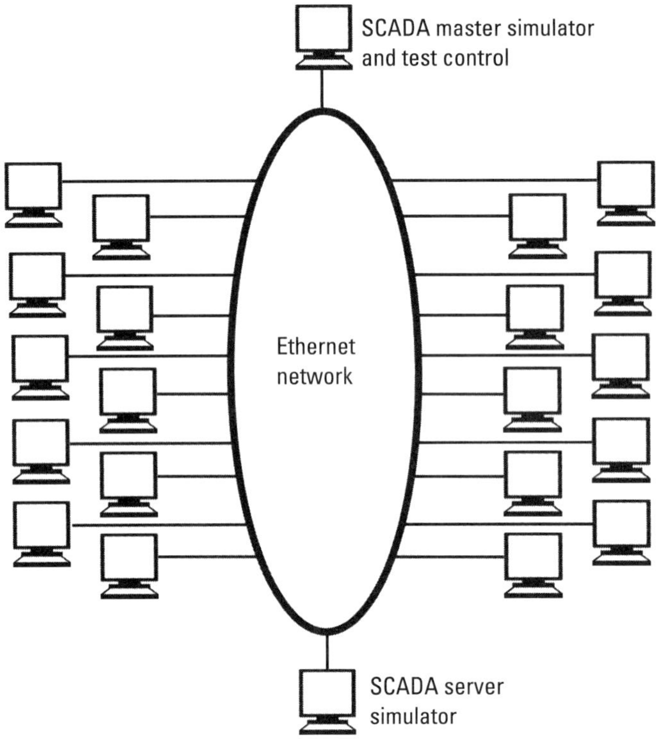

Figure 3.14 High-level test setup for Ethernet scalability test.

The computers were all Pentium class machines with the same desktop resources as used in the Profibus/Ethernet benchmark (see Figure 3.14). Computer C is the test control and responsible for generating the background SCADA and oscillography information retrieval from computer S. Computers 1 to 20 are configured to respond to the printer port trigger and to collect transactional results. The transactional results are then collected by computer C. The Ethernet infrastructure connecting all the computers was varied and included:

- *Generic Ethernet Hub (a very inexpensive hub) for 10BaseT.* The generic hubs were 8-port hubs that used an uplink ports to radially connect to three other hubs in order to support Ethernet connectivity for the 22 computers.
- *24-port 3com FMS Linkbuilder 10BaseT hub.*
- *3com Superstack-II 100BaseTX hub.* To achieve support for the 22 computers, a stack of two hubs was required. Stacked technology provides a high-speed backplane used to exchange Ethernet packets between the hubs instead of actual Ethernet CSMA/CD technology. The stack is also managed as a single entity.
- *10BaseT Cisco Catalyst 5000 Ethernet Switch.*

Each test campaign generated a base load of SCADA and oscillography data. For each test case, the number of computers generating requests was increased (e.g., 1–20). To generate statistically significant results, no less than 1,000 iterations were executed. For each iteration the base load was maintained as were the number of intentionally created simultaneous packet generations (e.g., computers 1–20 transmitting). Test synchronization was observed via an HP 1663c-32 channel logic analyzer. Network traffic validation and observation was provided via a Network General Ethernet Analyzer.

The SCADA traffic was generated using the same Ethernet connection-oreiented communication profile that was used previously. However, to be able to accurately test the UCA multicast technology, a connectionless profile was introduced and is shown in Table 3.8.

The MMS Protocol Data Unit (PDU) used over the connectionless profile was the InformationReport. The reception of the publication was to a configured peer node that also received the synchronization signal. The time recorder was the delta time between the reception of the synchronization signal and reception of the expected InformationReport.

3.3.3 Wondering What Happened

There are times when engineers are faced with a conundrum when the theoretical expectations are not shown by test results. When the benchmark tests were executed, the results showed some things that were unexpected. The unexpected results can be best demonstrated by looking at the resulting graphs of testing the performance of the system with no SCADA load.

To accurately reflect the performance of the Ethernet network, multicast processing within the actual computers needed to be removed from the experimental data. The results shown in Figure 3.15 are the adjusted mean value with a 3-sigma value added into the value. The 3-sigma addition was done to determine the probability of exceeding the 4-msec requirement. The graphs show that the results of the simulation were confirmed with the following:

- The best fit exponential curves (labeled as Expon) reflect the expected curve from the mathematical simulations. However, there is a slight offset between the projected and the actual results.

Table 3.8 Definition of First UCA Connectionless Communication Stack

ISO Layer	*Connectionless Profile*	*Connection-Oriented Profile*
7. Application	ISO 9506 Manufacturing Message Format	ISO 9506 Manufacturing Message Format
	Connectionless-ACSE (ITU X.237)	ISO/IEC 8649 and ISO/IEC 8650: Association Control Service Element
6. Presentation	Connectionless Presentation (ISO/IEC 9576-1)	Fast-Byte Presentation
5. Session	Connectionless Session (ISO/IEC 9548-1)	Fast-Byte Session
4. Transport	Connectionless Transport ISO/IEC 8072 and ISO/IEC 8073	ISO/IEC 8072 and ISO/IEC 8073
3. Network	ISO/IEC 8348 and ISO/IEC 8473	ISO/IEC 8348 and ISO/IEC 8473
2. Data Link	IEEE 802.2 LLC 1	IEEE 802.2 LLC 1
	IEEE 802.3 CSMA/CD	IEEE 802.3 CSMA/CD
1. Physical	IEEE 802.3	IEEE 802.3

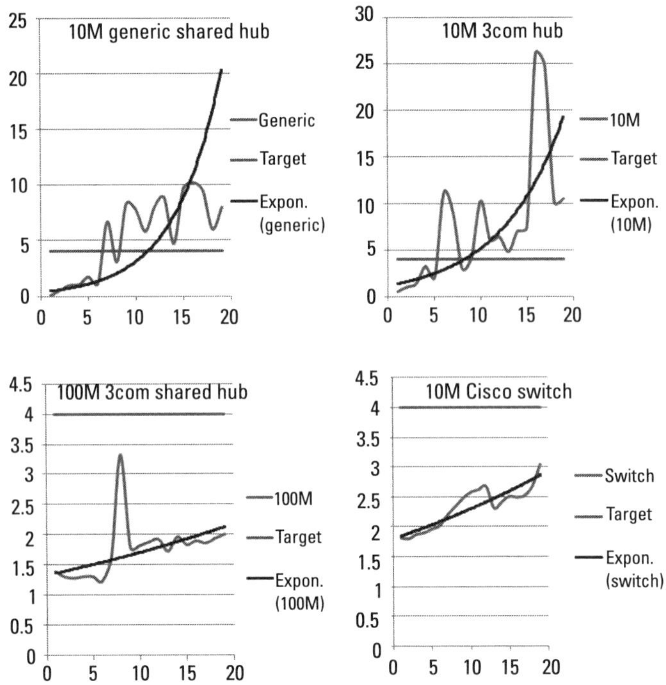

Figure 3.15 Adjusted Ethernet multicast performance with no SCADA load.

- The simulation results were confirmed for the 10M Shared Media/Hub technology in that the results showed that no more than nine collisions could be tolerated prior to exceeding the use case requirement.
- The simulation predicted, and the experiment confirmed, that switched Ethernet technology would be able to achieve the use case requirements.
- In all the results, there was an unexpected anomaly observed.
- The anomaly is best observed in the 100M Shared results but is actually present in all of the results except for the switched Ethernet results.

Figure 3.16 shows that at eight collisions, the performance was approximately 3.4 msec. However, performance improved with nine collisions. The theory of Ethernet and the simulations did not support this adjustment of performance at nine collisions. Additionally, the excursion was well beyond what the simulations predicted.

In many scenarios, the results would have been published without investigation into this anomaly since the results still showed that 100M Shared Ethernet could meet the requirements. However, good engineering practice dictated an investigation and the quest for understanding lead to an unexpected path and an encounter.

A detailed analysis of the results showed that regardless of the manufacturer of the shared hub technology, the anomaly was still present in the results. There were only three possible explanations once the hubs were ruled out:

1. *The computer software had a bug.* After thorough review of the code review, the test software was ruled out.

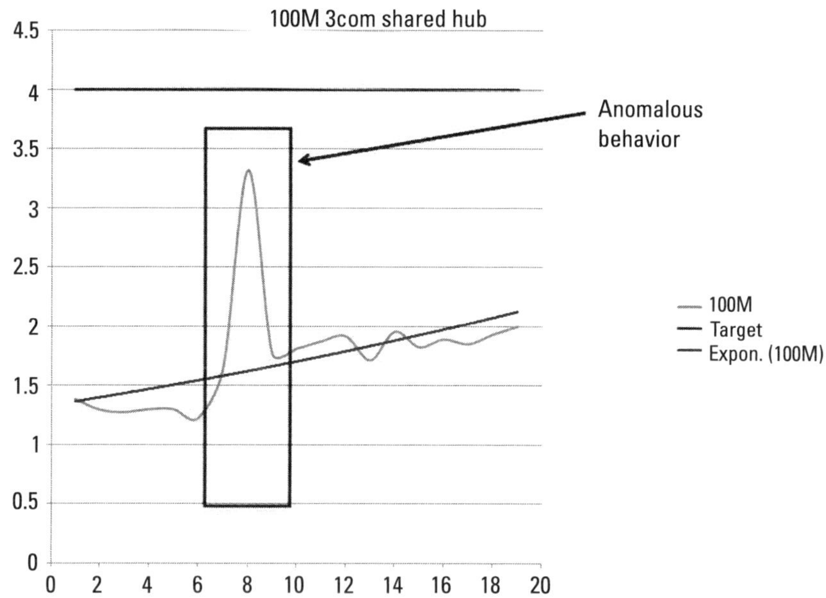

Figure 3.16 Observed anomalous behavior.

2. *The computers themselves could have been defective.* This was ruled out since all of the 20 computers were exhibiting the same problem. In retrospect, the computers should not have been discounted so quickly since they were all purchased from the same manufacturer at the same time and could have had the same manufacturing defect.
3. *The Ethernet cards could be defective.* It was this conclusion that was reached and investigated.

The problem was reported to 3com support and after a week of discussions the team was told that the anomalies were not being reported by any other users of the card. The team was skeptical that any other user, except those doing detail experimentation into the performance of Ethernet networks, would ever observe this problem. The team also concluded that the anomaly would not have impacted normal application usage of the cards and therefore the probability that it would have been reported was very small.

Driving a problem to ground sometimes leads to encounters that one never anticipated. Given that 3com support became a dead end, another path toward truth was needed. At the time, Bob Metcalfe was the president of 3com. He was also one of the foundational inventors for Ethernet. After discovering his email address through posts on other internet feeds (way before to Facebook or LinkedIn), the Ethernet results were packaged up and sent to him asking him if the 3com card results showed that they actually implemented Ethernet. An email response was never received from Mr. Metcalfe; instead 21 new 3com Ethernet cards appeared 1 to 2 weeks later. During retesting, the anomaly was greatly diminished.

Given the scope of the skepticism of the utility industry in regard to the performance of Ethernet, an understanding of what changed was needed. Once again, 3com support was contacted and the team was informed that the engineering change was not disclosable. At this juncture, the threat of reengaging Mr. Metcalfe was raised, and the engineering change was disclosed. It turns out that the testing had detected a buffer management algorithmic optimization that was not implemented 100% correctly.

With the explanation of the anomaly, simulation results and experimentation results confirming the simulations the industry conceded that Ethernet could achieve the requirements. As they say, the rest is history!

CHAPTER 4

Harmonizing IEC 61850 and IEEE TR 1550

The basis of the current IEC 61850 resulted from two organizations, IEEE and IEC, agreeing to try to reach consensus on a single international standard. The agreement put on hold the publication of UCA 2.0 as an IEEE standard and provided time for EPRI, U.S., and IEEE members to join the IEC 61850 activity within IEC to determine if there could be conceptual and technology transfer from what is now IEEE TR 1550 into IEC 61850.

Note that there was much angst on both sides of the technical discussion. During a meeting in Edinburgh, the question was asked about what would happen if consensus could not be reached. The response was politically incorrect and reflected the potentially ugly potential of two competing global standards in which neither would probably be globally accepted. This discussion galvanized the combined groups to work toward a single solution.

UCA 2.0 had two major documents whose core concepts were eventually accepted as tenets of the current IEC 61850. The documents were Common Application Service Models (CASM) and GOMSFE. Basic concepts in these documents are summarized as follows:

- CASM provided a set of abstract service definitions and base objects with which these services interact. In object-oriented terms, the objects would be classes with methods that can be invoked.
- CASM also provided a mapping of the abstract objects and services to concrete protocols. One of the concrete protocols was MMS (ISO/IEC 9506), which was inherited from the GM MAP initiative.
- CASM also provided multicast over Ethernet mappings (e.g., GOOSE).
- GOMSFE provided a set of object definitions based on the base objects that had semantic meaning for applications in the power domain. As an example, an object that represents the functionality and information related to a circuit breaker. Some of these definitions were large and complex leading to the GOMSFE objects being nicknamed "bricks."

The UCA/IEEE CASM model in Figure 4.1 represents an abstraction and extension to the services and objects that were defined by MMS (ISO/IEC 9506). The major elements were

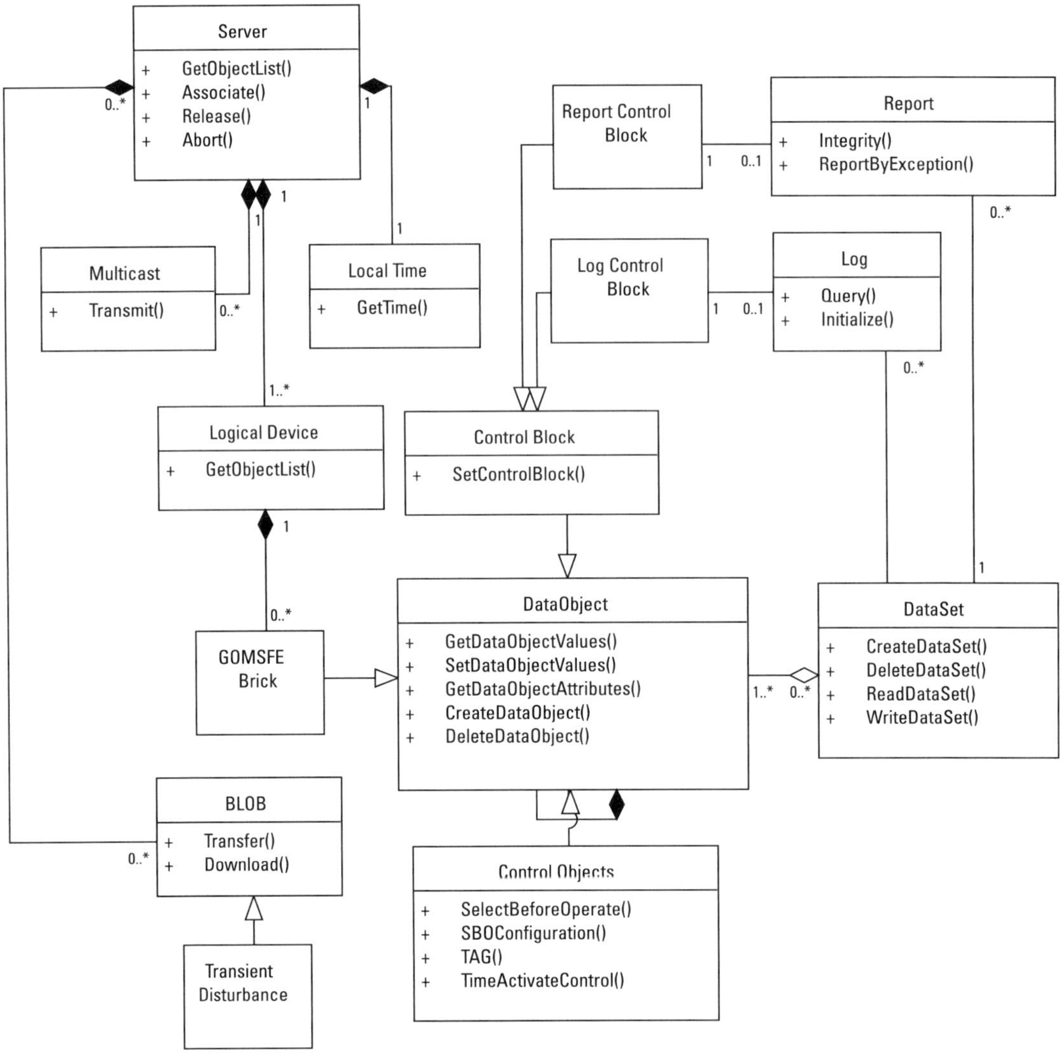

Figure 4.1 CASM abstract model.

- *Server:* This is an abstract object that represents a communicating node that provides information. The node can support connections and the ability to send information via multicast (e.g., GOOSE) messages. The ability to manage the connection to the server is represented by the following methods:
  - *Associate:* This service is used by a client to establish a connection to the server. In CASM terminology, this client/server connection would be referred to as an association.
  - *Release*: This service is used to gracefully terminate an established association.
  - *Abort:* This service is used to abruptly terminate an established association.
  - *GetObjectList:* This service is used to retrieve the names and types of other objects that exist within the server.

- *Local time:* A server must expose its local time, which may be obtained through the GetTime service.
- *Binary large object (BLOB):* A CASM BLOB can be used to transfer a file of Transient Disturbance information or to download new firmware.
- *Logical device:* This abstract object represents a collection of functions affectionately known as GOMSFE bricks.
- *GOMSFE bricks:* These large DataObject(s) consist of concretely defined functional semantics through the composition of DataObjects.
- *DataObject(s):* Besides the defined semantics there are specialized data objects to control the reporting and logging models through report and log control blocks. the services that manipulate the object are
  - *GetDataObjectValues:* This service allows the values of a data object to be queried by a client.
  - *SetDataObjectValues:* This service allows the values of a data object to be written (set) by a client.
  - *GetDataObjectAttributes:* This service allows the definition of a data object to be queried.
  - *CreateDataObject:* This service allows the remote creation of a data object.
  - *DeleteDataObject:* This service allows the remote deletion of a created data object.
- *DataSet:* This object represents a collection of data objects that would be used to be reported or logged. The services that manipulate the object are
  - *CreateDataSet:* Allows the remote creation of a dataset.
  - *DeleteDataSet:* Allows the remote deletion of a created dataset.
  - *ReadDataSet:* Allows the values of the data objects contained in a dataset to be retrieved with a single query.
  - *WriteDataSet:* Allows the values of the data objects contained in a dataset to be written with a single query.
- *Report:* This is an abstract representation of the ability to send the information contained in a dataset in an unsolicited fashion. The control block values determine if a particular report is enabled. The unsolicited information can be sent based on a time-based integrity (i.e., all the values are reported) or ReportByException (i.e., only send the values that have changed).
- *Log:* This is an abstract representation of what would typically be used for a sequence of event (SOE) recording. These recordings can be queried (i.e., read) or initialized (i.e., contents of the log emptied).
- *Multicast:* A server may have the capability to transmit multicast (e.g., GOOSE) information. The CASM multicast was not based on data objects; rather, the service sent a packet set of double bit statuses.

The precursor of IEC 61850 within IEC was focused on system engineering practices, distributed automation, and semantics leveraging IEC 60870-5 technology (e.g., indexes). When working toward a harmonized standard, this IEC focus caused several changes to the concepts in CASM and GOMSFE:

- The analysis of the GOMSFE bricks were determined to not be able to provide reusable or distributable functions. Therefore, the concept of a brick was transformed into smaller functional units that became known as logical nodes.
- In CASM, clients and servers communicated. In order to model and engineer distributed automation systems, the abstract concept that logical nodes exchange information was adopted.
- The need for discoverable semantics was agreed on. However, the multicast payload of UCA/IEEE did not have traceable semantics. Therefore, another multicast service was created based on the dataset construct. Since this new service was object-oriented, the CASM service was renamed from GOOSE to GSSE. (Note that this rename caused confusion in the industry and several

Table 4.1 Comparison of UCA CASM versus IEC 61850 Objects

Concept	*From IEC*	*From IEEE/ UCA*	*IEC 61850 Result*
Environmental	☑		IEC 61850-3
Project management	☑		IEC 61850-4
Communication interchange	☑		IEC 61850-5
Configuration language	☑		IEC 61850-6
Abstract services			IEC 61850-7-2
• Server		☑	• Server
• Logical device		☑	• Logical device
• Brick		☑	• Logical node
• Dataset	☑	☑	• Dataset
• Reporting	☑	☑	• Reporting
• Log		☑	• Log
• Multicast		☑	• GSE
▪ GOOSE		☑	▪ GSSE ▪ GOOSE
• Control	☑	☑	• Control
• Setting groups	☑		
• Time	☑	☑	Time
• BLOB		☑	Files
• Control blocks		☑	Report, Log, GSE, SettingGroup
GOMSFE Bricks	☑	☑	IEC 61850-7-3 and IEC 61850-7-4
Protocol mappings			IEC 61850-8-1
• Mapping to MMS		☑	• Mapping to MMS
• Ethernet for field bus		☑	• Ethernet Multicast
▪ GOOSE		☑	▪ GSSE ▪ GOOSE
Optical current transformer and potential transformer	☑		IEC 61850-9-2
Conformance testing			IEC 61850-10

integrationissues for those that did not understand that UCA/IEEE GOOSE was not the IEC GOOSE. Although technically correct, it was probably a mistake to reuse the name.)

- Time: The IEC concentration on integration, potentially across multiple time zones, forced the universal adoption of coordinated universal time (also known as temps universel coordonné or UTC).

The traceability of the IEC and UCA/IEEE harmonization can be seen in the Table 4.1.

A good standard is one in which everybody is dissatisfied. IEC 61850 represents a good standard and the harmonization and other work to produce the standard created other technical changes that are not represented in the table. The IEC 61850 model and services will be covered in detail in Chapter 8.

CHAPTER 5

Structure of the IEC 61850 Standard

Standards development is like the effort involved in developing a law from an original idea. It requires commitment and effort by many people, the ability to resolve technical and political disputes and to herd the cats toward a common objective. Many standards are local or national standards and the effort is immense to create a standard of this standing. The effort needed for the development of international standards pales in comparison to national standards.

International standard development inherits all the corporate and regional development issues as well as national interests of the participating countries. If this set of complications was not enough, there are typically overlapping standards organizations or groups within a standards group whose buy-in is needed to achieve the status of an international standard. The primary standards organizations involved in the IEC 61850 effort are shown in Table 5.1.

Of these four organizations, there are three different balloting methodologies required to reach the level of being a standard. ISO and IEC both vote based on a single vote being cast by member countries. IEEE balloting is on an individual basis. IETF has a different methodology from the others which involves standard developing and approval by either the Internet Architecture Board (IAB) or Internet Engineering Steering Group (IESG) governing boards. These differences in scope and rules create issues when it comes to coordination and synchronization of maintenance of the IEC 61850 standards.

IEC TC57 is truly a global organization the addresses key issues relating to the power industry in a global economy. Member countries include countries of every type of political system and therefore politics is not something that is discussed at these meetings. Political and theological friends and foes pull together to create and maintain IEC 61850.

At the time of this writing

- IEC TC 57 has 35 member countries.
- Working Group (WG) 10 has 25of the 35 member countries participating. There are 253 individual members that represent the 25 countries. Each country has a single vote.
- Working Group 15 has 21 participating countries represented by 110 individuals.
- Working Group 17 has 20 participating countries represented by 98 individuals.

Table 5.1 Primary Organizations Involved with IEC 61850

Organization	*Organization Name*	*Subgroups*	*Standard or Responsibility*
IEEE	Institute of Electrical and Electronics Engineers (www.iec.ch)	802	Ethernet and Ethernet Redundancy
		1588	Precision Time Sync Protocol
		PSRC*	TR 1550 (old EPRI UCA 2.0)
		PSRC	Power profile for time sync
		PSRC	Synchrophasor measurement methods
		PSRC	Security requirements for IEDs
ISO	International Standards Organization (www.iso.ch)	TC 184	Manufacturing Message Specification (MMS) Security framework
IEC	International Electrotechnical Commission (www.iec.ch)	TC38	Instrument transformers and switchgear
		TC57 WG10	IEC 61850 Core standards
		TC57 WG15	Cybersecurity
		TC 57 WG17	IEC 61850 Distributed Energy Resource (DER) functions and architecture
		TC 57 WG18	IEC 61850 hydroelectric functions
		TC 57 WG19	Coordination of standards within TC57
		TC 88	IEC 61850 wind power functions
IETF	Internet Engineering Task Force		Core communication technology including TCP/IP, directory services, core security technology

*PSRC (Power System Relaying and Control Committee) has had some of its groups moved into the IEEE PSCCC (Power Systems Communication Committee).

- Working Group 18 has 18 countries represented by 46 individuals.
- There are approximately 134 IEC 61850 Technical Reports, Standards, or Technical Specifications available for purchase from IEC. In many cases, the documents will be available in both French and English. If you are interested in ordering one or more of the documents, please help by ordering the electronic versions and not the paper version.

Instead of listing and explaining all 134 documents, the following eye chart format might help explain the philosophy and relationships behind many of the documents.

Figure 5.1 shows several different categories that the documents can be classified within:

- There is the expected introductory material category where the definitions and overview of the structure of the standard can be found.
- IEC 61850 is requirement driven and therefore some stand-alone documents can be classified as requirement-only documents. The documents specify environmental (e.g., temperature, humidity, surge withstand) project engineering processes for IEC 61850 systems and information exchange requirements. It is the information exchange requirements that drive the performance and time synchronization requirements that will be discussed later. The chart also

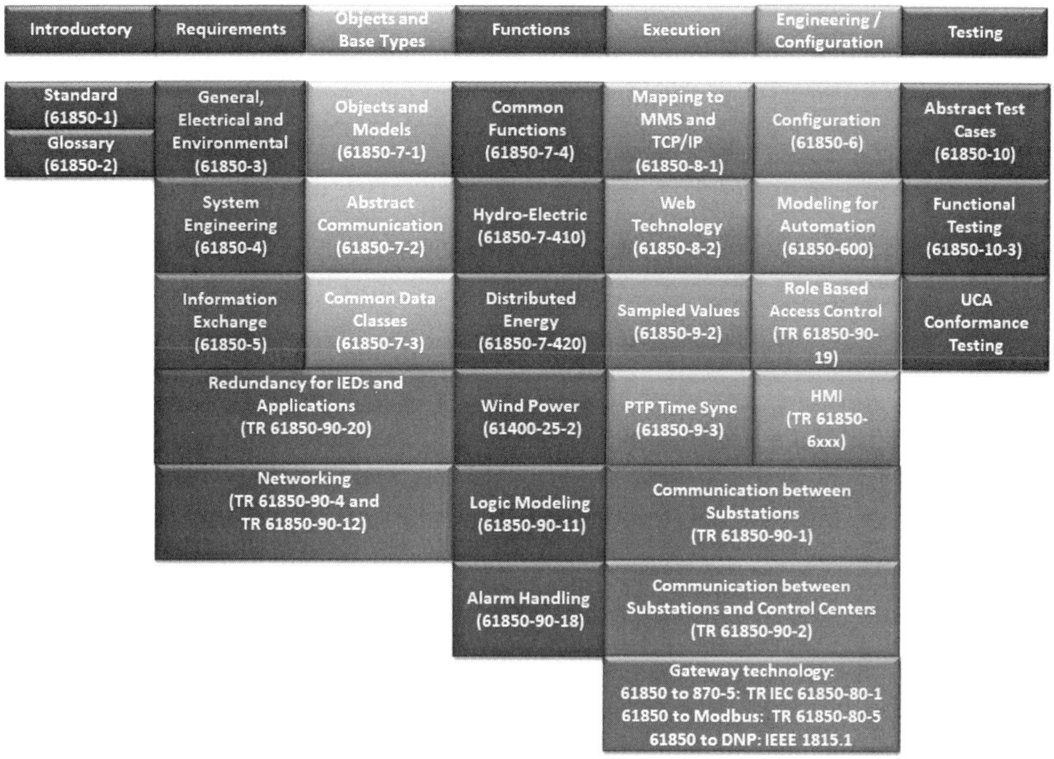

Figure 5.1 Overview of IEC 61850 standards.

shows that there are some TRs that include requirements as well as including information that could be classified in other categories. The specific documents in the chart detail how redundancy should be handled by IEDs and applications as well as network design guidelines. There will be more a little later about how IEC 61850 utilizes Technical Reports.

- The requirements drive a base set of object and type definitions.
- The requirements also drive the standardization of functionality that is needed to provide information exchanges to fulfill specific applications. Using IEC 61850 terminology these types of functions are typically Logical Node definitions.
- The abstract concepts of the previous categories must eventually be made concrete and implemented that eventually allow actual information exchange within IEC 61850 domains. The documents in this category detail how to implement the requirements, concretely. As an example, 61850-5 specifies the requirement for time synchronization, but does not specify the actual protocols to be used that meet the precision. The documents in the execution domain specify which protocols and options to be implemented to meet the requirements using NTP and PTP. There are other documents that provide details regarding implementation of information exchanges via TCP/IP, Ethernet, and web technologies. Tossed into this category are documents that specify architectures and technologies used to facilitate intersubstation information exchange as well as to control centers. Additionally, there are

specifications that detail the use cases, configuration, and implementation of gateways (e.g., mapping of information) from IEC 61850 to DNP, IEC 60870-5, and Modbus.

- The category of Engineering and Configuration has documents that specify concrete methodologies and technologies used to configure IEC 61850 systems, automation, role-based access control (e.g., security), and HMIs (e.g., displays).
- Any engineering solution must be tested. The testing category contains documents that provide abstract test cases for conformance testing, functional testing, and documents from UCA that codify conformance testing.

Except for the RBAC document, there does not appear to be many standards related to cybersecurity. Figure 5.2 correlates the relationships of the IEC 62351 security suite of standards and the suite of IEC 61850 standards.

IEC has several different categories of publications, each with different processes and time requirements for progression. The major documents are

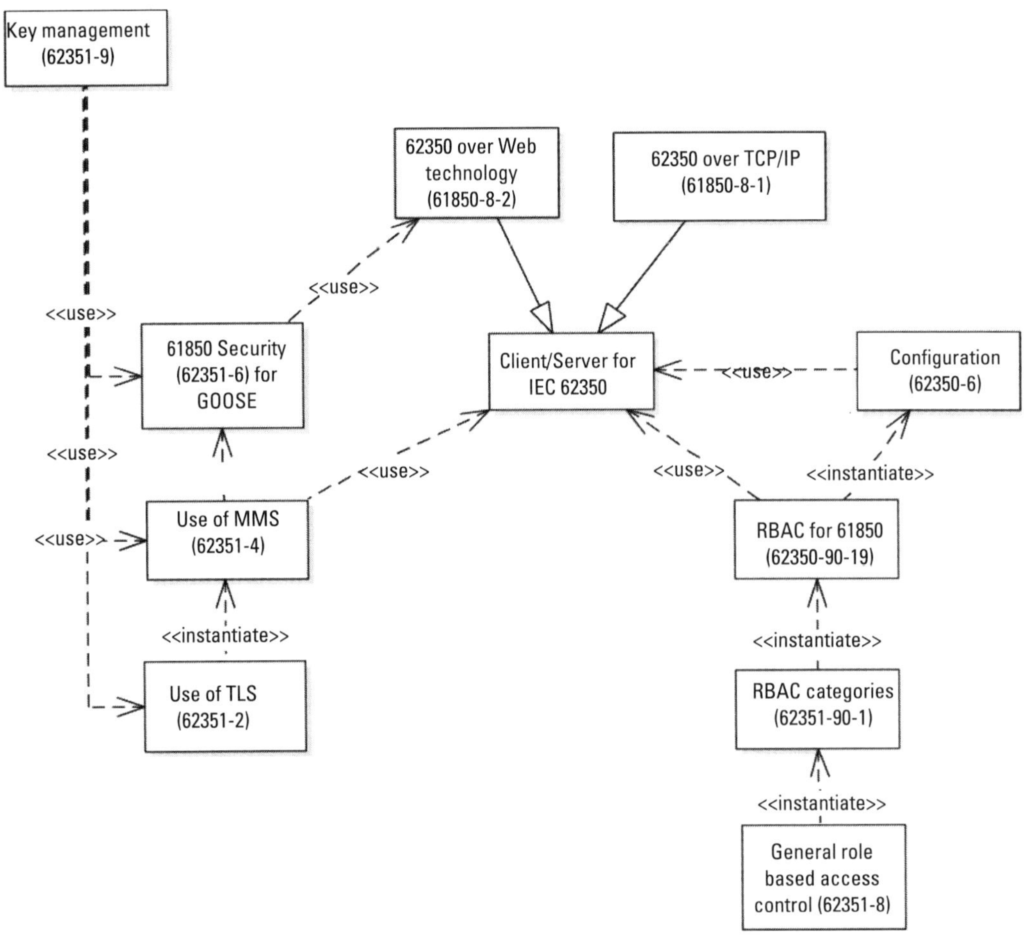

Figure 5.2 Overview of IEC 61850 related IEC security standards.

- IS is a normative document that was developed by consensus procedures and approved by the IEC National Committee members. These documents are intended to provide information that give implementation details that are, in many instances, the basis of conformance testing. These provide directives with the words "shall" and "must." They may also contain nontestable directives that utilize the words "may" and "should" (i.e., the choice is up to the implementer).
- TS: "Technical Specifications approach International Standards in terms of detail and completeness but have not yet passed through all approval stages either because consensus has not been reached or because standardization is seen to be premature." (From www.iec.ch.)
- TR: These documents are intended to be different from an IS and TS in that they are supposed to contain data, survey data, use cases, and other material that is not implementation-related. As such, IEC defines TRs as not being normative. However, the IEC 61850 working groups utilize TRs to gather technical changes that impact multiple IEC 61850 documents in a single document for approval prior to changing the impacted ISs. As such, the IEC 61850 community views these documents as potentially normative while IEC does not.

Table 5.2 summarizes the IEC balloting process as is defined in ISO/IEC DIR 1:2016. There are other processes within IEC, such as Publicly Available Standards (PAS), but to date the IEC 61850 working groups have not developed these types

Table 5.2 IEC Balloting Process

Document Type	*IS*	*TS*	*TR*
Proposal stage	New Work Item Proposal (NWIP)	New Work Item Proposal	Preliminary Work Item (PWI)
Preparatory stage	Working Draft (WD)	Working Draft (WD)	Working Draft (WD)
Committee stage	Committee Draft for Comments (CD)	Committee Draft for Comments (CD)	Not applicable according to IEC directives
Ballot length	2-4 months	2–4 months	
Enquiry Stage	Committee Draft for Vote (CDV)		
Ballot length	2–4 months		
Approval	2/3 NCs		
Approval stage	Final Draft International Standard (FDIS)	Draft Technical Specification (DTS)	Draft Technical Report (DTR)
Ballot length	2 months	2 months	2 months
Approval	2/3 approval or less than 25% disapproval	2/3 approval or less than 25% disapproval	Simple majority
Publication	IS published 1.5 months after FDIS approval and translation	TS Published 1.5 months after FDIS approval and translation	TR Published 1.5 months after FDIS approval and translation
Best case elapsed time	2 years	1 year	1 year
Revalidation or revision	5 years	5 years	At least every 5 years

of documents. The process of ISO and IEC are the same as is indicated by the ISO/IEC joint publication.

The IEC/ISO revalidation or revision process requires that each published final document is reevaluated to determine if the document should continue to be published as is, deprecated (e.g., removed as a standard/document), or requires a new edition. The use of TRs by IEC 61850 typically removes the TR from publication once the technical content of the document has been transferred into actual standards documents and this may occur whenever revision to the target standards are performed.

The IEEE is another critical standards organization for the foundational technologies on which some IEC 61850 standards are based such as Ethernet, security, and Precision Time Protocol. Additionally, IEEE has whitepaper/usage documentation creation capability that gives guidance to users and vendors in several areas regarding IEC 61850. However, the voting process is based on personal individual votes (not country-based) and the document creation process is different than IEC's.

The IETF)is another standards organization that develops de facto standards that are used by standards created by IEC, IEEE, and ISO. Some examples of these de facto standards are TCP, IP, and SSL/TLS. These standards are documented as Request for Comment (RFC) documents and may be referenced by other organizations that have more compliance and conformance requirements. The actual base process is documented in RFC 2026. This RFC has been amended several times and this amendment process represents a major difference between IETF and the other standards organizations. As an example, when an IEC standard is revised or amended, the same document number is used but a new date is indicated (e.g., IEC 61850-7-2:2010 versus IEC 61850-7-2:2003). IETF creates an entirely different document for the change (e.g., the base TCP RFC, RFC 793 is updated by RFC 1122, RFC 3168, RFC 6093, and RFC 6528). Due to the IETF change process, other organization must carefully analyze the changes and decide which in the chain of documents to use as their basis.

Originally, the IETF process was more rapid than those of ISO and IEC. However, this quickness created a potential wild west of draft RFCs and RFCs. Currently, the IETF process has slowed and there has been the introduction of Technical Standards and Applicability Statements. Primarily, IEC 61850 utilizes RFCs.

Table 5.3 shows the general IEEE process.

The IETF has several types of publications: RFC, TS, Applicability Statement (AS); and Best Current Practice (BCP). The processes for publishing these documents can be found in RFC 2026. IEC 61850 refers to IETF RFCs (TCP, IP, etc.).

The IEC 61850 standardization process attempts to maintain backward compatibility with previous versions of the standard. To accomplish this, each new version of other referenced standards must be analyzed. If the analysis reveals compatibility issues, those issues must be evaluated to determine if the revision should be adopted. This evaluation process is neither trivial nor 100% perfect.

In the specific case of Edition 1 and Edition 2 of IEC 61850, there are major compatibility issues that could not be addressed in a backward-compatible manner. A critical evaluation of the causes of this incompatibility, in many cases, was due to the flexibility allowed by the Edition 1 suite of standards. This flexibility causes interoperability and conformance testing issues that needed to be addressed within Edition 2. Edition 2, although still flexible, is much more constrained that Edition 1

Table 5.3 IEEE Standard Process

Document Type	*IEEE Technical Standard and Guide*
Proposal stage	A meeting is required to vote on the submission of a Project Authorization Request (PAR).
Approval ballot	Approval by the IEEE Standards Board (SASB) is required for the PAR. The SASB meeting occurs six times a year.
Committee stage	Internal committee document.
Committee approval	IEEE working group votes to progress the standard for approval. This requires a 75% approval. Upon approval, the document progresses to the main committee for approval.
	Once approved, the document is sent to out for sponsor ballot.
Ballot length	Sponsor ballot length has a maximum length of 180 days.
SASB Approval	The SASB meets every 2 months and may choose to approve or send the document back for revision.
Publication	Standard is typically published 1–2 months after the approval of the SASB.
Best case elapsed time	1–2 years

and deprecated some of the capabilities in Edition 1. The resulting incompatibility means that coexistence and migration strategies needed to be defined in Edition 2.

Conformance to a standard (e.g., IEC 61850) is tested through the development of the abstract test cases that need coverage and then develop the concrete test cases with expected results. Conformance testing is primarily based on the mandatory aspects (e.g., where the standard states "shall") of a standard. Conditional test cases are developed should a implementation declare that it supports an optional capability. If the specified behavior or result is a "may" in the standard, then multiple expected results must be planned for. Originally, IEC 61850-10 provided the abstract test cases for IEC 61850. However, there are many more concrete conformance tests developed by the UCA International User Group (IUG).[1] The concrete tests now cover GOOSE, SV, and SCL.

The UCA IUG IEC 61850 TPWG is responsible for the development and maintenance of the concrete test cases. However, in many instances there may be multiple interpretations of a specific part of the standard. When this occurs, a Technical Issue (TISSUE) is submitted to IEC for resolution. The result of the resolution is used to update the next version of the standard as well as providing the clarity required for the test case. The current TISSUE database may be found at http://iec61850.tissue-db.com/default.mspx. The maintenance of the TISSUES is being transitioned to IEC and will have a different URL in the future. Conformance testing tests a system under test (SUT) against a reference implementation. Therefore, such testing should be viewed as part of the vendor of the SUT quality assurance process.

One might think that two conformant SUTs would be interoperable; this is not the case since different IEC 61850 capabilities could be implemented by each SUT. Therefore, it is possible to be conformant and not interoperable. Conversely, it may be possible to be interoperable and not conformant if both SUTs implement in a

1. For more information see: http://www.ucaiug.org/.

similar nonconformant manner. Interoperability testing is performed by testing information exchange between multiple implementation, none of which is a reference implementation. The UCA IUG is not only responsible for conformance test cases, but also for test lab accreditation and SUT certificates. It has also been performing large-scale interoperability tests biannually since 2011.

Most utilities specify that SUTs must be conformance certified. However, they stage their own interoperability tests to ensure that the integrated system performs as desired.

CHAPTER 6

Read Before Proceeding: Use of UML in This Book

A picture is worth a thousand words. To express the relationships and concepts found in IEC 61850, this book utilizes a Unified Modeling Language (UML) and images. IEC 61850 has several different classifications of objects depending on which IEC 61850 standard is being read. The same name can be used by different parts of the standard to express slightly different aspects that can best be explained as abstract, configuration, and instantiation. Many of the IEC 61850 makes use of UML or UML-like diagrams to express some concepts. This book extends those concepts using UML. However, many of the readers of this book may need a brief tutorial on how this book utilizes UML. To understand what is being expressed in other parts of this book, please take the time to read this chapter. For more detailed information regarding UML, see https://people.eecs.ku.edu/~hossein/Teaching/Fa13/810/Readings/UML-diagrams.pdf. If a book is desired, a list of books can be found at: https://modeling-languages.com/list-uml-books/.

6.1 Classes, Attributes, Operations, and Multiplicity

UML classes represent the definition of characteristics of objects. There are two typical types of characteristics that are graphically represented in a UML Class: attributes and operations.

As an example, a building has certain characteristics such as size, number of windows, number of external walls, number of doors, and more. This information could be represented by Figure 6.1, which shows the name of the definition (e.g., class) named Building. The definition contains the characteristics of size, number of doors, number of windows, and number of external walls. Each characteristic, expressed as shown, is a UML attribute. The main parts of an attribute definition are

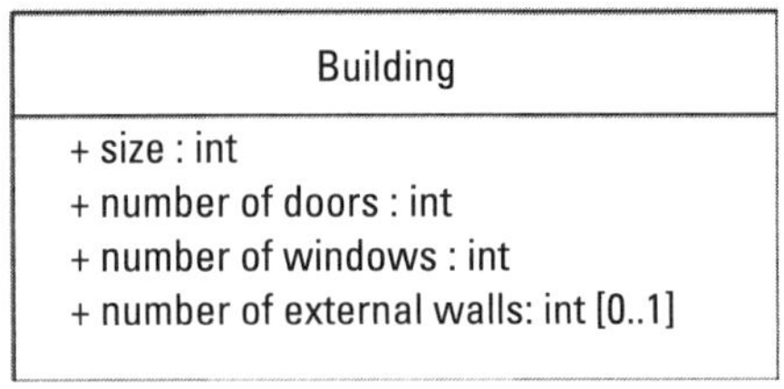

Figure 6.1 Building class with attributes.

- The name of the attribute (e.g., size, number of doors).
- The datatype whose value would be used to define the concrete definition of the attribute when an actual building is built. As an example, an outhouse might have a single door (e.g., a value of 1), no windows (e.g., a value of 0), four external walls, and a size of 10 square meters.
- A specification of how many values of a specific attribute must be specified in an actual building. This is called UML multiplicity. Attribute multiplicity is typically restricted to indicate if the attribute is mandatory for a definition (e.g., multiplicity of 1..1) or optional (e.g., 0..1). The 1..1 multiplicity is not displayed as that is the default. The optional characteristic of number of external walls is shown as [0..1].

Operations are methods that are used to interact with a particular class, or in this case, a building.

Figure 6.2 shows the operations with the "( )". These represent ways that can be used to interact with a building. It is possible to build, tear down, or clean a building. The example is a gross simplification and there may be many others. An example: How many external walls does the building in Figure 6.3 have? Does it have a roof? All things that need to be considered when designing a model.

6.2 Generalization

UML specialization graphically represents that a definition is based on another definition and inherits all the characteristics of the parent definition.

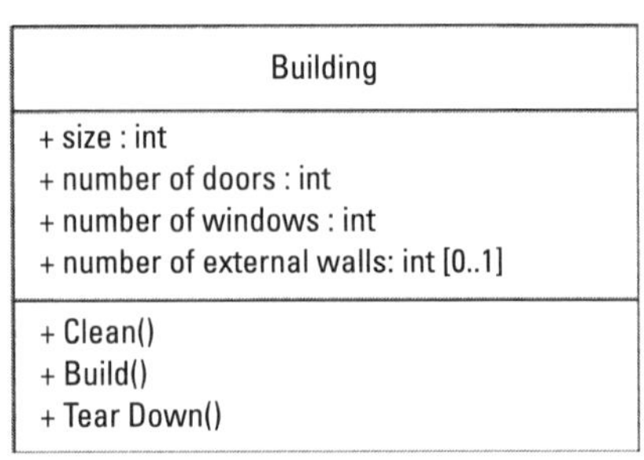

Figure 6.2 Building class with attributes and operations.

Figure 6.3 Geodesic dome. (Image courtesy of Biodomes, http://www.biodomes.eu/.)

Figure 6.4 shows various types of buildings: commercial, residential, and for the purposes of this book, outhouse. Outhouses come in two different varieties: plumbed and pit. If you have ever been camping, it is probable that you have

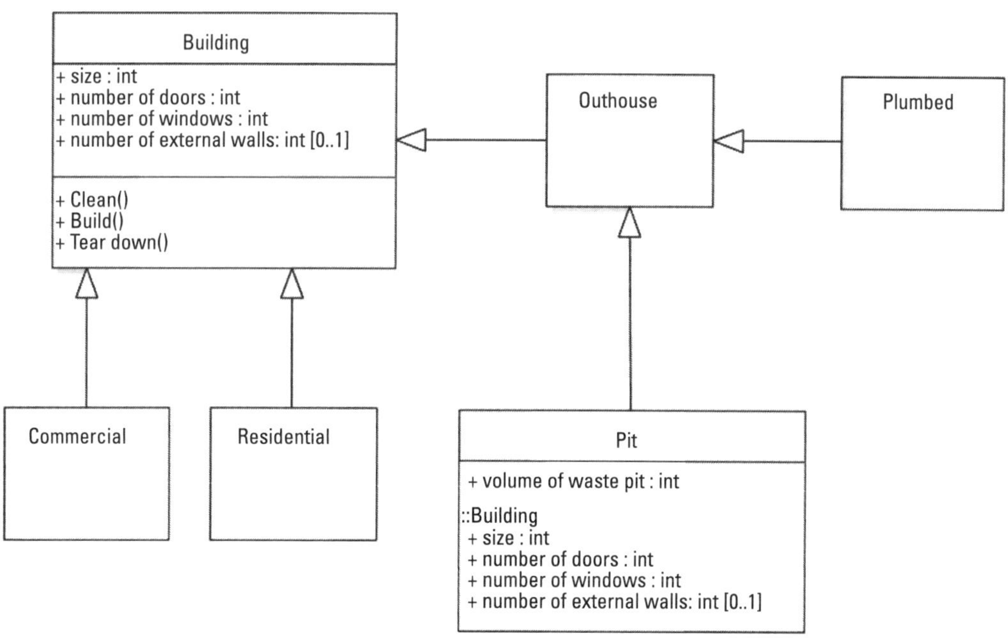

Figure 6.4 Various types of buildings: UML generalization.

experienced both. The relationships shown by the arrows in the diagram are symbols of generalization. Generalization means that the pit type of outhouse inherits all of the attributes and operations from building and outhouse. Since outhouse has no additional attributes or operations, the definition of a pit outhouse are shown to include those inherited from building.

6.3 Association, Composition, and Aggregation

The concepts of association, composition, and aggregation are similar in concept but represent different persistent patterns. When thinking of a building, we imagine that it contains rooms and furniture. If one does a total teardown of a building, all of the rooms, walls, windows, and doors are removed. This is the equivalent of a database cascade delete. The representation of this type of relationship is depicted as UML composition (a filled black diamond).

However, a room may have either built-in furniture or regular furniture. In a building that is sold or torn down (e.g., removed), the regular furniture can be moved out whereas the built-in furniture is also destroyed or sold. The relationship that a room can contain furniture that is not to be sold as part of the room/building is a relationship known as UML aggregation. This is represented by a nonblack diamond.

Both aggregation and composition can be considered as a type of containership (e.g., a building contains rooms). Figure 6.5 also shows that a room must belong to a single building (e.g., the multiplicity of 1 at the black diamond) but a building may have multiple rooms but must have at least one room (e.g., the 1..* at the opposite end of the black diamond). This type of multiplicity declaration also applies to aggregation and associations.

Whereas aggregation and composition can be thought of as a containership relationship, associations are more of a "has" relationship. However, there is no parent/child relationship in an association and therefore no cascade delete is represented using association in the UML diagram.

However, there is a differentiator that allows a modeling choice between aggregation and association: an associated class can be used to represent information

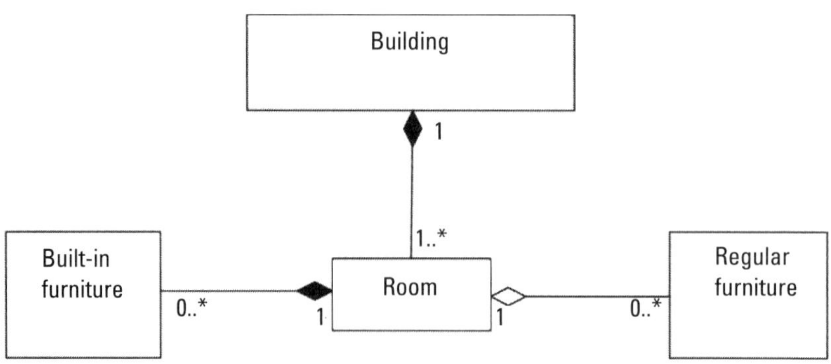

Figure 6.5 Rooms and furniture containment: UML composition and aggregation.

about multiple instances of other classes. Figure 6.6 shows that multiple buildings can share the same district. The sharing aspect of association is not able to be accomplished with aggregation or composition. Associations represented by the line in the diagram have multiplicities similar to aggregation. The diagram expresses that a location can be associated to multiple buildings and that a building can have multiple types of locations. However, a location is not required to be associated to a building; rather, a building must have at least one location.

6.4 Dependency, Instantiation, and Stereotypes

The concepts of dependency and instantiation can be demonstrated through the use an example of creating a building, as is shown in Figure 6.7.

When the Sydney Opera House was being considered, there was a conceptual drawing. The blueprints, or construction instructions, were based on the conceptual drawing. The building we know today was created using the instructions to build the concrete opera house we see today. In terms of UML, these types of relationships could be diagramed as shown in Figure 6.8.

Figure 6.8 shows that a blueprint is dependent on the drawing (e.g., concept). The actual building is a concrete representation of the blueprint (e.g., instantiates the blueprint). The use of dependency and instantiation allow the relationships to be expressed. In the example, the same object name is not reused; however, this is not the case with the IEC 61850 standards. Additionally, there are models for each relationship that are pseudoindependent of each other as is demonstrated in Chapter 7.

Figure 6.9 shows that there are definitions of the same or similar object classes in both IEC 61850-6 and IEC 61850-7-2. IEC 61850 7-2 has the abstract

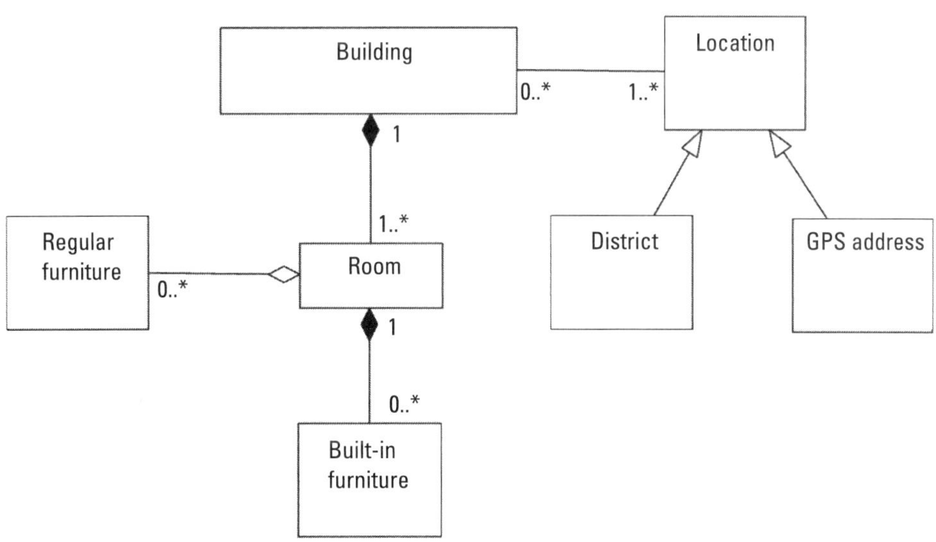

Figure 6.6 Multiple buildings can share the same location: UML association.

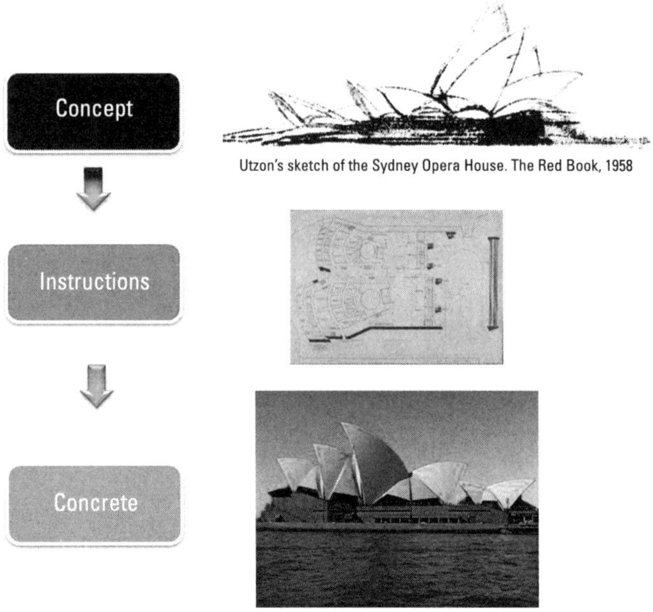

Figure 6.7 Dependency and instantiation example. (Images used under license from Shutterstock.com.)

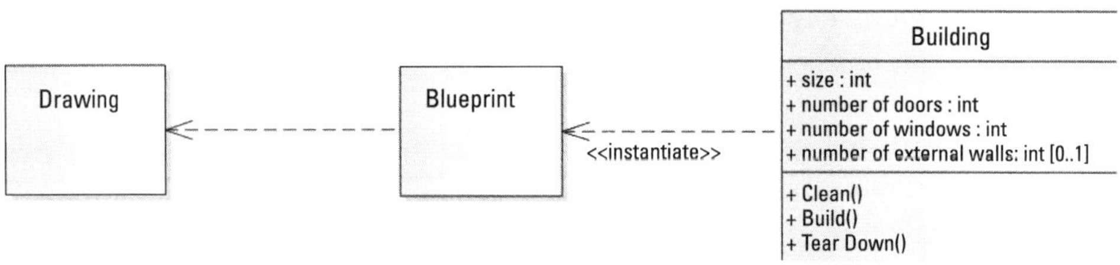

Figure 6.8 UML representation of dependency and instantiation.

definitions of these objects whereas IEC 61850-6 is used for configuration. IEC 61850-6 objects are therefore dependent on the abstract definitions in other parts of the IEC 61850 standard.

An analysis of the service access point definition shows that IEC 61850-7-2 is general, and IEC 61850-6 provides a mechanism to specify the instantiation of a particular service access point as an access point. The access point clarifies that if the access point represents an IED, clock, or router. However, it is not possible to communicate to an actual device using IEC 61850-6 as this part is used to configure IEC 61850-8-1-based devices. The server construct of IEC 61850-8-1 is another standard, ISO/IEC 9506, which is known as MMS. It is the implementation of an MMS-based device that instantiates the actual objects and allows information to be exchanged.

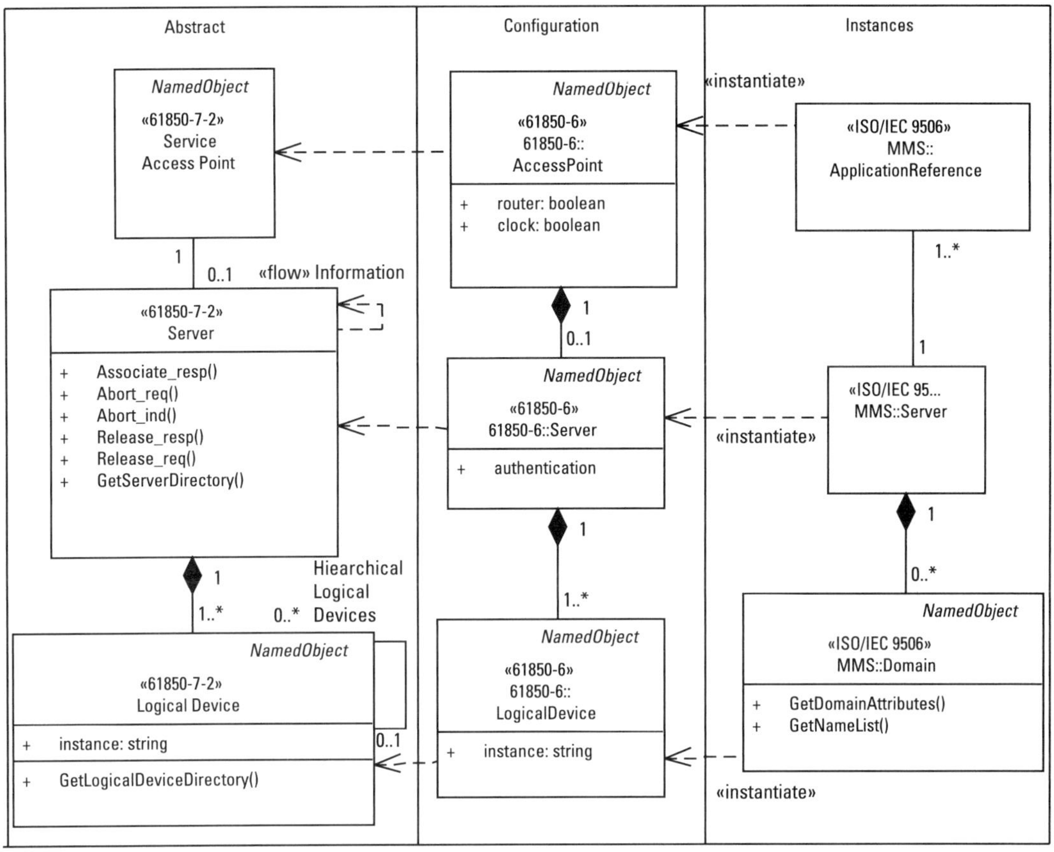

Figure 6.9 Relationship example of IEC 61850 abstract, configuration, and instances.

6.5 Stereotype

The IEC 61850 standards typically concentrate on server functionality, and client functionality is left to implementations that must interact with servers in a conformant manner. Since there is no real abstract definition of client in IEC 61850-7-2, this book introduces the object. The UML stereotype (e.g., <<Book>>) is used to represent that this class/object is an artifact introduced by the book. Additionally, the stereotype nomenclature is also used to indicate the primary standard that is defining the particular use.

6.6 UML Cheat Sheet

Table 6.1 shows the graphical representation of the UML artifacts utilized and a definition for those artifacts.

Table 6.1 Quick Cheat List for UML Use within this Book

Graphical Representation	*Usage*
<< >>	This graphical representation is used for a UML stereotype. There are typical UML stereotypes (e.g., enumerations). However, this book extends the use of stereotypes to indicate the primary standard(s) from which the class definition can be found.
◆	The black diamond shape is used to indicate that a class is used to aggregate objects/classes at the opposite end of the diamond. This is a composition of the whole class at the opposite end of the diamond and therefore represents the equivalent of a parent/child relationship. The parent is the class that has the diamond.
	The other aspect of the filled diamond is that if the instance of the parent class is deleted, the children of the composition shall also be deleted.
◇	The nonblack diamond represents aggregation. It is similar to composition except it that if the parent is deleted, the child is not deleted.
———	The solid line represents a UML association. As an example, the ISO/IEC 9506 server may have 1 or more (e.g., 1..* ApplicationReferences). However, a specific ApplicationReference can only represent one server. The numeric (e.g., 1..*) are known as UML role multiplicities.
	The actual role name can be provided in the diagram but is typically determined by the class relationships themselves. As an example, the role name for the (1..*) role is Server. ApplicationReference. This naming convention applies to associations, composition, and aggregation.
Numeric	The numeric(s) represents the multiplicities (i.e., how many instances of a specific class) can be used for a given relationship or role.
Role name	The actual role name can be provided in the diagram but is typically determined by the class relationships themselves. As an example, the role name for the (1..*) role is Server. ApplicationReference. This naming convention applies to associations, composition, and aggregation.
◄------ <<instantiate>>	The dashed arrow represents that the class at the opposite end of the arrowhead marked with "instantiates" defines the information from the class at the arrowhead end in a more concrete fashion.
	As an example, the ISO/9506 domain instantiates the configuration provided by the IEC 61850 logical device.
◄------	The dashed arrow represents that the class at the opposite end of the arrowhead marked with "depends" on the definition of the class at the arrowhead end.
	As an example, the definition of an IEC 61850-6 logical device is derived, or dependent on, the definition of the IEC 61850-7-2 logical device.
△ (arrow with triangle head)	This symbol represents a generalization. As an example, a GOOSE publication is a type of publication and inherits the attributes and operations of the class at the triangle end of the generalization.
	Generalizations can also appear as names in the upper right-hand corner of the class. As an example, the IED class is a generalization of the NamedObject class.
class	A class is a metadata definition (i.e., type) of an object. Classes are typically defined in terms of attributes and operations.
class operation	UML operations are typically used to represent methods that have input parameters and return information. Within the construct of this book, they are used for this purpose and also to represent abstract and concrete services that are used to control or exchange information.
	As an example, the client has a operation through which it can request an association (e.g., Associate_req()). "( )" indicates that it is an operation.
class attributes	Attributes are the equivalent of information of a class. They have a type and a multiplicity. The attribute value can also have a default value.
	As an example, the attribute cbName is a type of ServiceSettingKind and is optional since the multiplicity is [0..1]. The book uses the UML initial value to represent the default value if the value is not given. In this example, the default value for cbName is "Fix."

CHAPTER 7

Integration Patterns

Users of IEC 61850 devices and application utilize concrete instance of function, objects, behavior, and protocols. The protocols utilized to exchange function and object information are based on two different integration patterns:

1. *Client/Server*: The interaction of clients and servers are based on what IEC 61850 defines as a two-party association. This exchange pattern allows a single client to interact with and receive information from a single server. This pattern typically allows a client to interact with IEC 61850 objects through abstract services. The behavior of the interaction of the services with objects defines the various abstract models in IEC 61850.
2. *Publish and Subscribe*: This integration pattern allows a single server to send information to other entities without knowledge to which entities the information is being delivered.

7.1 Client and Server

Our daily lives are filled with invisible examples of client and server integration and information exchange patterns. The use of client server patterns allows individuals to make a personal phone call, query a database, search the internet, or browses a web page. As an example, consider web browsing.

A web server may provide information to multiple web browsers. As the name implies, a web server is a server. A web browser is a client. The browser issues an HTTP request to the web server. For the request to reach the server, there are several fundamental requirements:

Both the web browser and the web server must be connected to a communication network.

The web browser must have knowledge of the URL of the web browser (e.g., www.google.com). However, www.google.com is shorthand for an address lookup in the same way as a name in your phone's contact list is a lookup for a phone number. Depending on where you ask to connect to Google, an address for Google might be an IP address of 173.194.198.103.

Depending on the use of http:// or https:// different TCP ports are utilized. HTTP requests implicitly go to TCP port 80 whereas https requests utilize port 443.

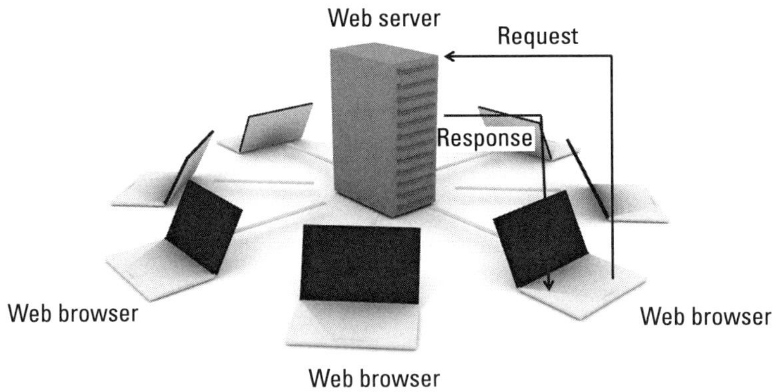

Figure 7.1 Web browsing. (Image adapted from Freeimages.com.)

The combination of the IP Address and TCP port allow the web browser request to be routed to the appropriate application, which is the web server. The web server, even if it can't find the information being requested, will return a response to the web browser unless network communications is interrupted. The web server can provide information to multiple clients and therefore can support multiple client/server exchanges.

The name web server implies that it is a computational platform (i.e., a server). In reality, a web server is a set of software that is hosted on a computational platform that provides access to information through a service based on http or https. Therefore, the web server should really be called web server service. The advent of SOAP and web service messaging provides an even more appropriate name, web service endpoint. A computational resource, either physical or virtual, can provide access to multiple different software services such as email, video streaming, database access, and more. Access to these software services are provided through a specific set of network interfaces that provide connectivity to the communication network(s).

The IEC 61850 client/server integration pattern is abstract and can be represented in UML, as shown in Figure 7.2. Note that the same abstract pattern is equally applicable to the web browsing example as a client issues a request and the server responds to the request.

The binding of a client/server in order to fulfill a request/response transaction is what IEC 61850 defines as a two-party application association (TPAA) and is represented by Figure 7.3.

As with web browsing, an IEC 61850 server can respond to requests from multiple clients and a client can interact with multiple servers. The abstract TPAA

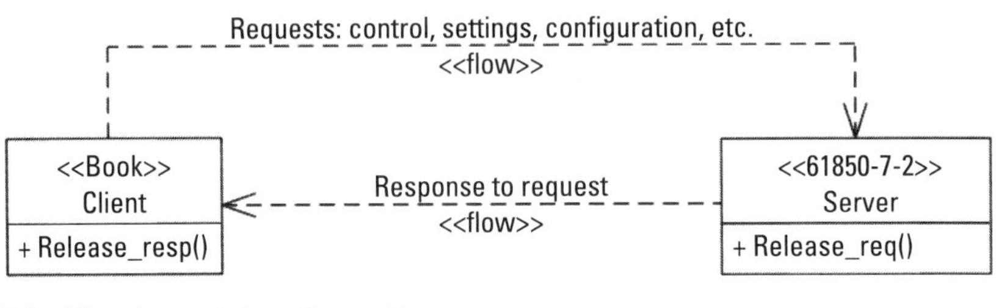

Figure 7.2 Client/server integration pattern.

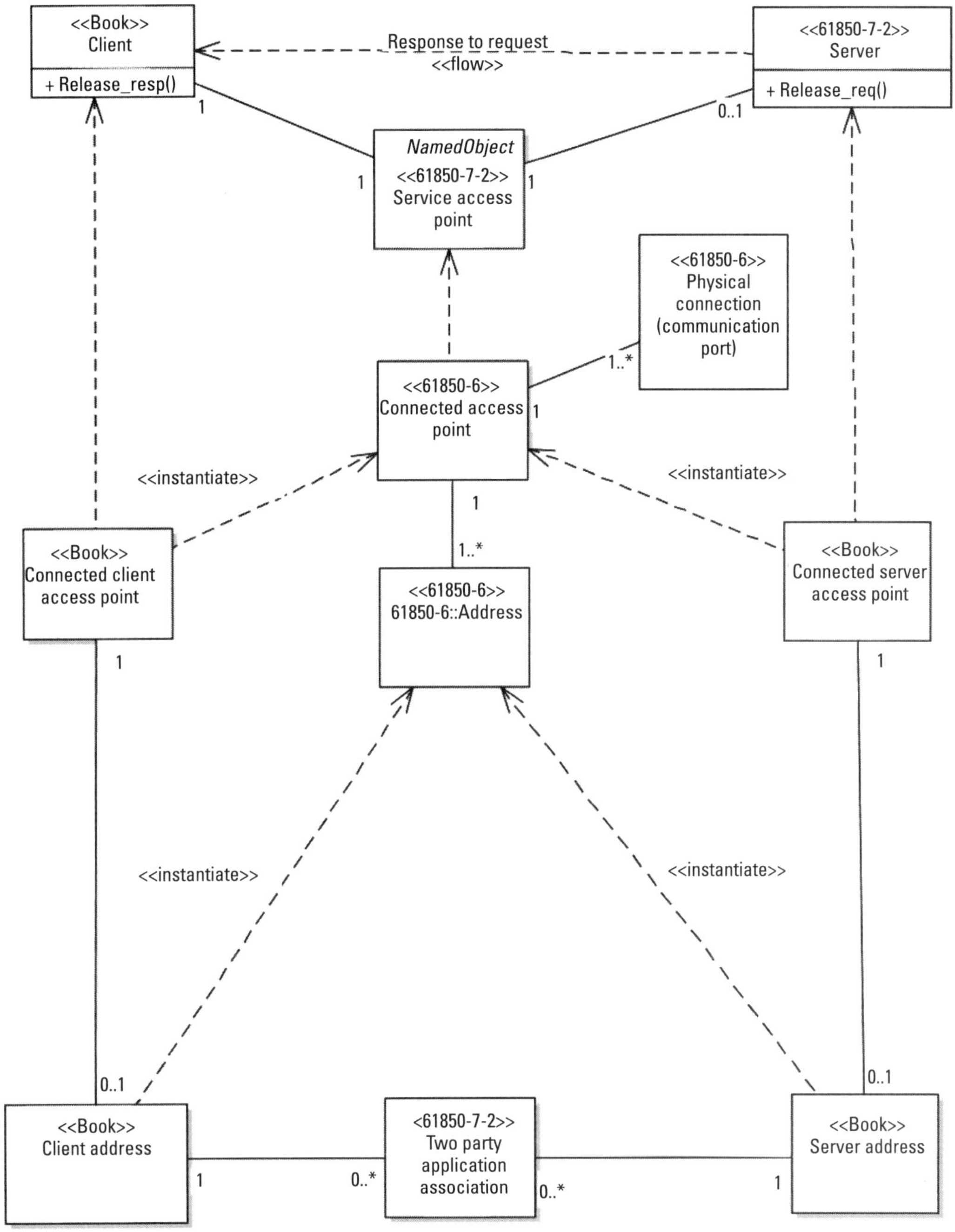

Figure 7.3 Two-party association model.

model defines the binding of a single client and server to allow an exchange of information between those entities.

Client and servers have an access point that represents the communication interfaces that provide access to or for a set of services (e.g., email, http, IEC 61850 client/server, and SNTP). However, IEC 61850 has abstract definitions of a service access point, which is concretely represented as a connected access point (connected

AP), which has the defined relationship to specific network interfaces known as a physical connection. A connected AP, just as with a computation server, can have multiple network interfaces that are bound together to provide communication redundancy.

The diagram shows a connected client access point and a connected server access point. These represent the fact that both a client and server have connections to the communication network. However, as in the example of web information exchange, the client and server must have at least one address. In order to establish a TPAA, a client using a specific address initiates a request to a server bound to a specific address. It is therefore the binding of the specific addresses that forms an association.

An example of the abstract concept can be illustrated by a telephone call. Somebody needs to have a voice conversation with another person. The initiating caller's (i.e., client) phone has a phone number (i.e., client address) and a connection to the telephone or cellular network (i.e., communication port). The caller dials the phone number (i.e., server address) of the party to which they desire a conversation. Once the connection between the two parties is established, the TPAA exists.

One might question the use of association versus connection. It is true that many TPAAs utilize connection-oriented protocols (i.e., TCP) to achieve application-level information exchanges. IEC 61850 specifies several protocols that provide TPAA, but one does not utilize TCP and is not connection-oriented. The IEC 61850 protocols that utilize the TPAA construct are

- There are two standardized methods or time synchronization specified by IEC 61850: Simple Network Time Protocol (SNTP) and Precision Time Protocol (PTP).
  - SNTP utilizes the connectionless UDP protocol. Although SNTP can use publish and subscribe, it is constrained within IEC 61850 to be unicast and therefore a two-party association.
  - IEC 61850-9-3 provides precision time synchronization through the use and extension of the IEEE 1588 PTP.
- IEC 61850-8-2 utilizes the Extensible Messaging and Presence Protocol (XMPP), which is a broker-based technology. XMPP clients can be either an IEC 61850 client or server. The XMPP clients connect to the XMPP server, but the IEC 61850 client and server can still establish a two-party association at the application layer in order to exchange information.
- IEC 61850-8-1 utilizes direct connections between the IEC 61850 client and server. However, there are multiple layers of connection establishment and not just at the Transport layer. Although there are connections at multiple communication layers, the two-party association represents the combination of all of these connections as they are required for IEC 61850 client/server information exchange.

Therefore, a two-party association (see Figure 7.4) is not the same as a connection because it represents the ability to exchange information between applications.

The two-party association supports two different exchange patterns as defined by IEC 61850. These are request/response and unsolicited information delivery.

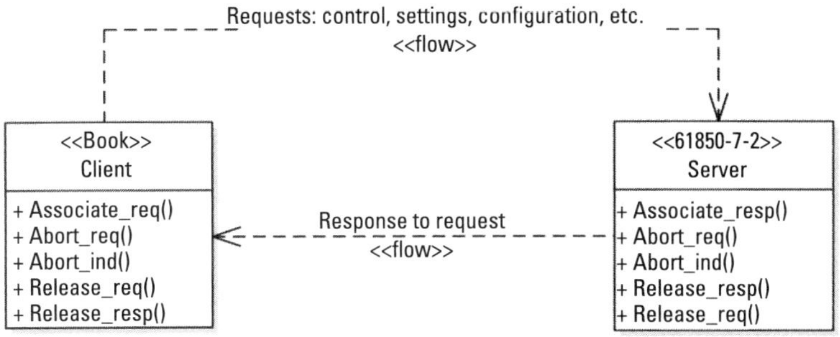

Figure 7.4 Two-party association exchange patterns.

From a high-level terminology perspective, a request is a call for certain information in the server and the response is the server providing that information. However, in some instances it is confusing to understand if the Request is being sent to the network or received by the server.

To minimize this confusion, or maybe introduce some additional confusion, ISO introduced the concept of a four-legged exchange pattern for request/response. The definitions of the primitives are found in Table 7.1.

The two-legged unsolicited primitives are similar in function and definition to the four-legged primitives except the request is issued by the server of information and the indication is received by the client. The real difference is that there is no response generated by the client. IEC 61850 has two behaviors that utilize the unsolicited exchange pattern: reporting and controls.

To establish a two-party association, IEC 61850 specifies that the server supports the abstract services of associate, abort, and release shown as UML operations

Table 7.1 Four-Legged Primitive Exchange Definition

Primitive	*Typical Abbreviation*	*Definition*
Request	_req	The client application packet that is calling on the server to provide specific information into the communication network.
Indication	_ind	The delivery of the client request to the server. To date, all the communication profiles within IEC 61850 deliver the same application protocol packet that was sent to the network by the client to the server. The indication is a demand for the server to perform processing or work.
Response	_resp	Once the server performs the processing and collects the information for the client, it packages the information into a response packet and sends it to the network.
Confirmation	_conf	The delivery of the server response to the client. To date, all the communication profiles within IEC 61850 deliver the same application protocol packet that was sent to the network by the server to the client.

Note: To date all the communication profiles within IEC 61850 deliver the same application protocol packet that was sent to the network to the receiving entity. However, with the emergence of Enterprise Service Bus (ESB) technology, this may not be the case in the future since ESBs typically have message transformation capability and therefore the protocol packet may not be the same between a request/indication or response/configuration pair.

in Figure 7.5. The IEC 61850-7-2 standard does not detail clients, but the clients issue the requests and the server responds.

The state transitions for a connection-oriented client/server TPAA is shown in Figure 7.6.

The state machine shows the transitions of the two-party association. An associate_req creates a pending association. A server may either accept the association request with a positive response (i.e., Associate_resp+) or deny the establishment of the association through a negative response (i.e., Associate_resp-). A client can request that the association be terminated through the issuance of a Release_req. The server may either accept the request to terminate the association (i.e., Release_resp+) or deny the request (i.e., Release_resp-). If the Release_req is accepted, the two-party association is terminated and is nonexistent. An Abort_req is similar to the Release_req but it may not be refused as is really a termination of the two-party association with extreme prejudice. Additionally, an Abort_req or Release_req may be issued by either the client or server.

The communication profiles that implement these state machines within IEC 61850 are IEC 61850-8-1, IEC 61850-8-2, and to a lesser degree SNTP. Both IEC 61850-8-1 and IEC 61850-8-2 utilize ISO 9506 as the application profile although the protocol is binary encoded in one standard and XML encoded in the other. IEC 61850-8-1 client/server utilizes over TCP/IP and over XMPP, which uses TCP/IP. The actual ISO 9506 services that are utilized to provide the two-party association services are

- Associate is provided by the ISO 9506 service of Initiate. The Associate_req is provided by Initiate_Request. Associate_resp+ or Associate_resp- is provided by Initiate_Response and Initiate-Error, respectively.
- Release is provided by the ISO 9506 service of Conclude. The Release_req is provided by Conclude_Request. Release_resp+ or Release_resp- is provided by Conclude_Response and Conclude-Error, respectively.

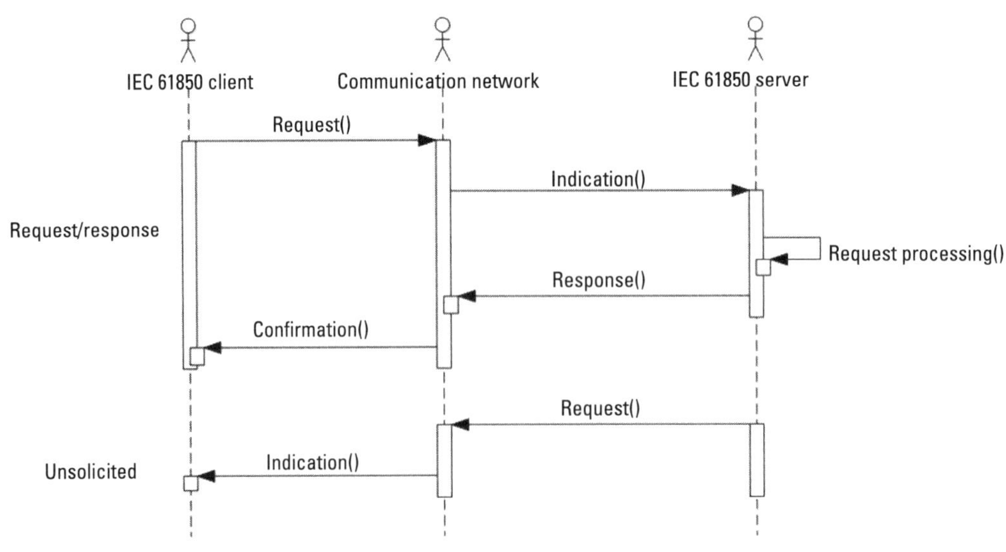

Figure 7.5 Client/server showing two-party association services.

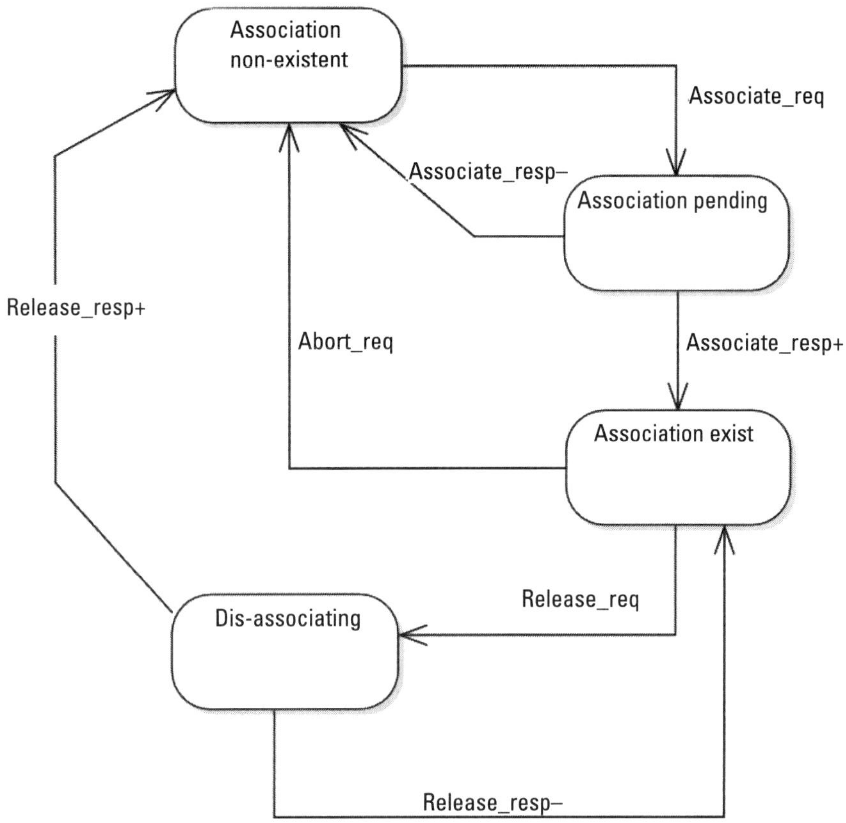

Figure 7.6 Two-party association state machine.

- Abort is provided by the ISO 9506 service of Abort.

Since SNTP is not connection-oriented, the two-party association is transactional in nature. However, the association is provided by the peer poll. The Association_req+ is provided by a Poll_req. The Association_resp+ is provided by the Poll_resp. The Association_resp- is typically provided by a lack of response. Since the two party association is transactional, there is no equivalent to Release and Abort in NTP.

The contents of the addressing for client/server exchanges is configured based on the communication profile that is being used to instantiate the exchange pattern. The configuration of this addressing information is found in the Communication Section of the following SCL and is related to an IED that is defined within the section known as the IED section. The interrelationships of the abstract, configuration, and instantiation are show in Figure 7.7.

The figure depicts a subtle difference between the configuration of an IED and an IED that is configured. IEC 61850 provides configuration through the use of an XML file. To be able to communicate to a physical or virtual computational resource the configuration must to loaded and activated by the resource. A simple example of this is the configuration of an IP address of a computer. Many companies pre-assign IP addresses to computers (e.g., the configuration assignment). Without

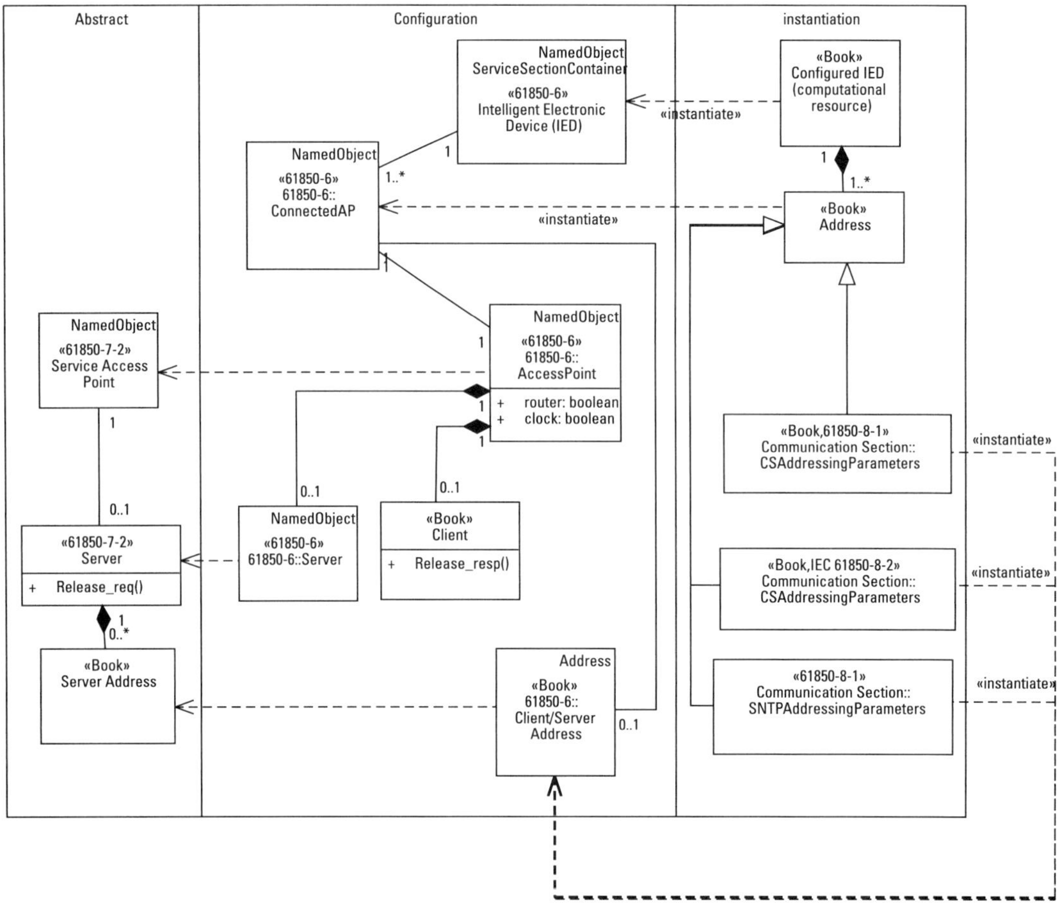

Figure 7.7 Addressing for client/server.

the configuration of the host address of the computer, no network based external communication to the computer can be achieved. It is the act of configuring the resource that allows communication and thus a "Configured IED."

The configuration aspect of the model also provides more clarity in that an AccessPoint can provide connectivity to a Client, Server, or both a Client and Server in the device. An AccessPoint is bound to a single ConnectedAP which contains the addressing information relevant to that specific AccessPoint. An IED can have multiple network connections which can provide multiple AccessPoints.

IEC 61850-6 provides many of the definition aspects required for addressing configuration. Many of these are driving by the addressing actually required to configure actual IEC 61850 devices. Therefore, the specific TPAA related addressing is shown in the instantiation domain and Figure 7.8.

Figure 7.8 shows the current addressing configuration information that can be configured. All TPAA addressing utilizes IP addressing either directly or indirectly. The Jabber ID provides an aliasing for IP addressing like URLs for web servers. An example of the XML serialization of the configuration information is shown in Figure 7.9. The serialization of the XML contents is defined by the XSD found in IEC 61850-6. The configuration example, shown in Figure 7.9, is not a full serialization.

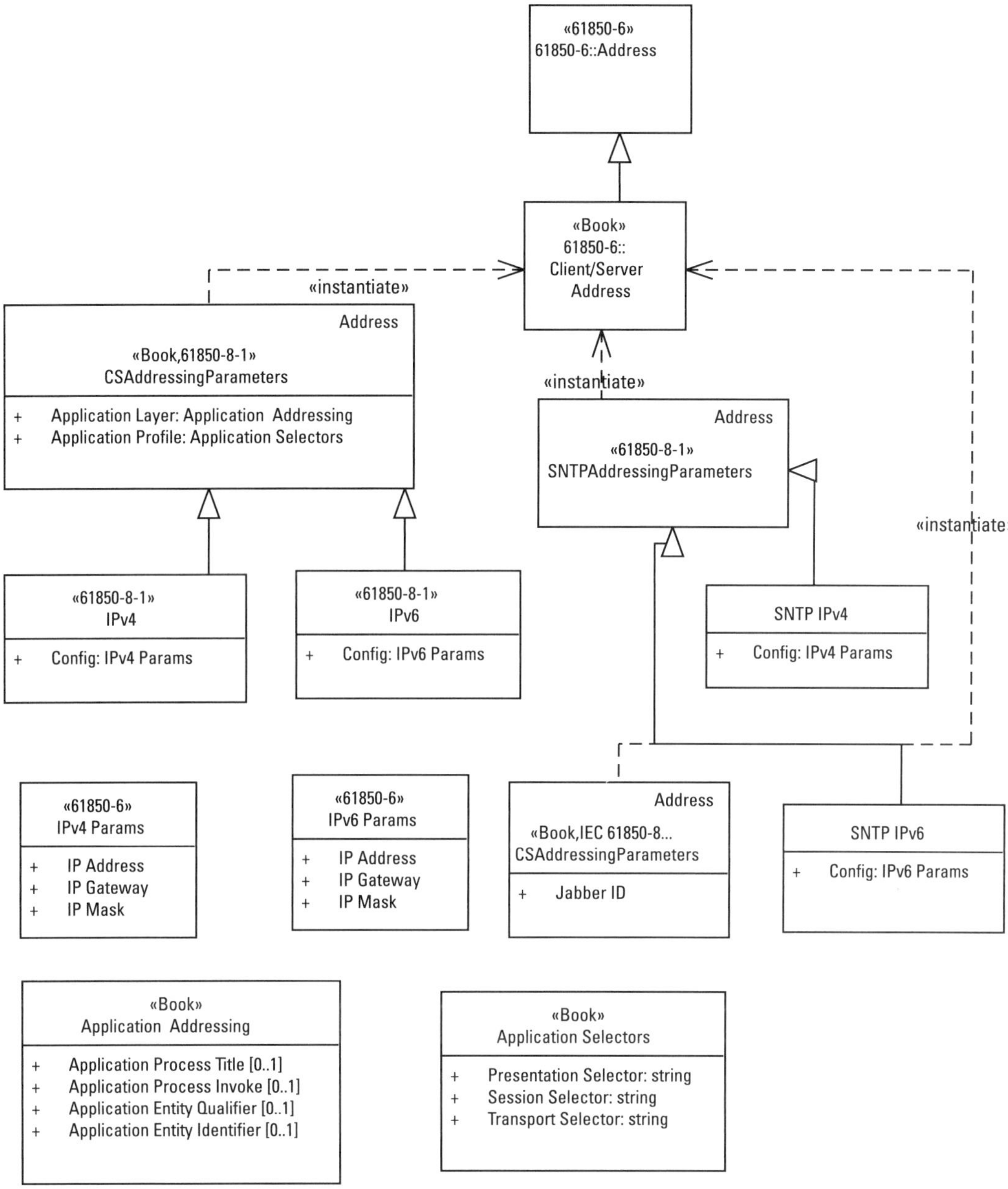

Figure 7.8 TPAA IEC 61850 addressing.

The <Address> production varies based on the configuration of a device/access point that is IEC 61850-8-1 or IEC 61850-8-2. Additionally, both IPv4 (i.e., type="IP") and IPv6 addressing is supported. For IEC 61850-8-2, the only element in the <Address> is type="JID" (Jabber ID). For SNTP, only the IP related elements would be present and the clock functionality would be indicated. A representation of the configuration parameters for client/server addressing follows.

```xml
<?xml version="1.0" encoding="utf-8"?>
<SCL>
    <Communication>
        <SubNetwork name="SubNet1" type="8-MMS">
            <ConnectedAP iedName="SampleIED" apName="IPAccessPoint">
                <Address>
                    <P type="IP">192.168.8.101</P>
                    <P type="IP-GATEWAY">192.168.1.1</P>
                    <P type="OSI-TSEL">0001</P>
                    <P type="OSI-PSEL">00000001</P>
                    <P type="OSI-SSEL">0001</P>
                    <P type="OSI-AP-Title">1,3,9999,23</P>
                    <P type="OSI-AE-Qualifier">23</P>
                </Address>
            <ConnectedAP>
        </SubNetwork>
    </Communication>
    <IED desc="IEC 61850-8-1 IED" name="SampleIED" >
        <AccessPoint name="IPAccessPoint">
            <Server>
            </Server>
        </AccessPoint>
    <IED>
</SCL>
```

Figure 7.9 Example of client/server addressing in SCL configuration.

7.2 Publish and Subscribe

We do not realize how many publish and subscribe or producer and consumer integration patterns have penetrated our lives. Facebook, LinkedIn, smartphone notifications, and Twitter are all examples of publish and subscribe patterns. In all of these, the producer of the information being delivered has little if any information with regard to whom the information is being delivered. This type of integration pattern is defined to be loosely coupled. Since the publisher doesn't know to whom to send the information, a pub-sub integration pattern includes a message delivery service that provides message routing, typically what is known as a topic. Examples of topics are

- *Twitter*: One of several ways to receive information through Twitter is following of a hashtag. The hashtag represents the topic subscription and any information posted to that "#tag" will be delivered.
- *Facebook*: Facebook allows the delivery of friends' information that is posted. It is the act of establishing Facebook friendship that establishes the topic. Information posted by a Facebook friend will be delivered.
- *LinkedIn*: In a similar pattern to Facebook, LinkedIn allows a linkage to be established with other people or organizations. It is the establishment of this linkage that establishes the topic. Information posted will be delivered to the other accounts that have linkages.
- *Smartphones*: Smartphones allow the enabling and disabling of notifications on an application-by-application basis. If the smartphone owner enables the application's ability to deliver a notification, the application itself subscribes to the information source from which the information is to be delivered. The topic of the subscription is determined by the smartphone application.

A producer and consumer message delivery service is typically referred to as a message broker that receives the message/topic combination and based on topic subscriptions forward the message/topic to the subscriber of that topic. There are two different architectures of message brokers: radial and distributed.

A radial architecture is also known as hub and spoke and is shown in Figure 7.10.

Subscribers use a subscription service to build the topic routing table of the message broker. When the publisher sends a message that includes the subscribed for topic, the broker will deliver it to the appropriate subscriber. In the figure, Subscriber 1 and 3 have subscribed for Topic1. The publishers send a message including Topic1 to the broker and the broker then delivers the message to Subscriber 1 and 3, but not Subscriber 2.

A distributed architecture has cooperating brokers that use topics to route from broker-to-broker and then to the subscribing application. A distributed architecture typically has a broker per node, as shown in Figure 7.11.

It is significant that the application-level information delivered to the subscribers is the same regardless of the message delivery architecture implemented. Users don't care about the architecture; they care about the information being delivered.

Topic-based routing is not the only routing mechanism that brokers can provide. There is another mechanism that augments topic-based routing called content-based routing. The difference between topic and content is that content routing typically requires parsing of the message payload. Thus, efficient publish/subscribe information exchanges have the topic external to the application message payload.

There are typically two other characteristics that are controlled by message brokers: quality of service (QOS) and message persistence (MsgP). QOS typically includes the priority of the delivery of the information of the message payload. MsgP determines the period that an undelivered message is stored and attempted to be delivered. In many situations, there are different queues with different MsgP configurations (e.g., message time allowed to live) before it is purged from the broker's queue. The MsgP configuration is typically determined by the application utilization of the messages in the queues. Critical real-time automation systems

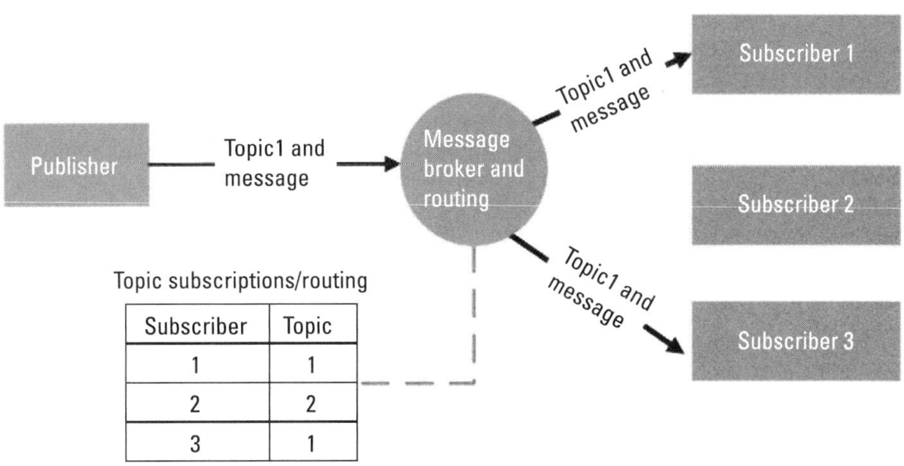

Figure 7.10 Example of topic-based routing and radial publish/subscribe.

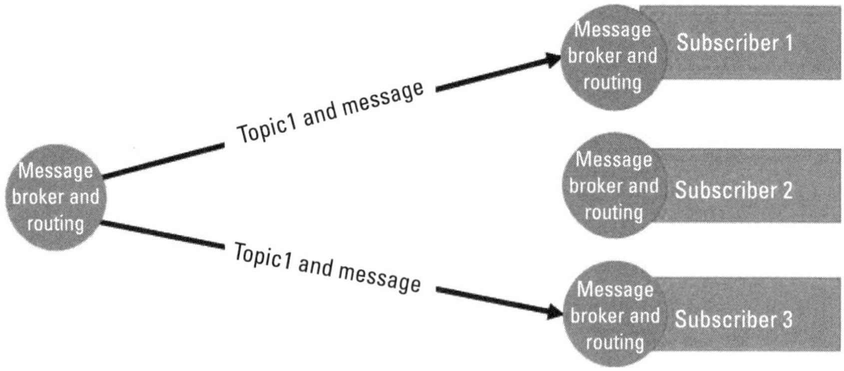

Figure 7.11 Example of topic-based routing and distributed publish/subscribe.

would typically have a very small MsgP (i.e., approaching zero) since delivery of old information could cause misoperation.

The IEC 61850 publish and subscribe integration pattern utilizes all of the aforementioned broker characteristics in that it

- Utilizes a subscription service;
- Utilizes topic-based routing;
- Utilizes content-based routing;
- Provides QOS configuration capability;
- Message queue persistence is determined by the communication profile and IEC 61850 service combination.

The details of the actual implementation of these characteristics are predicated on the specific protocol mappings. Two IEC 61850 information exchanges utilize the publish and subscribe integration pattern: GOOSE and Sampled Values. The UML representing the GOOSE exchange is shown in Figure 7.12.

There are constraints placed on the pattern by IEC 61850:

- A subscription may be posted by either an IEC 61850 client or server;
- Only an IEC 61850 server may publish information as it is the entity that has the information that needs to be provided;
- An IEC 61850 client-only implementation is only allowed to be a subscriber;
- IEC 61850 servers may be both a publisher and subscriber;
- Therefore, the integration pattern can allow information to be exchanged between servers and clients as well as servers to other servers.

The actual IEC 61850-7-2 concept of a multicast association represents the binding of a publication to one or more subscriptions. The creation of this one-to-many binding is done through a subscription mechanism. Although the pattern for GOOSE and Sampled Values subscriptions are the same, they vary slightly. Therefore, Figure 7.12 defines GOOSE-specific publications and subscriptions.

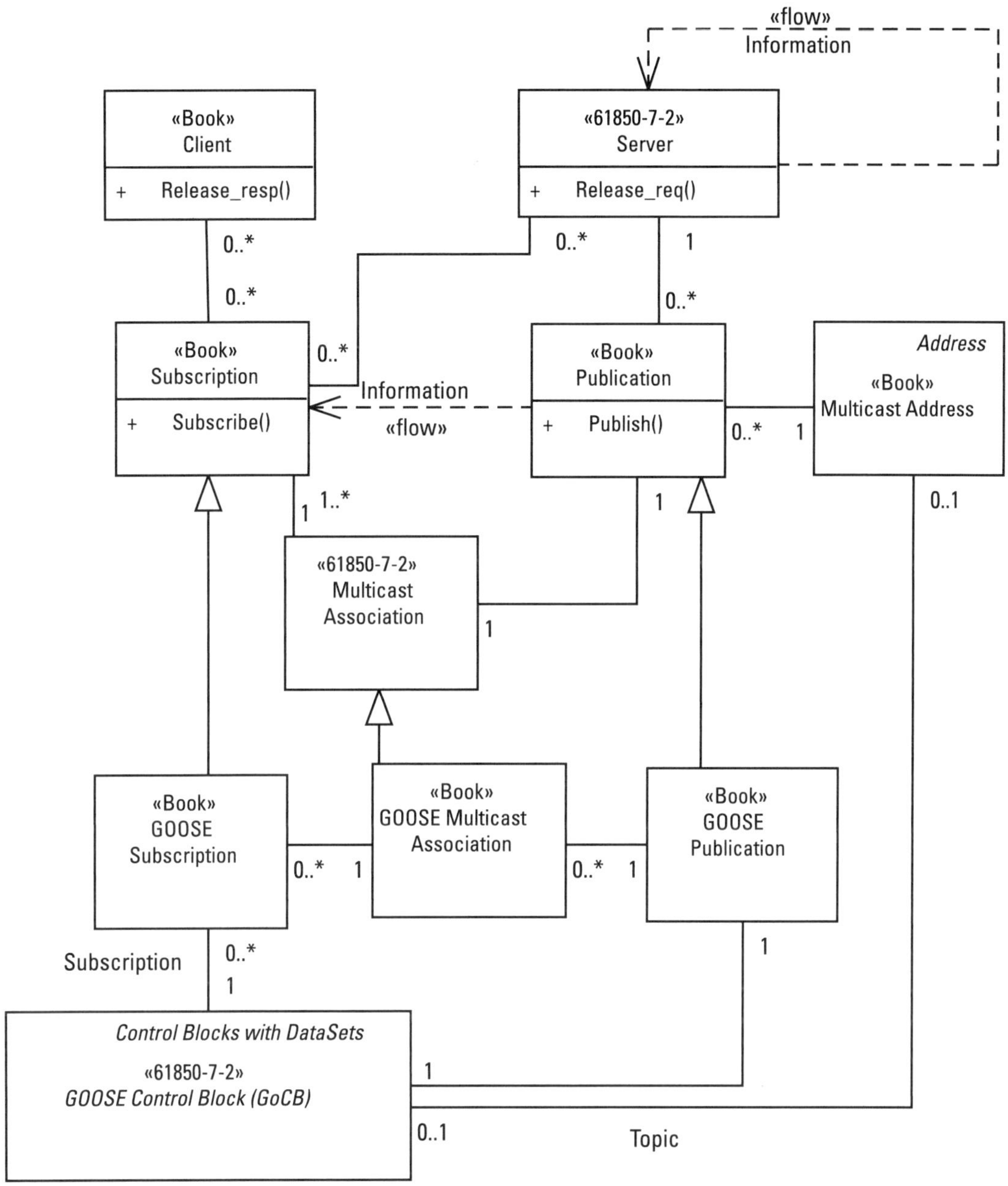

Figure 7.12 GOOSE information exchange.

GOOSE and Sampled Values, as specified by IEC 61850-8-1 and IEC 61850-9-2, can be routable and nonroutable (e.g., Layer 2). The addressing information and some of the contents of the control blocks, in the SCL file, varies based on these distinctions.

The routable variants of the addresses can support IPv4 or IPv6 even though Figure 7.13 does not explicitly show this definition.

The IEC 61850 subscription service is typically provided through the configuration expressed in a System Configuration Description (SCD) file. The actual mechanism involves the use of what is known as a control block in IEC 61850.

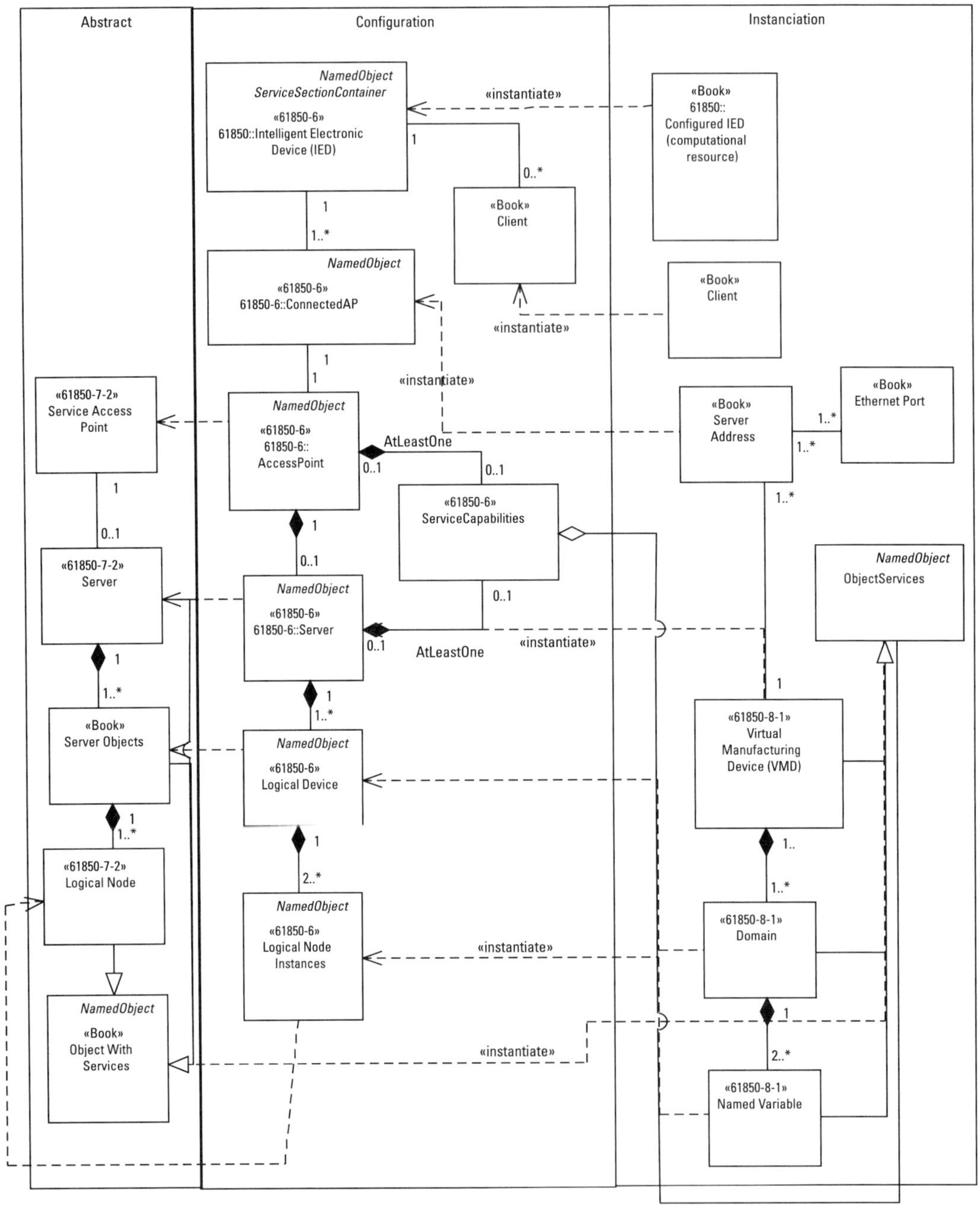

Figure 7.13 GOOSE and Sampled Value addresses.

The GOOSE subscription mechanism is provided using subscription syntax in the GOOSE control block, as shown in Figure 7.14.

The broker architecture for GOOSE and Sample Value, shown in Figure 7.15, is a hybrid of radial and distributed architectures.

IEC 61850 GOOSE and Sample Value publish and subscribe topic routing is performed by the communication network using multicast addressing and content filtering is based on the implementation receiving the message that was routed to it

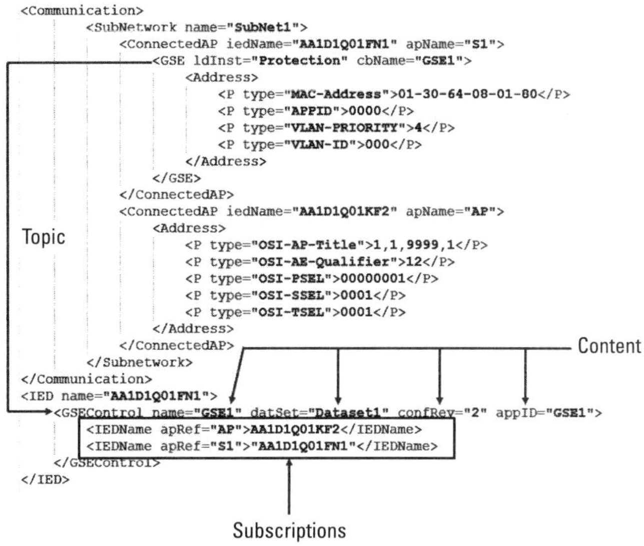

Figure 7.14 Example of Layer 2 GOOSE addressing in SCL configuration.

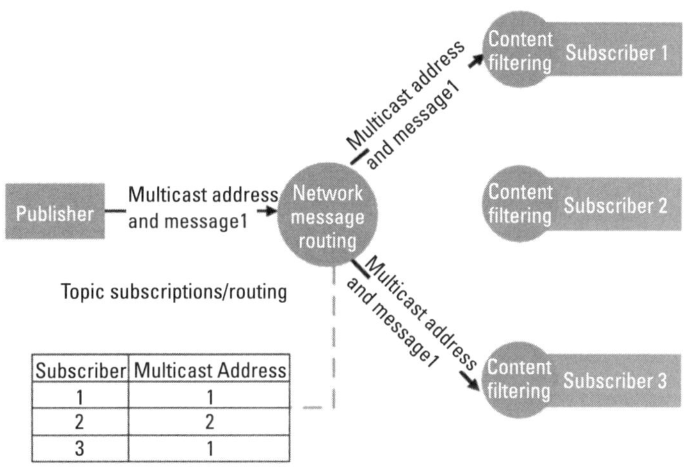

Figure 7.15 GOOSE and Sampled Value broker architecture.

based on the topic. The content filtering allows for the producers to publish with to the same topic (i.e., multicast address), have it delivered to the subscribing implementation, and have the subscriber decide on the appropriate processing. This allows for the same topic to be utilized by multiple publishers and create topics for a specific set of information deliveries and potentially redundancy. It is not difficult to envision that GOOSE and Sampled Values will eventually be conveyed, in a standardized manner, through a true message broker/ESB technology. A prototype of this construct was created to allow global integration of a GOOSE application for the 2017 IEC 61850 UCA International Interoperability Test.

Figure 7.14 shows the items that provide the opportunity for GOOSE content-based filtering. These are

- *Control Block Name (name)*: In a GOOSE message this information is embedded as an object reference that includes the full path reference of the control block. This reference includes the name of the logical device. Filtering on this information allows the differentiation of sources of the message that are using the same topic.
- *Data Set (datSet)*: In a GOOSE message this information is embedded as an object reference that includes the full path reference of the data set. This reference includes the name of the logical device and logical nodes. The data set reference name defines the actual information that is being delivered in the message. Filtering on this information allows the differentiation of sources of the message that are using the same topic as well as determining if the information being delivered is what is expected.
- *Configuration Revision (confRev):* The value of this item is embedded in the GOOSE message and allows the subscriber to determine if the contents of the data set have been modified from what was expected.
- *Application ID (appID)*: The value of this item is embedded in the GOOSE message and allows the subscriber to filter the delivered message based on the distributed application that the message is participating in.

Sampled Values has different content filtering enabling information embedded in the message.

The communication network intermediate systems (i.e., switches and routers) provide the quality of service and message queue persistence. The quality of service is defined by the VLAN-PRIORITY parameter value and is enforced by the intermediate system of the communication network (i.e., Ethernet switches). The network also provides limited message persistence through buffering on egress buffering. By design, the MQP approaches zero since GOOSE and Sampled Values is designed for automation.

CHAPTER 8

Basic IEC 61850

There are several different aspects of IEC 61850. There is the abstract model (e.g., IEC 61850-7-2), which consists of models that define behavior, objects, and services used to interact with the abstract objects (see Figure 8.1). There is the ability to provide configuration information (e.g., IEC 61850-6) and to instantiate the configuration (e.g., IEC 61850-8-1 or IEC 61850-8-2).

An IED can consist of either zero or more servers or clients. Servers provide server objects that define abstract services that are used to interact with the abstract objects.

The actual instantiated configured IED has objects as well. However, there is a mapping between the abstract server objects and the instantiated objects. The instantiated objects provide concrete communication services. The mapping of the abstract objects and services to instantiated services is provided by documents known as Specific Communication Service Mapping (SCSM) document (see Table 8.1). The IED, based on its implementation of objects and client services, including subscriptions, exposes its service capabilities in a configuration known as Service Capabilities.

The Service Capabilities represent the instantiated device's declaration of what its capabilities are, and they are shown in Figure 8.2. Service Capabilities can be exposed on an IED or per access point basis.

The Client Capabilities represent a set of declarations utilized by the IEC 61850 engineering process to allow the engineering tools to understand if the client functionality can subscribe to GOOSE (goose), subscribe to GSSE (gsse), receive and control buffered (bufReport) or unbuffered (unbufReport) reporting, reading of Log information (readLog), and the ability to subscribe to Sampled Values (SV).

The IEC 61850 abstract server objects provide access to several different types of abstract objects, as shown in Figure 8.3.

IEC 61850 is about allowing communication and information exchange with and between intelligent electronic devices (IEDs). An IED is a physical entity that is given a unique name within the configured system.[1] It must have at least one communication media connection (a port) through which to communicate (e.g., an Ethernet connection). The media connection can be used to support one or more IEC 61850 servers. The server is the entity through which all communication with an IED occurs. The IED has the following objects:

1. This is a requirement from IEC 61850-6.

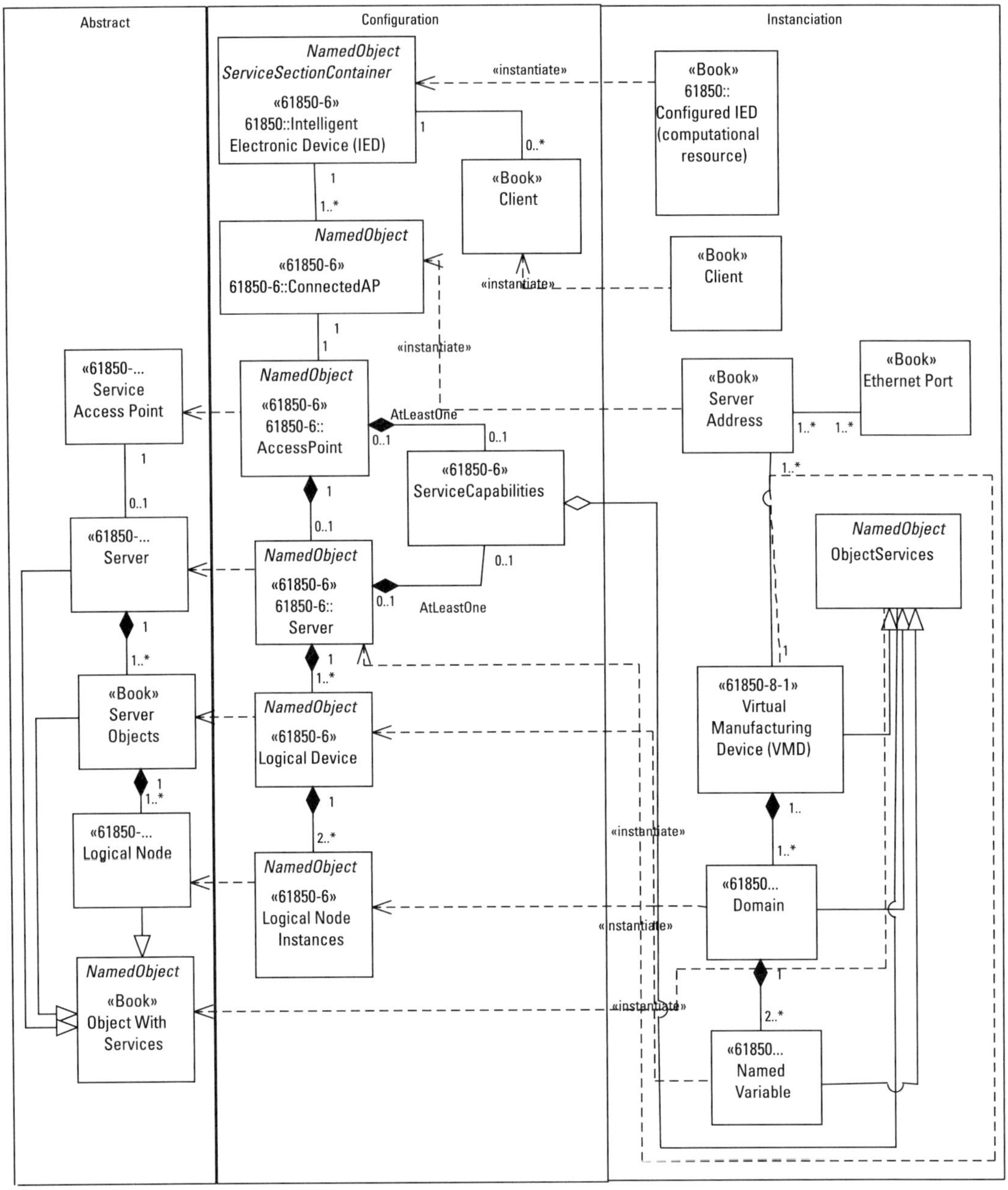

Figure 8.1 IEC 61850 object and service relationship.

Table 8.1 Mapping of Abstract Server Services

Service Capability	*Abstract Server Services*		*SCSM*	
	Object	*Service*	*Object*	*Service*
DynAssociation	Server	Associate	VMD	Initiate
		Release	VMD	Release
		Abort	VMD	Abort
			ProtocolMachine	Reject
GetDirectory	Server	GetServerDirectory	VMD	GetNameList

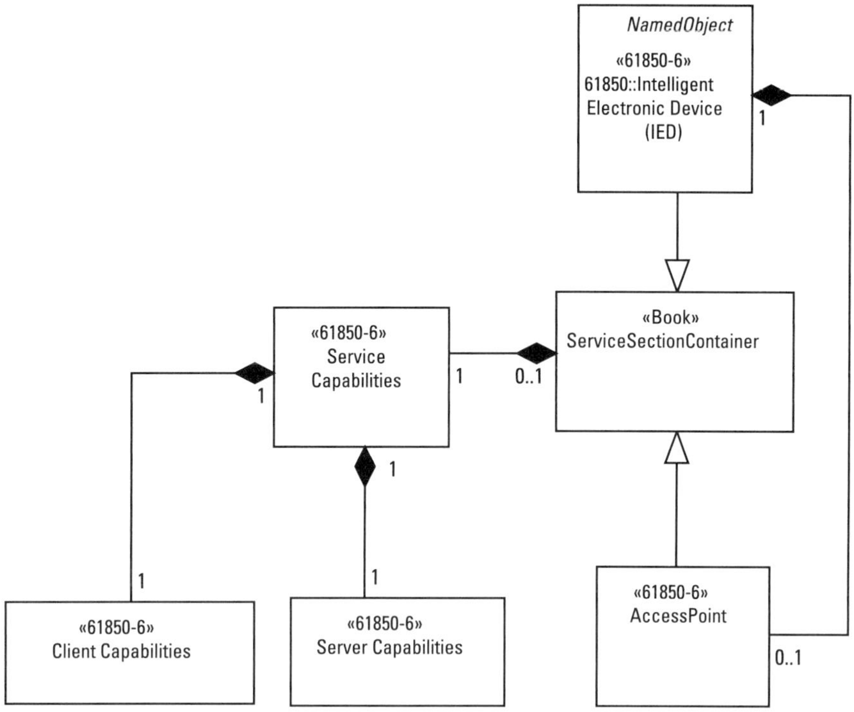

Figure 8.2 IED Service Capabilities.

- *Server:* This is an abstract concept that represents a set of communication endpoints that provide access to other IEC 61850 objects. See Figure 8.3 for further detail.
- *Logical device*: A named object that is an aggregation of other IEC 61850 resources or objects. There must be at least one logical device in a server. See Section 8.2.1 for further details.
- *Logical node (LN):* Represents specific functionality that provides information that is to be exchanged. The IEC 61850 communication concept allows logical nodes to exchange information with other logical nodes. A logical device must have a minimum of two logical nodes. These nodes represent the functionality to control the resources in the logical device (LLN0) and one that represents the physical IED (LPHD). See Section 8.2.2 for further information. LNs are the equivalent to the UCA GOMSFE Bricks but are much more granular.
- *Data object (DO):* Represents a named sub-object of a LN that provides specific semantics and exchange capability.
- *Data attribute (DA):* Represents a named sub-object of a DO that provides specific semantics and exchange capability.
- *Data set:* A data set is a collection of named information that is a constrained set of information of either the data object or data attribute type. The constraint that creates the subset is known as a functional constraint (FC). The constrained subset of a DO is known as functionally constrained data (FCD).

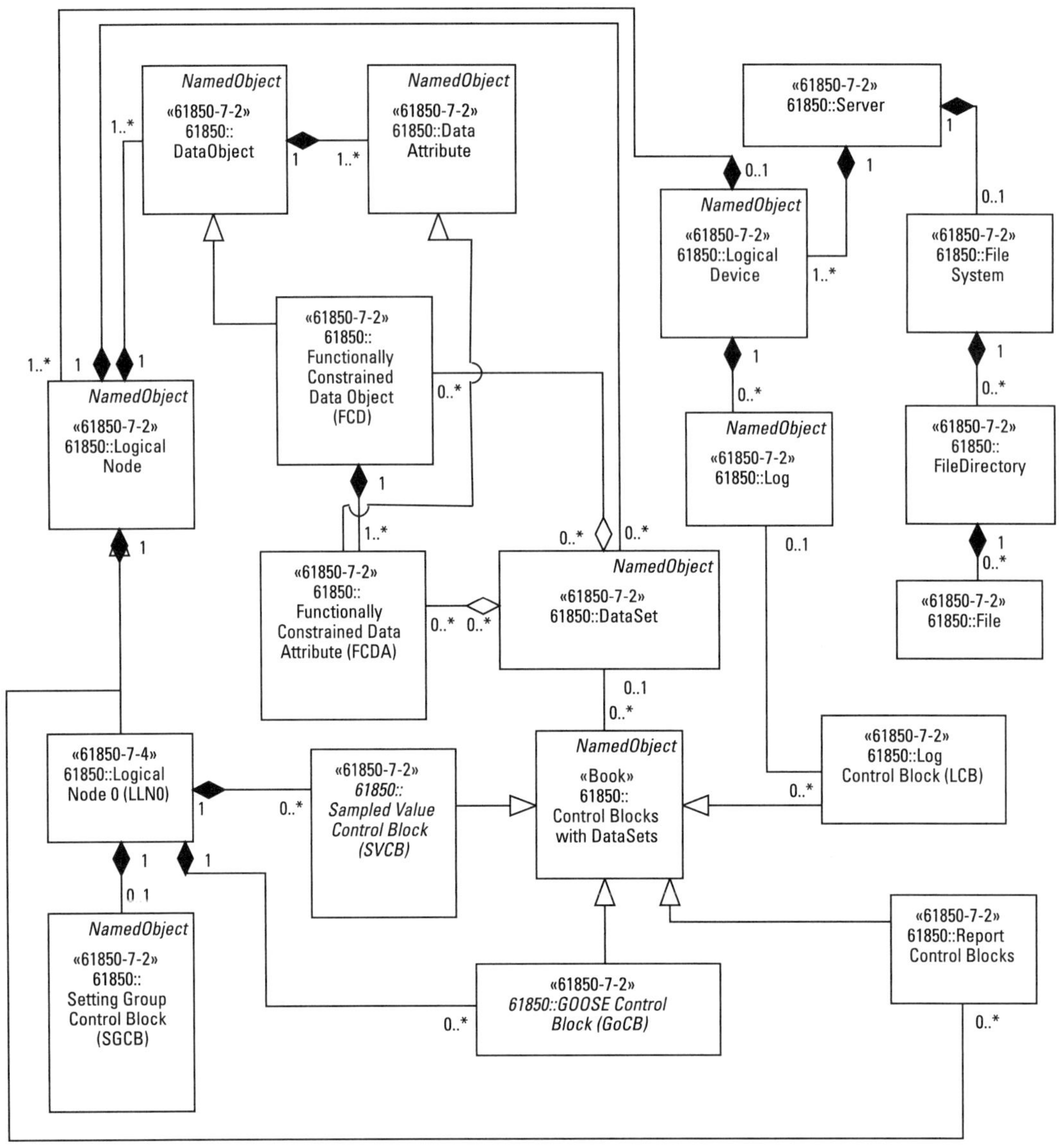

Figure 8.3 IEC 61850 server objects.

The constrained subset of a DA is known as a functionally constrained data attribute (FCDA).

- *Functional constraint (FC):* Are characteristics that are assigned to specific attributes' data attributes.
- *Control blocks:* Control blocks represent the ability to control some aspect of the logical node or logical device. The setting group control block controls the Setting Group Model and does not utilize a data set within the model. All other control blocks utilize a data set in order to determine which set of information is to be exchanged or logged. Each control block with data sets has a detailed model provided. See Section 8.2.2.4 for more detailed information regarding control blocks.

Each of these objects will be detailed and explained in the following sections.

8.1 Intelligent Electronic Devices

The vernacular of IEC 61850 defines an IED as an intelligent electronic device. However, this simple definition masks the complexities that need to be understood when using the moniker of IED. First and foremost, an IED must be able to communicate via a communications medium. A computer using LEDs to indicate status would not fulfill this definition as it is providing information visually, even though it is digital in nature. Take the same device and allow it to communicate with other devices via Modbus, DNP, IEC 61850, SEP 2.0, or IEC 61850 and it becomes an IED. It is the ability to exchange information to other devices and entities that is at the core of an IED.

In most non-IEC 61850 protocols, the only integration or exchange pattern is that of client and server. Modbus utilizes the terminology of master to represent client functionality and slave to represent server capability. In a similar fashion, DNP uses the terminology master and outstation. Client and server represent functional capabilities and both capabilities can be contained in the same device. In many instances, a device that contains both capabilities may be referred to as an RTU, proxy, or gateway. Likewise, devices can contain a client-only or server-only capability. Only server capability can expose information to other devices.

There is an additional integration pattern of publish and subscribe that is possible within the context of IEC 61850. As discussed as part of the integration pattern discussion, only devices with server capability can be publishers. Otherwise the subscription capability can be associated with any functionality. A subscriber would not need to be either a client or a server if not for the configuration methodologies specified in IEC 61850-6. There are two types of publish and subscribe exchange protocols within IEC 61850: GOOSE and SMV.

The declaration of the functional capabilities is defined within IEC 61850-6 as ServiceCapabilities as shown in Figure 8.4.

The functional capabilities of the client or server are exposed in SCL as part of an access point or server definition. The access point service capabilities can be used to express the client capabilities and be used to differentiate server capabilities on a per access point basis (e.g., one access point may support only GOOSE and another access point supports server services). The service capabilities express capabilities such as GOOSE publication and subscription is supported as well as how many publications or subscriptions can be supported. The detailed description of Service Capabilities can be found in Section 10.2.

Based on the functional capabilities, devices can support different applications. Most applications can be categorized into SCADA, automation, and synchrophasor. Many of these applications may be with or without the use of digital electronic sensors.

8.1.1 IED and Their Applications

IEC 61850 IEDs can be utilized for a wide range of applications. The following sections provide high level information regarding typical applications.

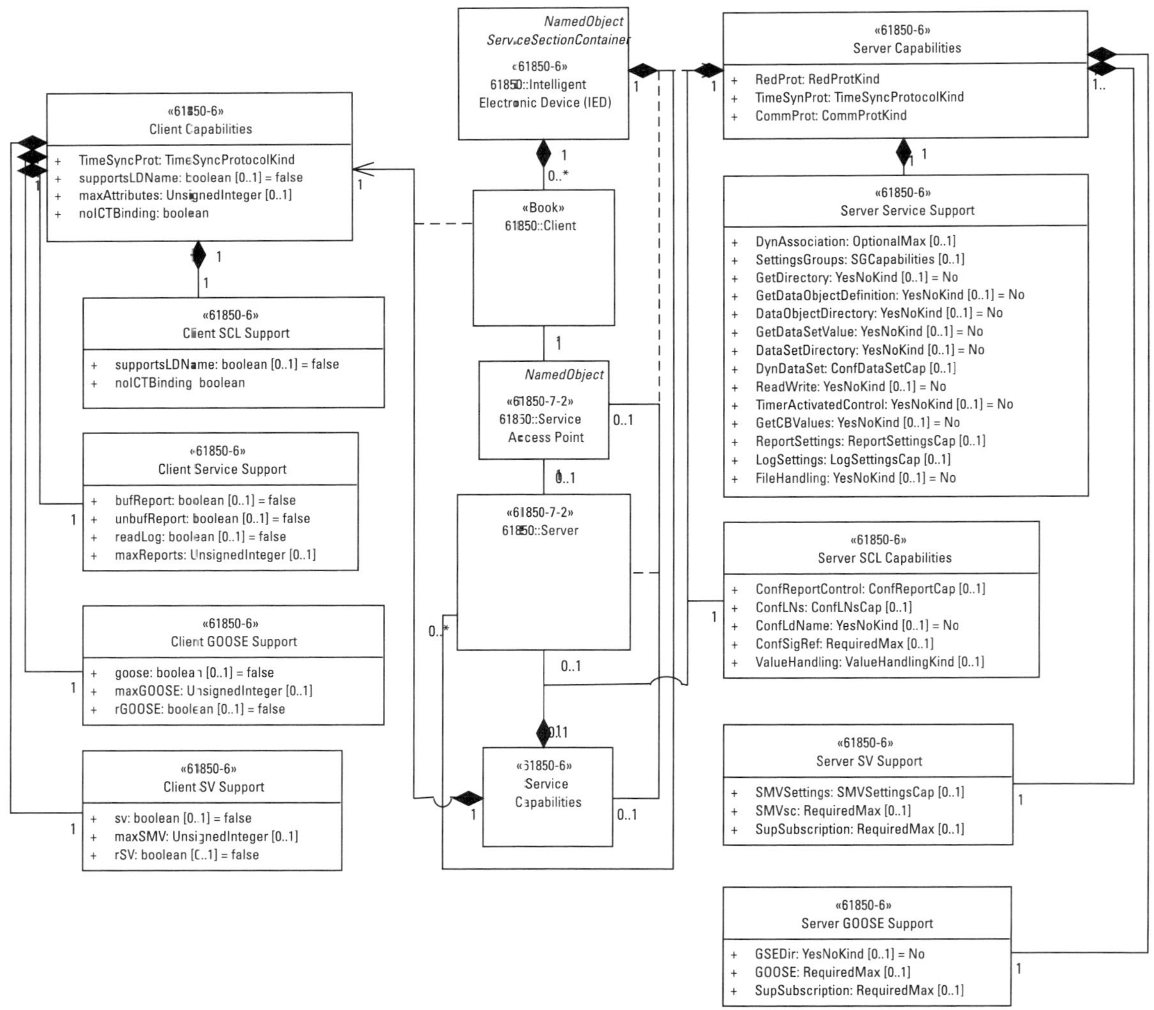

Figure 8.4 UML Definition of IED service capabilities.

8.1.1.1 SCADA Applications

There are two typical SCADA application architectures: direct connection to an IED or the use of a proxy or gateway.

Figure 8.5 shows some of the typical functions (e.g., logical node) functions that might be supported for a SCADA application. Noteworthy is that Client functionality typically requires at least one logical node. Servers contain other functions.

The diagram also shows that an RTU consists of both a client and a server with information flowing internally between the client function and the server function. The client function acquires information from other IEDs and the server function proxies or repackages that information for another device's consumption.

8.1.1.2 Automation Applications

Automation applications can utilize the SCADA construct where clients interact via control commands to perform control. However, what differentiates DNP control applications from IEC 61850 automation applications is the integration of the high-speed peer-to-peer exchanges provided by GOOSE. It is rare that proxies are utilized with automation applications at the substation level even though an RTU

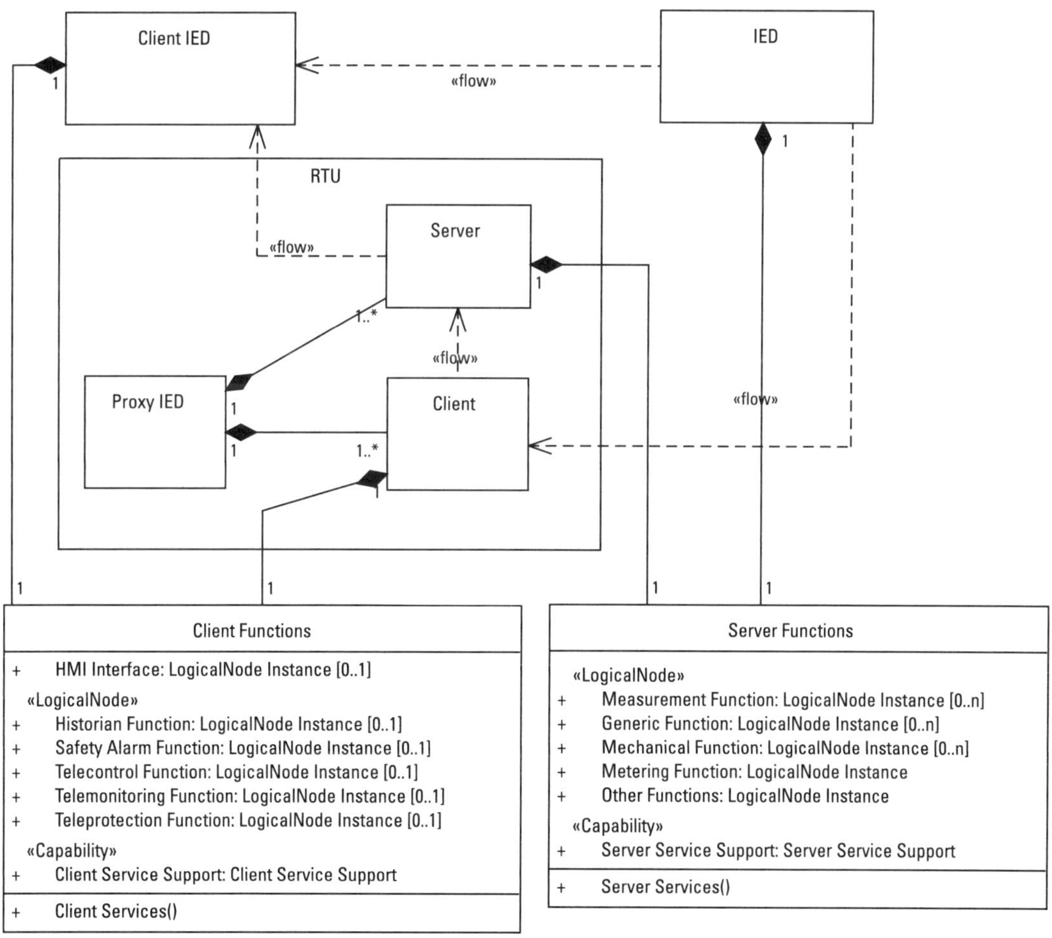

Figure 8.5 UML representation of typical SCADA applications.

can provide the automation logic. Thus, the UML for automation applications does not reflect the use of proxies.

An automation application consists of IEDs and automation IEDs. Figure 8.6 asserts that in most automation applications the deployed IEDs have client and servers; however, this is not necessarily true.

8.1.1.3 Synchrophasor And Sampled Value Applications

The original design of Sampled Values was designed to replace the analog distribution of current transformer (CT) and voltage/potential transformer (VT/PT). The analog distribution mechanism can best be described as shown in Figure 8.7.

The high voltage is monitored by a PT. The measured voltage is converted from the primary voltage (e.g., 300 Kv) to a lower secondary voltage in a similar fashion to a step-down transformer. The secondary voltage is conditioned and calibrated.

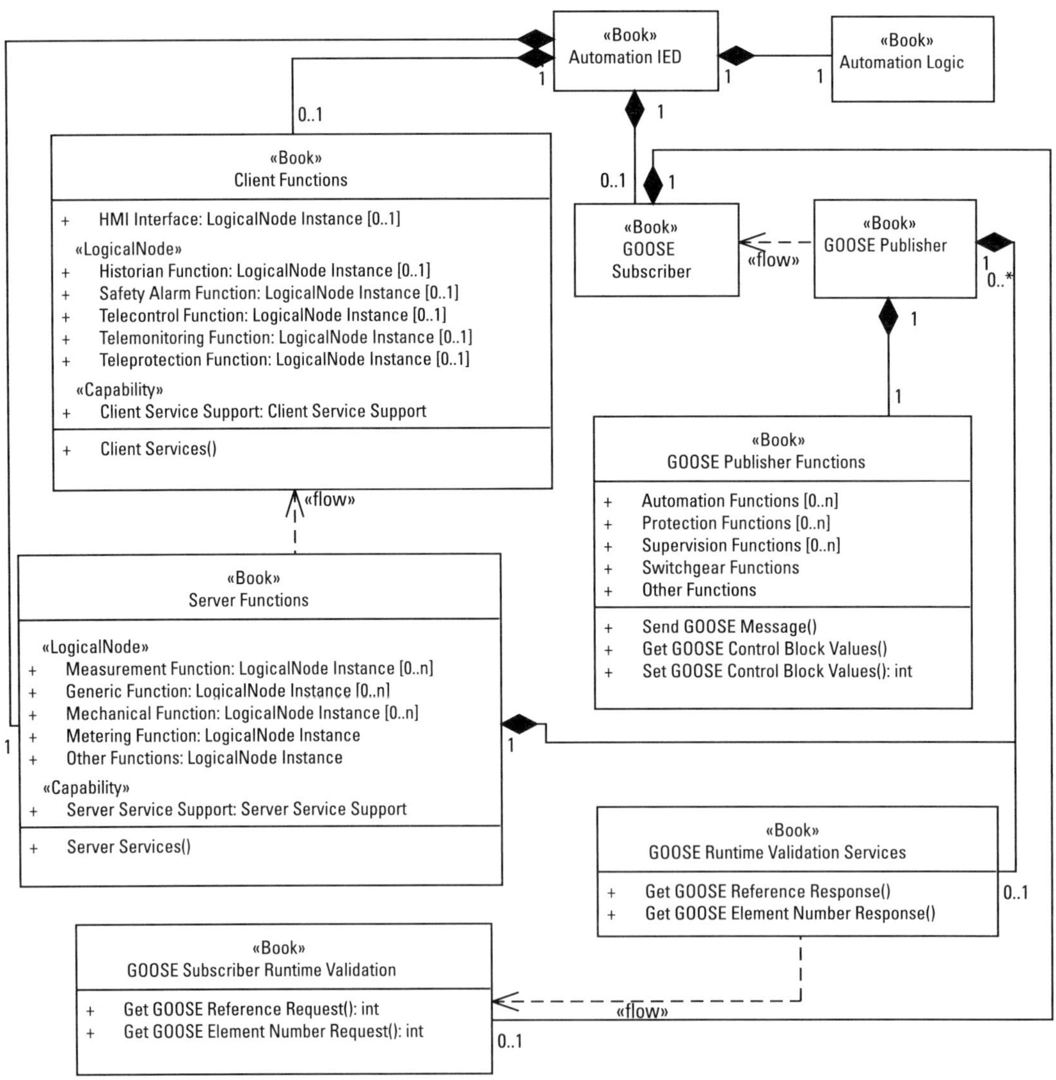

Figure 8.6 Typical GOOSE automation applications.

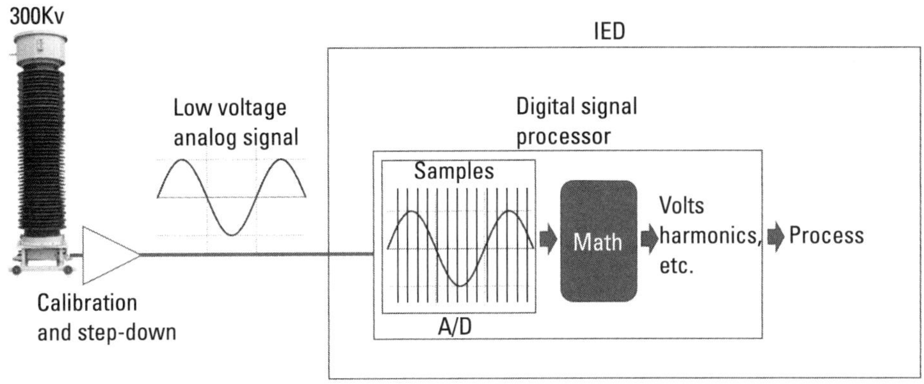

Figure 8.7 Sampled Values—analog processing. (Adapted image courtesy of PACWorld.)

This calibrated secondary voltage is then distributed by copper wires to the inputs of the IEDs. Depending on the length of the distribution additional compensations may be required (e.g., temperature, impedance). This makes it difficult to share the low voltage signal over disparate distances and different environments. Enquiring minds might ask why not just add additional PTs; the answer is that the cost of transmission level PTs is high so it is cost-prohibitive to do so. Additionally, analog VTs are large and require much space since they must be concerned with allowing electrical arcs to ground. That is why the stack (e.g., the disks on the PT) is large and gets larger for higher voltages.

The modern IEDs typically utilize a digital signal processing[2] (DSP) that performs an analog to digital (A/D) conversion, scaling, conditioning, and math to produce the digital information that is required to be processed within the IED. The sampling rate is tightly controlled within the DSP and variance in the sampling rate would introduce potential errors. Key to the precision of the process information is an accurate time synchronization typically a 1 PPS source such as IRIG-B or GPS.

The advent of optical CTs/PTs changed the paradigm. The optical CT/PT distributes some of the signal processing as shown in Figure 8.8.

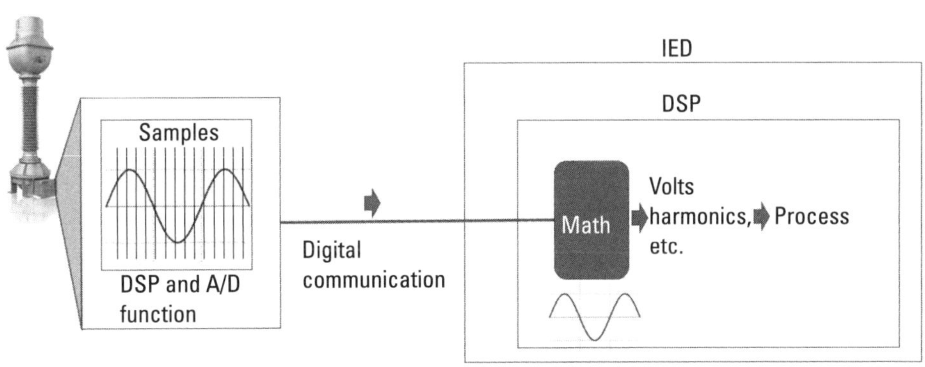

Figure 8.8 Sampled Values—optical processing. (Adapted image courtesy of PACWorld.)

2. A DSP is a special purpose micro-processor optimized for signal processing.

An optical CT/PT contains local A/D conversion and a digital communication interface. The digitized samples are transmitted through a digital communication interface to an IED where a DSP is used to produce the process information that the IED needs to perform its function. The first optical CT/VTs had a proprietary digital interface where a vendor of both the IED and optical sensor could control the entire distributed process. A standard for digital information exchange was needed and that produced IEC 61850-9-2 which is commonly referred to as Sampled Values (SV).

Since a standard allows a mix and match of optical sensor with IEDs and the sampling of the analog values is no longer under the control of the IED, the SV standard needed to somehow provide the time at which the sample was taken and the sampling rate in the digital stream. This information was designed into the SV protocol.

In 2003 there was a major power outage in the Northeastern United States. The National Electric Reliability Corporation (NERC—www.nerc.gov) attempted to analyze the cause of the blackout. NERC discovered that it was near impossible to correlate the information provided by the various utilities due to in-precise time stamping and different algorithms which provided the process values to those utilities. This represented a major issue for post event analysis and caused the North American Synchrophasor Initiative (NASPI—www.naspi.org) as well as an initiative within IEEE to develop the required standard for algorithmic production of synchrophasor values and the digital communication of those values. The initial standard combined the sampling/algorithm with the digital communications in IEEE C37.118. When this standard was presented to IEC for joint logo (e.g., to be both an IEEE and IEC standard), IEC rejected the joint logo concept since the IEC SV protocol had the ability to convey the synchrophasor values produced by the C37.118 algorithm. Based on this feedback, the IEEE C37.118 standard was split into IEEE C37.118-1.

The C37.118-1 algorithm relies on precise time synchronization, as sampling is synchronized to the top of each second. It is this precision and algorithm that allow the synchronized phasor values to be usable over long distances, referred to as wide area as would have been needed for post event analysis of the blackout or prior detection of the grid becoming unstable.

The advent of the synchronized measurement technique allows for phase differences to be used to determine real-time phase shifts, grid oscillations, and frequency shifts across the United States transmission system as shown in Figure 8.9.

The differences can be plotted (e.g., the vectors of magnitude and angle). The foretelling of grid instability is typical when the angular difference between values starts increasing. Not only can these measurements be used across the United States, but also within the transmission and distribution networks of local utilities. The values are becoming integral in Wide Area Monitoring Systems (WAMS) and Wide Area Monitoring Protection and Control (WAMPAC) systems.

IEC 61850-9-2 SV already had the communication capability to carry these types of vector quantities. However, the L2 SV was not easily routable and did not have sufficient security[3]. The need for routing, and the use of GOOSE in WAM-

3. IEEE C37.118-2 had no security either. Work is ongoing at IEEE to address this deficiency.

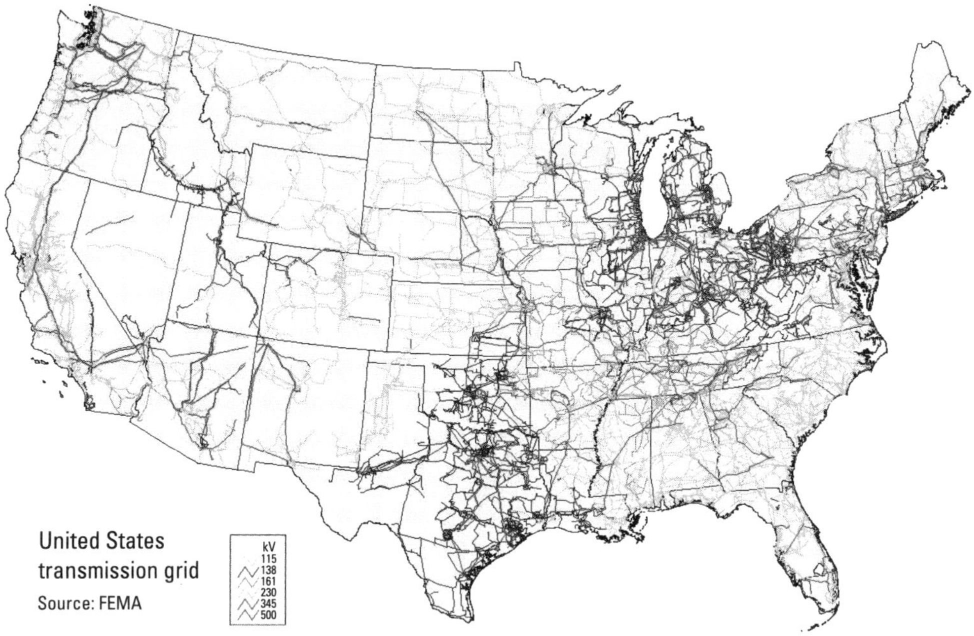

Figure 8.9 Transmission grid of the United States. (Source https://commons.wikimedia.org/wiki/File:UnitedStatesPowerGrid.jpg.)

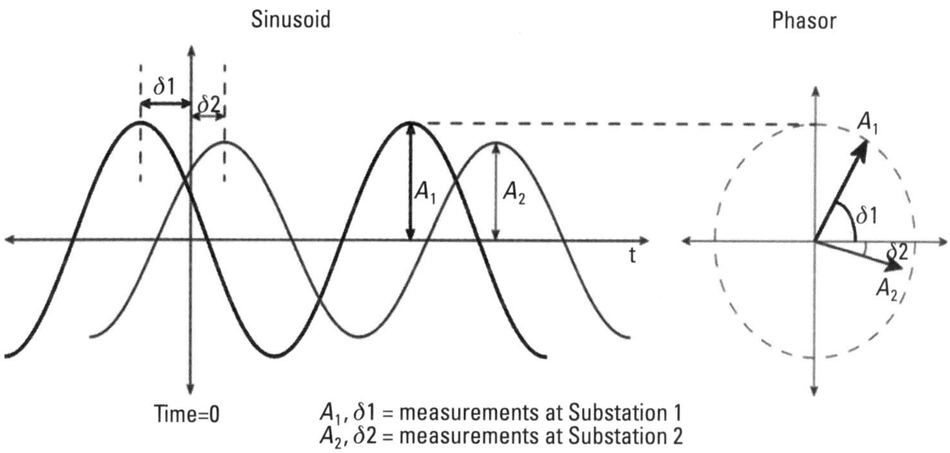

Figure 8.10 Sampled Values—synchrophasor plot. (Image courtesy of PACWorld.)

PAC situations, was the reason for the development of the Routable GOOSE and Routable SV.

The sampling rate for SV or Synchrophasor projects is widely different. For CT/PT information it is typically 80-256 samples/cycle (e.g., 4,800–15,360 samples per second). Synchrophasor applications sample 30-120 samples/second. Due to the sampling rate, it is difficult to distribute the CT/PT information over a WAN. The synchrophasor sampling rate does not restrict the transmission via WAN.

The R-SV can not only be utilized for WAMS and WAMPAC, but also can be used to assist in load flow and state-estimation[4] applications.

The use of SV, see Figure 8.11, requires a publisher of information and one or more multiple subscribing IEDs. It is typical that these IEDs are themselves IEC 61850 Servers and use the SV information to drive other functions and the GOOSE messaging required for automation and protection.

8.1.2 Naming of IEDs

There are two different IEC 61850 naming conventions for IEDs. Functionally, one could think of these as being user assigned which is known as location assigned. The selection of the naming convention affects the naming of logical device objects (see Table 8.2). User assigned is as it sounds; a user just enters the name of the IED. However, most utilities have a naming convention that includes the system in which the device/IED is located. Although a user can manually assign the name to the IED, most IEC 61850 System Configuration Tools (SCTs) can provide automation and thereby remove errors. The location naming is directly related to use in the functional naming for logical devices.

IEC 61850 has specified a specific hierarchy of objects in a substation and other systems (e.g., Lines and Plants). The hierarchy is shown in Figure 8.12.

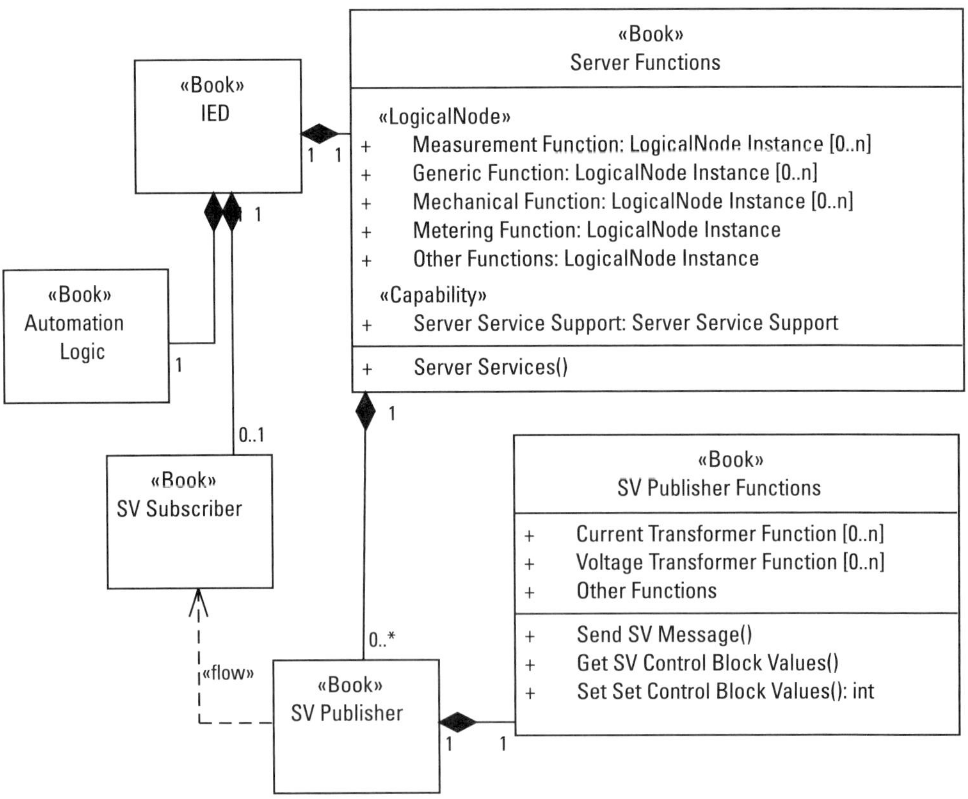

Figure 8.11 Synchrophasor and SV application.

4. See https://www.nerc.com/comm/PC/Synchronized%20Measurement%20Subcommittee/NERC_SMS_EPRI_Synchrophasor_State_Estimation.pdf

Table 8.2 Attributes of Logical Devices

UML Attribute	*SCL XML Attribute*	*Purpose*
instance	inst	This is a required attribute that must have a non-null value. If the configuredName attribute does not have a value, the exposed name of the specific logical device will be the contentation of the name of the IED and the value of instance.
description	desc	A textual description of the logical device. Typically, it describes the high level functionality provided by the logical device. It can be changed during the engineering process.
configuredName	ldName	Provides an alias to the logical device name created through the concatenation of the IED name and instance. It is this name that will be exposed via IEC 61850 if configured.

Note: There are constraints within IEC 61850-6 that require the resulting names of logical devices to be unique within the scope of the system (e.g., SCL configuration file).

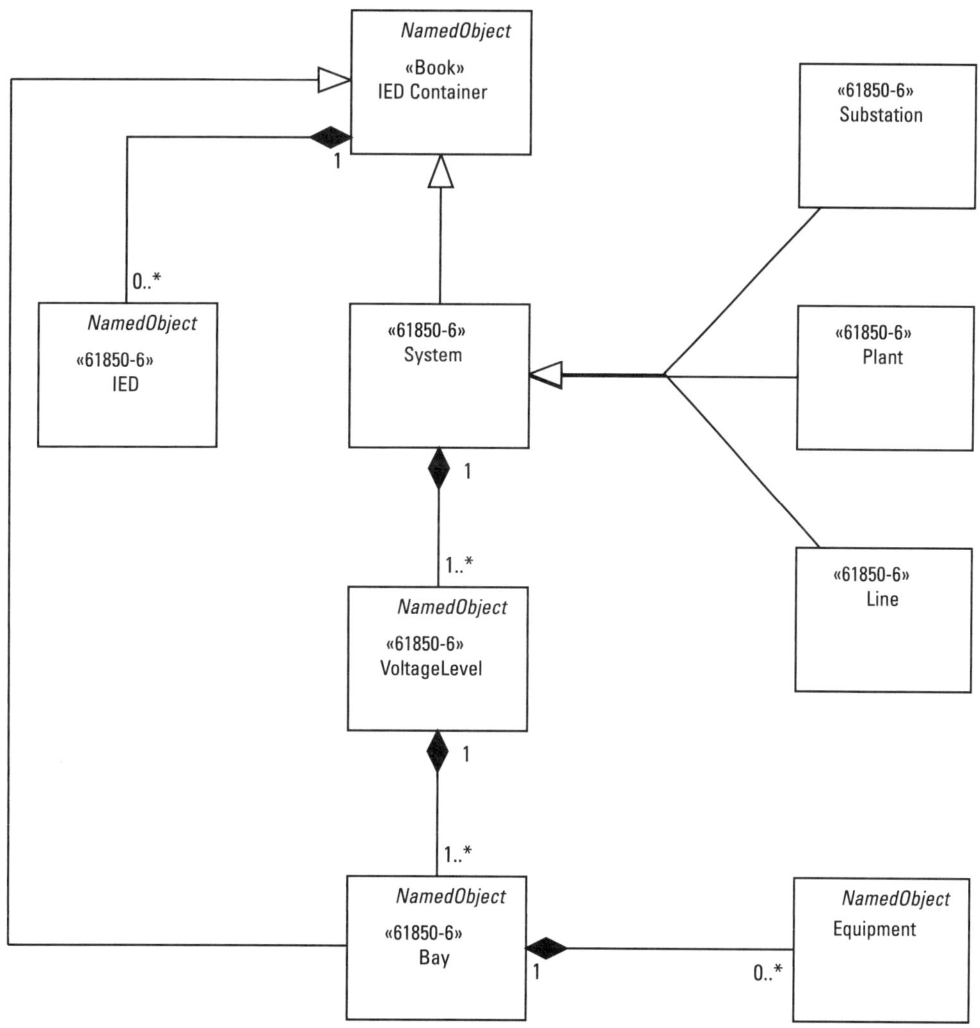

Figure 8.12 IEC 61850 system hierarchy and IED placement.

IEDs are physical and can be installed at the system (e.g., in a substation, on a line, or within a plant) or within a bay (see Figure 8.12). Automation within the tools will concatenate the names of the objects in the hierarchy where the IED is installed. As an example, an IED installed in a hierarchy of:

- Substation whose name is "Airport".
- VoltageLevel whose name is: "VL1".
- Bay whose name is "Line1".

The resulting prefix of the IEDName would be AirportVL1Line1. Since multiple IEDs can be installed within the same location of the hierarchy, users typically need to append a unique identifier. The final name would be part of the logical device functional name that is placed in the configured name attribute of its SCL configuration (see Table 8.2).

8.1.3 When Is an IED Not a Physical Box

Up to this point, the analogy has been that an IED is a physical box and that an IED might contain multiple IEC 61850 servers, with access points as shown in Figure 8.13. However, with the advent of virtualization, we can now have a single physical computer that hosts one or more virtualized computers. Each of these virtual computers has its own health, status, and functionality.

To determine if a physical box is one or more IEC 61850 IEDs, there is a simple litmus test. If a single physical box has multiple health indications (e.g., multiple LPHD logical nodes) that can have different health indications, this would indicate that for each different potential health indication, there would be a different IED represented in the SCL configuration file. There is a well-known exception which is when there is an IED that is being a proxy for other IEDs.

8.1.4 Access Point

An IED can have one or more access points. Each access point represents a single communication interface. A single access point may only be bound to a single IP address and subnetwork. This binding is found in the communication section of an SCL file and is known as a connected access point (ConnectedAP). A communication interface can consist of multiple Ethernet connections that act as a single interface by providing redundancy.

Simplistically, each access point may provide a different set of services. in general, an access point may be either client only (e.g., a consumer of information) or a provider of information (e.g., server of ServerAt). The relationship of services to access point is shown in the Figure 8.15. Specific details on the overall service definition can be found in this figure.

8.1.4.1 Service Capabilities

The service capabilities are expressed in XML format. The services associated with two-party associations are shown as server service support.

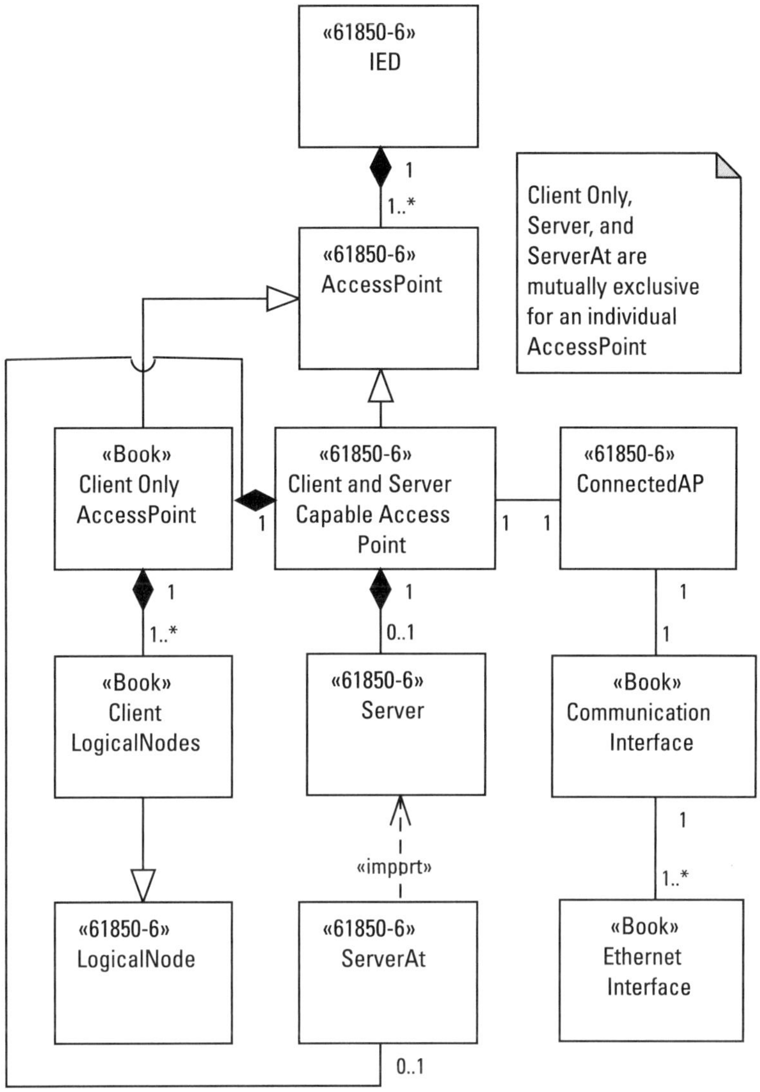

Figure 8.13 IED and access point functionality.

A reject is issued when the ISO 9506 protocol is violated and typically causes the client to issue an abort. The protocol can be violated by attempting to utilize a service that was not negotiated, a message whose size exceeds the negotiated size, and others. Errors, such as requesting data that is not available does not cause a reject, rather it causes specific error responses.

8.1.4.2 Server Access Point

IEDs can be client, server, or both depending on the type of application into which they are being applied.

The abstract model of a server is found in IEC 61850-7-2. Servers contain objects that are externally visible remotely using the IEC 61850 set of communication services, see Figure 8.17. The abstract model associates many communication

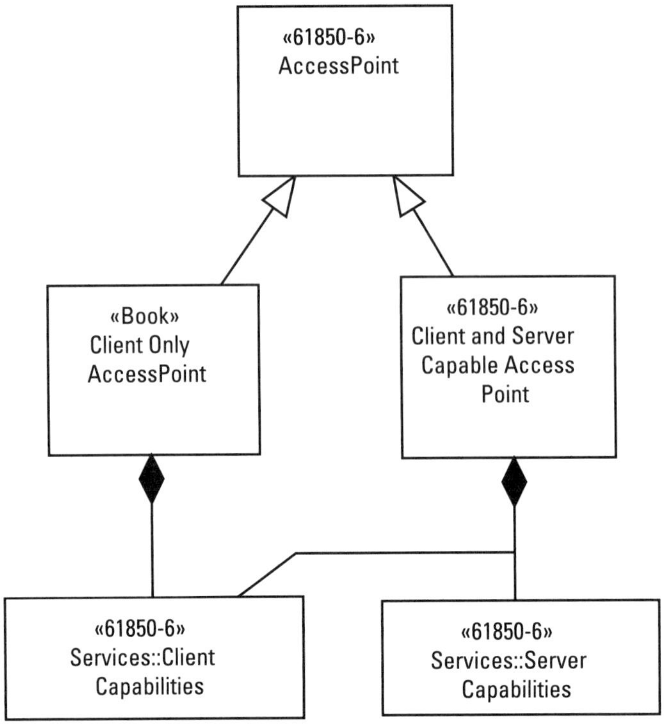

Figure 8.14 Access point and Service Capabilities.

services with specific objects. Configuration exposes these communication capabilities represented in ServiceCapabilities so that the system engineering process can have the information exposed in a manner that can be computationally processed without human intervention and without an actual device being online and communicative.

Instantiation of a communicative server is dependent on the protocol being utilized to provide IEC 61850 services. The mapping of the abstract services to a specific protocol is provided in a Specific Communication Specific Mappings (SCSMs), as shown in Table 8.3. A SCSM is responsible for providing the mapping from the abstract services and objects to one or more concrete communication protocols and concrete objects. This book concentrates on the use of ISO 9506 as the protocol utilized and the mappings in IEC 61850-8-1 and IEC 61850-8-2 SCSMs.

The IEC 61850-8-1 and IEC 61850-8-2 SCSMs use the same objects and services although the underlying encoding and transport of the communication services are different. In both, an abstract server is mapped to what is known as a virtual manufacturing device (VMD).

Figure 8.18 shows the abstract types of objects that a server may contain.

A server must contain at least one object known as logical device. This object provides the ability to group functions that are known as logical nodes. Logical nodes consist of more than one data object. Each data object consists of more than one data attribute.

Constraints on data objects and data attributes are used to create functionally constrained data objects (FCDs) and functionally constrained data attributes (FDCAs). FCDs and FCDAs can be aggregated into a list known as a dataset (e.g., a set

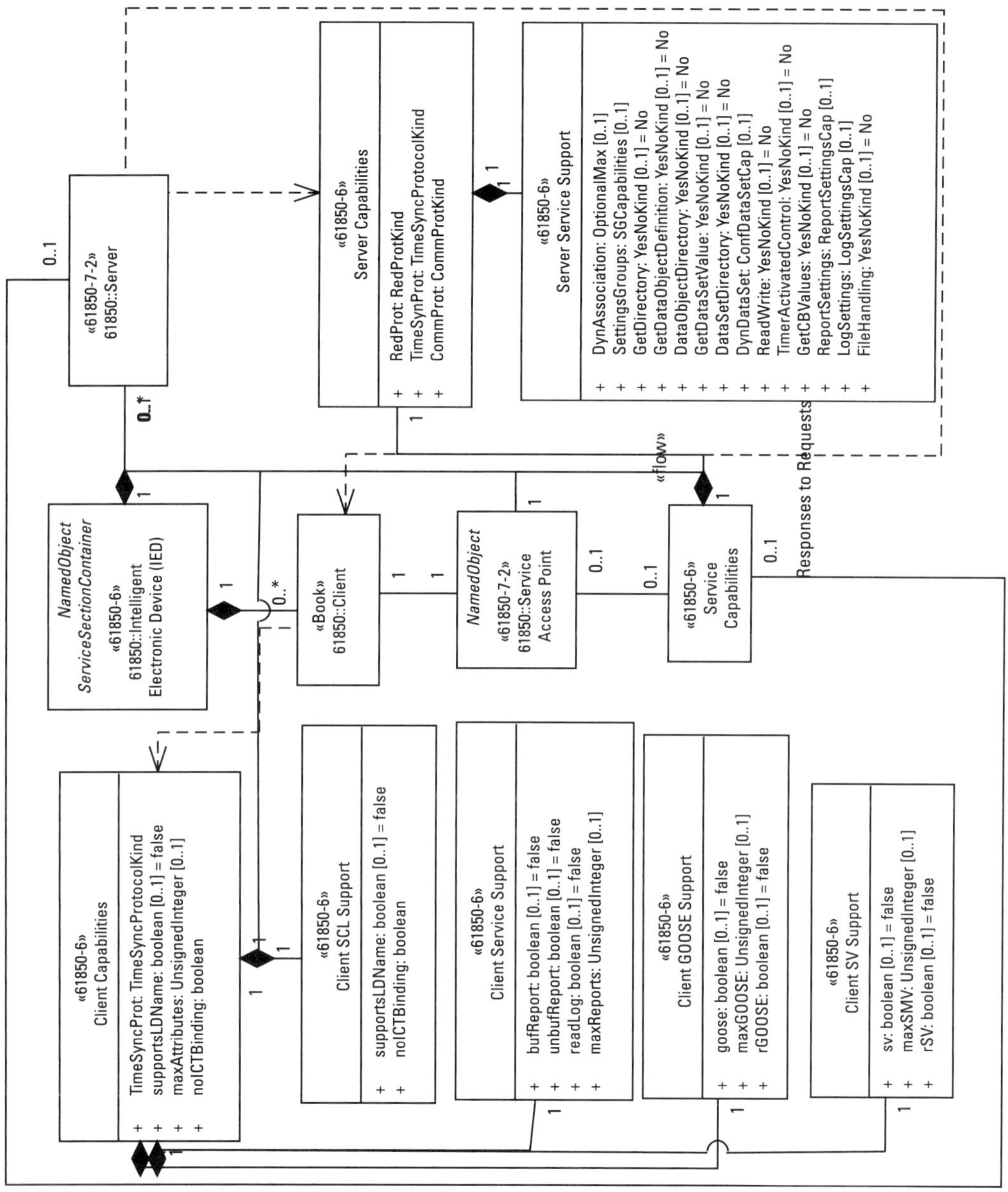

Figure 8.15 IEC 61850-6 Service Capabilities. The IEC 61850-7-2 abstract server object's services and their SCSM mappings are shown in Figure 8.16 and Table 8.1.

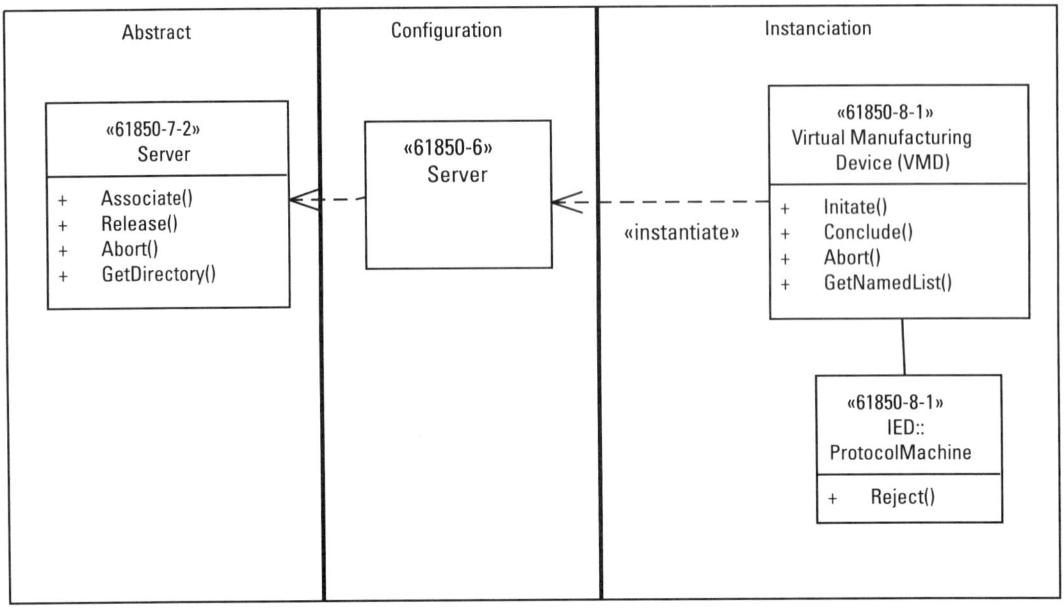

Figure 8.16 Abstract server services to SCSM mappings.

of FCDAs and/or FCDs). These datasets can be referenced by several different types of control blocks. Many of these control blocks have been previously discussed. However, log control block and its associated log object are introduced here. There is one type of control block that does not utilize the concept of a dataset and it is known as the setting group control block.

Each object has its own set of services, attributes, and configuration methodologies. Each object will be discussed in the following chapters.

8.1.4.3 ServerAt Access Point

A ServerAt represents an object clone of a server. The basic construct of ServerAt provides the ability to use a different host address to allow access to the objects defined in the referenced server. This provides the ability to provide communication path redundancy to a single IED and achieve greater system resiliency.

The ServerAt construct also provides the ability to provide specialized access/publications of the server objects through a different communication interface. As an example, IEC 61850's logical architecture separates station bus and process bus functionality leading to an implementation strategy that two-party exchanges and control of SMV control blocks is through a different access point and communication interface (e.g., LAN) from that over which the actual sampled values are published. The use of ServerAt can also allow different GOOSE messages to be published through different access points and communication interfaces.

8.1.4.4 Client Only AccessPoint

It is rare to encounter a client only access point as most IEDs have both client and server capability. The typical exception to the rule is a SCADA front-end processor

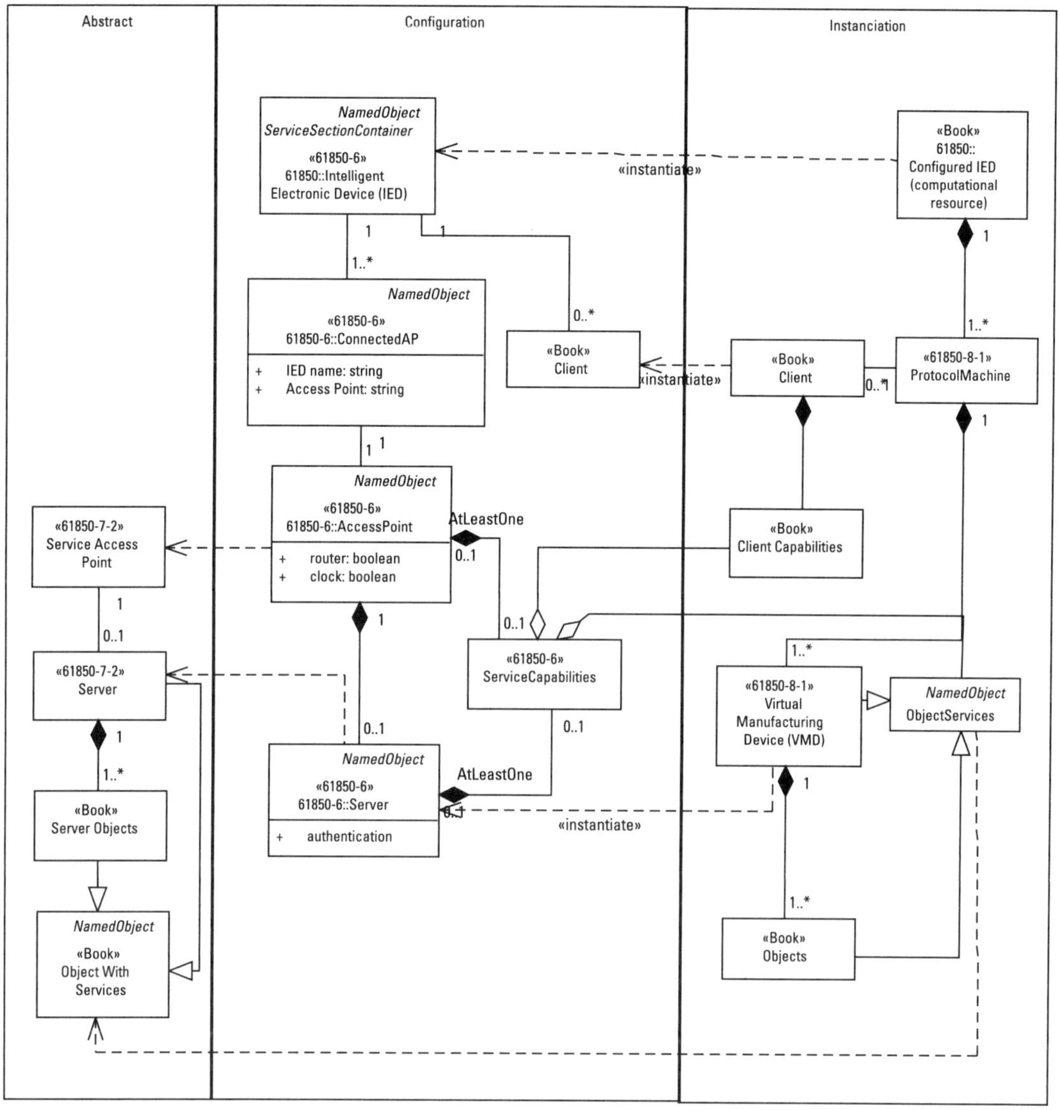

Figure 8.17 General server model.

Table 8.3 Association Abstract Services versus SCSM Services

Abstract Server Services	*SCSM*	
	Object	*Service*
GetServerDirectory	VMD	GetNameList
Associate	VMD	Initiate
Release	VMD	Release
Abort	VMD	Abort

or human-machine interface (HMI). These systems need the ability to be configured to reserve reports and subscribe for GOOSE or sampled values, but do not

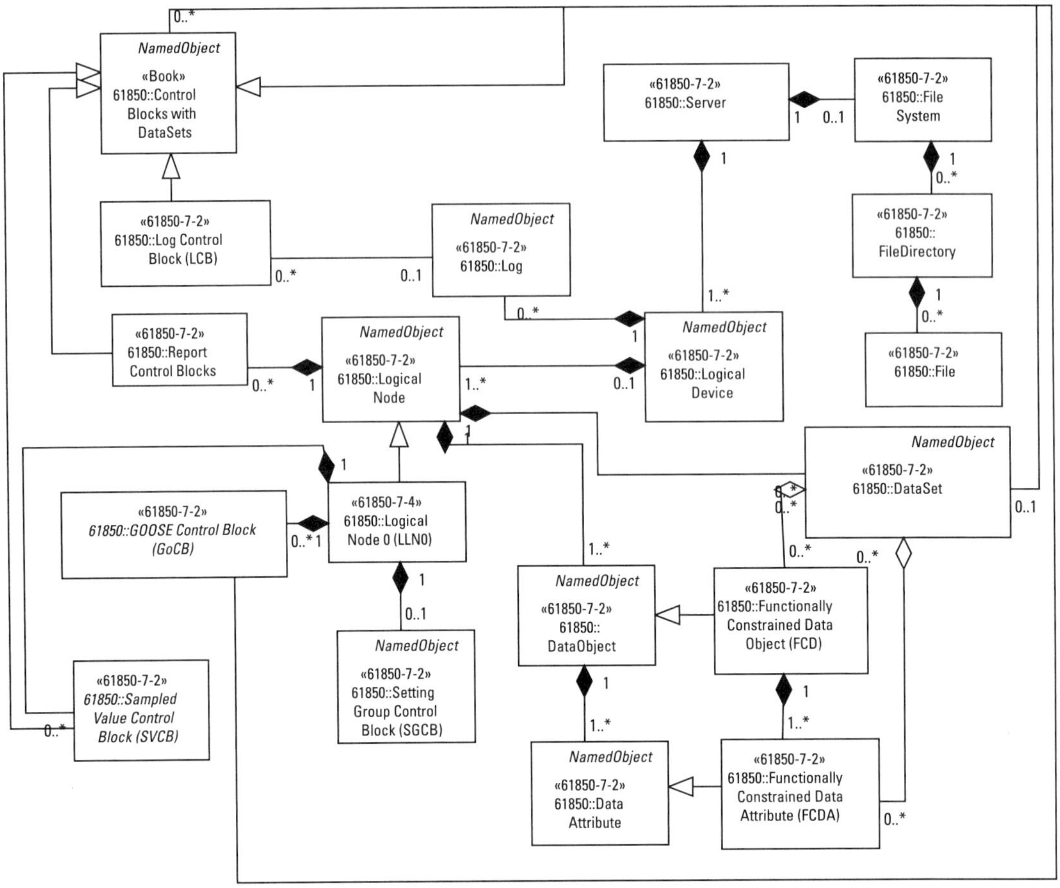

Figure 8.18 IEC 61850-7-2 server and object abstract definitions.

publish or provide information via IEC 61850 services. However, as explained on 9, these applications can interact with IEC 61850 IEDs via two-party communication services.

A client only access point contains one or more functions (e.g., logical nodes) that represent client functionality. Typically, a client only function represents a capability such as a historian or archive, HMI, alarm functionality, and certain teleprotection functions. Within IEC 61850, these are designated as logical node classes in the interfacing and archiving logical node group. More information on logical nodes can be found in Section 8.2.2.

8.2 Logical Device and Logical Nodes

8.2.1 Logical Device

A logical device (Figure 8.19) is an object that provides the ability to group a set of functions known as logical nodes. Each logical device must contain a logical node that contains nameplate information and therefore must contain an instance of LLN0. The IEC 61850 instantiation of the LLN0 definition is LN0. The physical device health is also typically present in a logical device and it is represented by an

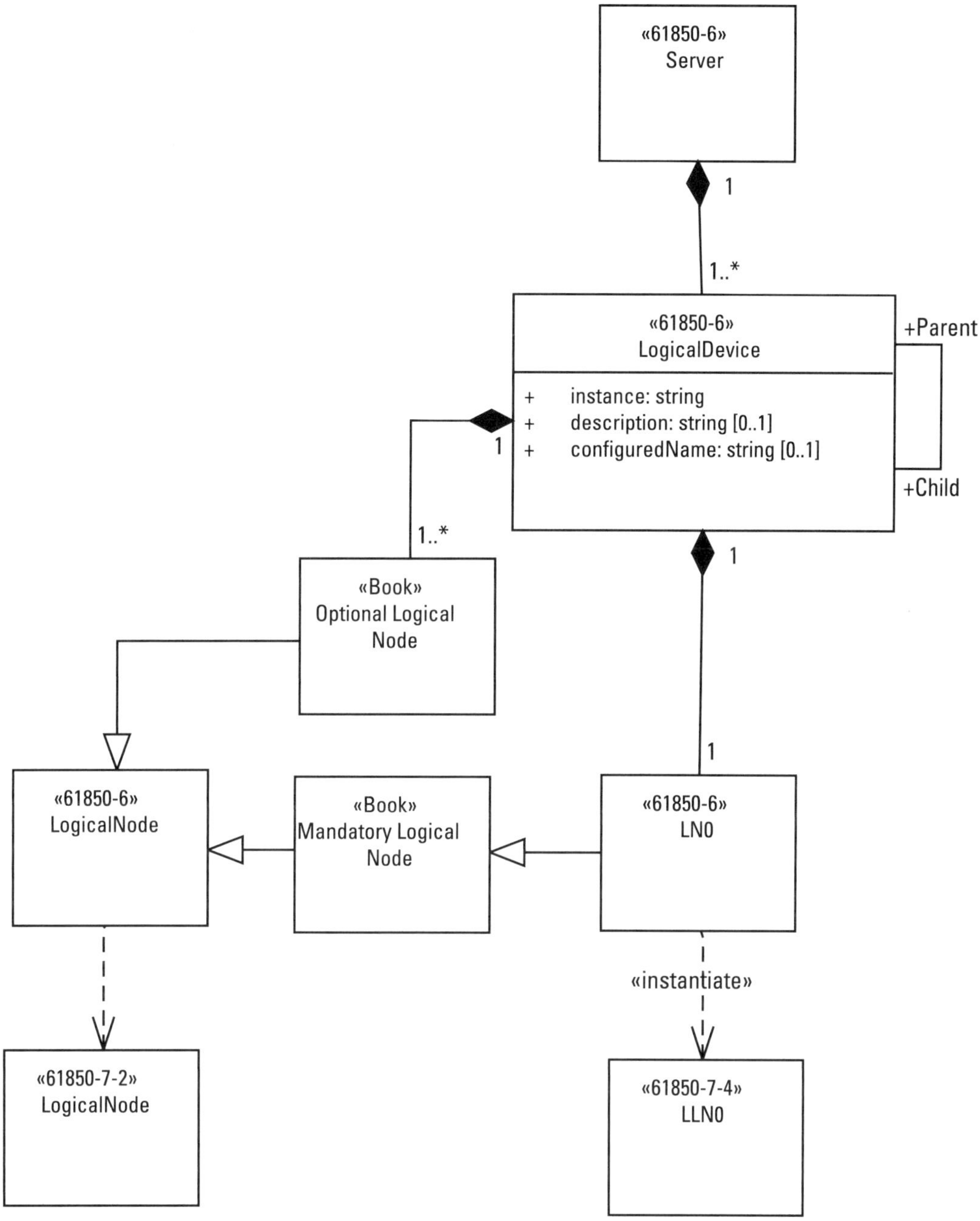

Figure 8.19 Logical device UML relationships.

instance of LPHD. Besides LN0 and LPHD, and to be useful for IEC 61850 applications, there must be other logical nodes in the logical device (see Figure 8.20). Typically, these additional logical nodes are from other logical node class groups beside the general group (e.g., Lxxx). Many logical devices will also contain an addition general logical node to represent the physical device nameplate and health information. Logical devices may form a hierarchy that allows an LLN0 in the parent logical device to control the mode and behavior of the children in the hierarchy.

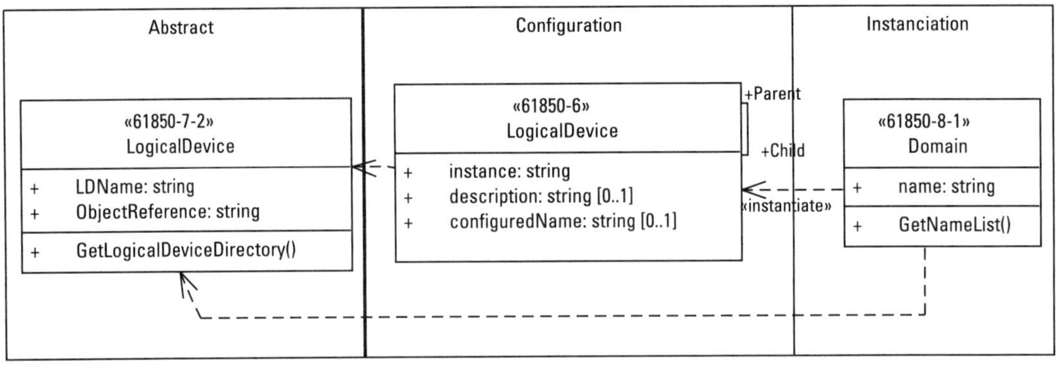

Figure 8.20 Logical device UML.

The abstract LDName is instantiated by the domain name attributed. That value of the domain name is determined by either the configured value of instance or configuredName. The ObjectReference is either the concatenation of the IED name with the instance or the configuredName. The method GetLogicalDeviceDirectory is instantiated by the MMS GetNameList service.

The attributes of a logical device are shown in Table 8.2. The table shows the UML attribute, the actual SCL attribute used to serialize the UML, and a description of the purpose of the attribute.

The ability to change the preconfigured name or instance of a logical device is controlled by ConfLDName Server SCL Capability. If true, the values of the initial provided for instance and configuredName may be redefined during the IEC 61850 engineering process. The resulting name of a logical device forms the basis for all ObjectReferences within that logical device.

8.2.1.1 Logical Device Hierarchy

There is an explicit mechanism to express the hierarchy of logical devices within an IED. The hierarchy is used to determine the hierarchy of management of certain functions within an IED. The design of the logical device management hierarchy was based on the hierarchy of functions and subfunctions that exists for logical-nodes within a logical device.

As an example, it is possible to put an individual logical node into test mode, all logical nodes within a logical device into test mode (e.g., setting the Mod in LN0), but there is no obvious mechanism to command all logical devices in an IED into test mode. The ability to command all logical nodes into test mode is one of the management functions that the hierarchy of logical devices allows (see Figure 8.21).

All logical nodes, except for the logical node that monitors the physical health of a device (e.g., LPHD), has an ability to have its mode and therefore its behavior controlled through controlling the Mod DataObject. Inherently, this means that the mode of an overcurrent function (e.g., PIOC) or measurement function (e.g., MMXU) can be controlled by direct interaction with the logical node itself. However, all modes of logical nodes within a logical device can be set through control-

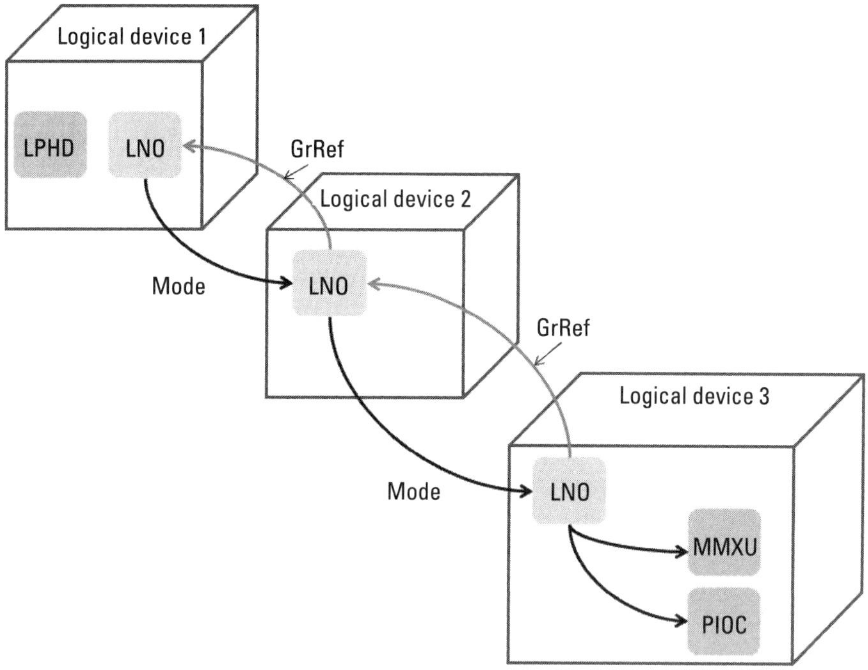

Figure 8.21 Hierarchy of mode control.

ling the Mode of the LN0 that is contained within that specific logical device (e.g., logical device 3).

The management hierarchy is predicated on the design pattern of logical nodes. The logical device hierarchy is a hierarchy created between LN0s that are instantiated in different logical devices. The method of indicating the hierarchy is through a value in one LN0 set to the name of the logical device that contains the mastering LN0. Figure 8.21 depicts that the LN0 is logical device 3 and is controlled by the LN0 in logical device 2. It also shows that the LN0 in logical device 2 is controlled by the LN0 in logical device 1. It is the data object GrRef in the instances of LLN0 (e.g., LN0) logical node that creates the logical device hierarchy. The mechanism of having the Mod information changed in a logical device hierarchy within the same IED is a local issue and is not subject to standardization, but is typically some type of internal messaging or data sharing.

IEC 61850 also has the potential to have a hierarchy of IEDs that control the behavior of other IEDs within the system. In the typical case there is a bay controller coordinating the behavior of the IEDs within its substation bay. The logical device hierarchy is reused to allow a hierarchy to be created based on GrRef values that specify a LN0 that is in a different IED. Since the exchange of the Mod from the controlling logical device must be communicated over the network, and therefore the exchange mechanism must be configured. This is typically accomplished through the logical node ExtRef construct.

Within an internal IED hierarchy, the logical device that has no GrRef value must have an instance of the physical health functions (e.g., LPHD). Other logical devices may also have instances that mirror that LPHD with the exception of the logical device acting as a proxy for another logical device.

8.2.1.2 RTU and Proxies

Data aggregation and proxying concepts were developed as soon as digital computers were developed and are still prevalent in many aspects of business today. The basic need for aggregation arises from the need to easily control access to information, minimize the number of remote entities that a client application needs to support, and automation. Data warehouses[5], data cubes[6], databases, and web portals are all examples of proxied information and information aggregation.

One could argue that the first SCADA system in 1912 was the first utility centric data aggregator as it presented information to humans from multiple substations, even though there were no computers in 1912. As computers and digital communications began to permeate the utility industry, there were mechanical relays and analog I/O. These nondigital sources of information needed to be converted into the digital world. An RTU in the utility industry started as the equivalent of a programable logic controller (PLC) in the industrial domain. However, as automation became more complex, there was a need for more digital devices in the substation. The proliferation of devices was due to a couple of different reasons: proliferation of I/O beyond the capability of a single RTU and the need for resiliency in automation and protection. The evolution and history of RTU functionality is evident in the hardware and communication interfaces that the typical RTU supports. Although RTUs still have the capability to utilize hardwired I/O, there has been an inevitable march forward to a communication centric functionality. The RTUs of today can digitally communicate with backend devices along with utilizing hardwired I/O.

The first RTUs offered upstream communications to a primary and secondary SCADA system, the current generation can provide upstream links to SCADA, maintenance, and other systems. As with PLCs, RTUs evolved to be register centric. Therefore, information from the backend IED registers, acquired by digital protocols, are mapped into the register map of the RTU. It is the register/index map of the RTU that is exposed to the other clients, see Figure 8.22.

This figure depicts the typical methodology that RTUs employ to provide mapping to information to the backend IEDs (e.g., IED 1 and IED 2). It shows that the IEDs can differ in the type of information used (e.g., analog inputs/outputs and well as digital inputs/outputs). It is typical that not all of the information from each IED is aggregated into the RTU. As an example, only the first four of the analog inputs from IED 1 are mapped into and exposed by the RTU.

As with data warehouses, RTUs provide the ability to decimate information from the backend devices. However, using register-based technology means that the registers accessed through the RTU are not the same register assignments that are in the IEDs. Thus, accessing the same information from the IED requires access to a different register than if accessed through the RTU.

IEC 61850 is object oriented and semantically rich. As such, the IEC 61850 standard IEC 61850-7-1 defines a methodology that allows IEC 61850 RTUs to be created instantiating logical devices that represent the logical devices of the backend IEDs. There are two approaches to creating a proxy based on:

5. For more information see: https://en.wikipedia.org/wiki/Data_warehouse
6. For more information see: https://en.wikipedia.org/wiki/Data_cube

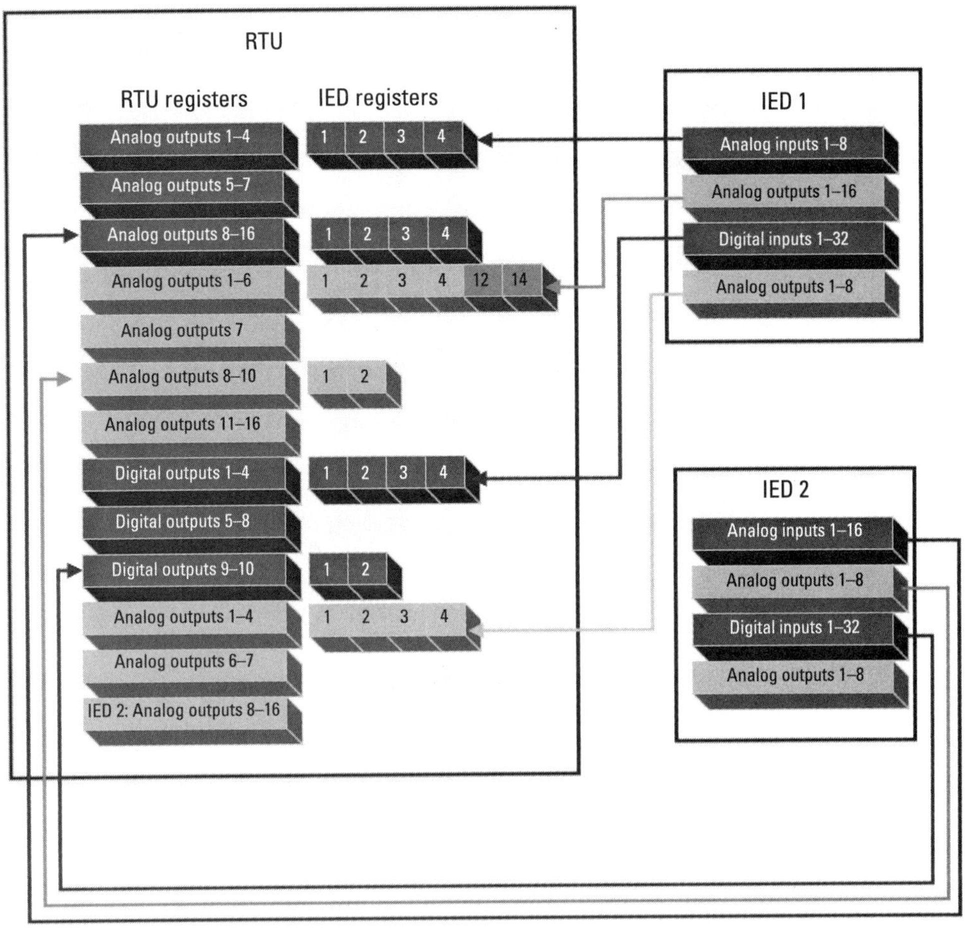

Figure 8.22 Example of non-IEC 61850 RTU register aggregation.

- Creating proxy logical devices whose name is inherited from the proxy's IED name value (e.g., IED naming). This type of naming creates the equivalent to a new instance of logical devices since the data object names will be different.
- Creating proxy logical devices whose name is the same as the logical device names from the backend devices. This approach allows the information to be proxied using the same names as would be used to access the back-end information but through a different communication address. Since the names of the information will typically be the same, this is the preferred naming approach.

The foundation of an IEC 61850 RTU is that a proxy (Figure 8.23) should present the same objects and names of the information in the back-end IEDs. If a logical device name is the same as another name in the system, it represents the logical devices of the same name assumed to contain the same (although potentially decimated) objects of the back-end IEDs.

Figure 8.23 shows two IEDs that the RTU is aggregating (e.g., IED1 and IED2). Each IED has two logical devices (e.g., inst="_1" and inst="_2"). However, IED2 has a logical device with the ldName attribute set to a value of 'ABC'. In order to

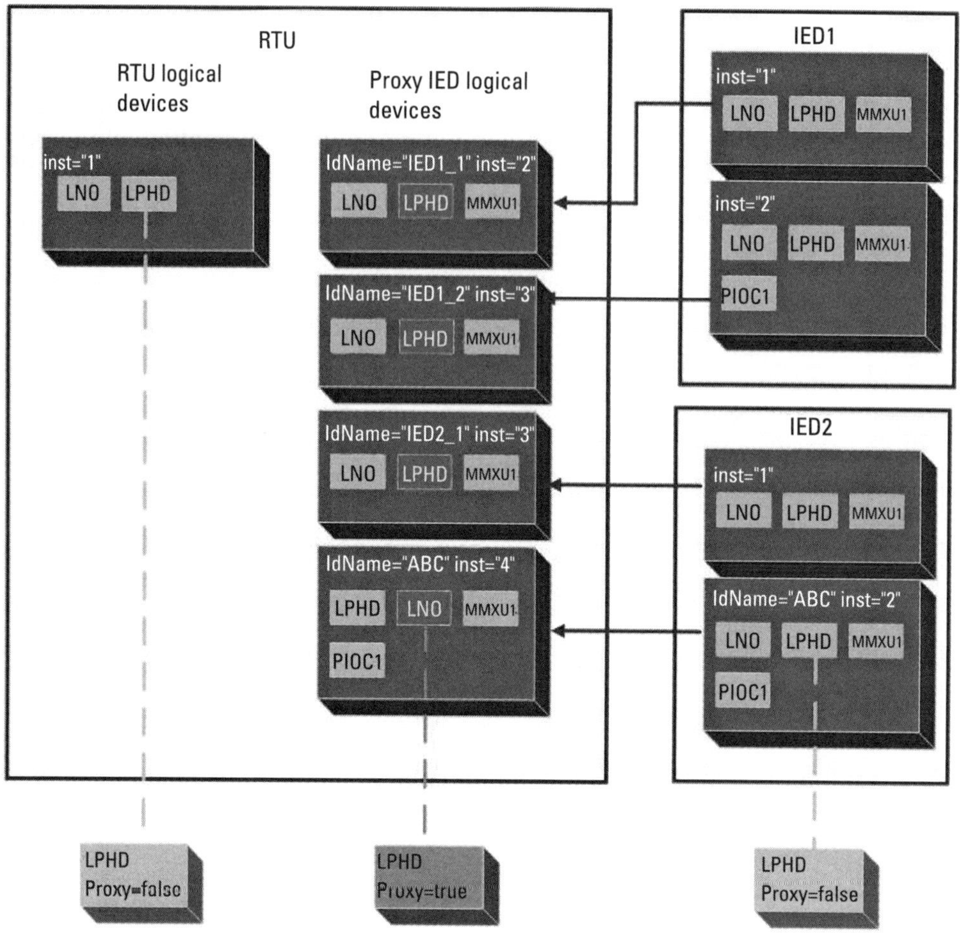

Figure 8.23 Conceptual model of IEC 61850 RTU or proxy.

aggregate the IEDs using the same logical device names, the SCL configuration of the RTU must utilize the configuration of the ldName attribute. It is worthwhile to note that the "inst" attribute values need to be different from the back-end IEDs since it is not allowed to have duplicate values of "inst" within an IED (e.g., the RTU). In order to create the same logical device name as those in the back-end IEDs, there are a couple of simple rules:

- An RTU must have a logical device that contains at least the nameplate and health information of the RTU IED. The figure depicts this as a logical device whose "inst" value is "1".
- If the back-end ldName attribute has value (e.g., not ""), the RTU should be configured to use the ldName attribute value of the back-end IED. The figure shows an example of this type of configuration in the depiction of the aggregation of IED2's logical device "inst" whose value is 2. The figure shows that the IED2's ldName for that instance (e.g., "ABC") is configured in the RTU, but the instance number was changed.

- If the ldName attribute is empty or nonexistent, the RTU should be configured to use a ldName whose value is the concatenated value of the IEDName of the back-end IED and the value of "inst". The figure shows and example of this type of configuration in the depiction of the aggregation of IED1's logical device "inst" whose value is 2. The resulting ldName value in the RTU is IED1_2.
- Within the RTU logical devices that are proxying backend IED logical devices, the LPHD Proxy status value must be set to "true".
- It is also true that the proxied logical devices may contain all, some, or decimated object definitions from the back-end IED logical devices. This is shown by the proxy logical device ldName= "IED1_2" where the PIOC logical node of the back-end device is not represented.
- Based on constraints in SCL, duplication of the same value of logical device name is not allowed within the same SCL access point declaration.

Figure 8.24 is an example of how the configuration of a proxy mimicking the same back-end logical devices might be represented. The "ldName" attribute of the LDevice in the proxy IED is the same value as the name would be for the IED whose name is "T60" with the logical device instance or "Master". If functional naming was utilized in the T60 IED (e.g., ldName had a non-null value), the proxy's ldName would be the same value. Additionally, the figure shows how the proxy is indicated within the LPHD of the proxy logical Device.

The concept of a proxy can be used to create an information aggregation capability that may be required to minimize the number of SCADA connections. However, the proxy construct can also be utilized to decimate backend information, transform the backend information, or create new information based on logic or calculations within the proxy.

```
<IED desc="UR" configVersion="7.60" manufacturer="GE Multilin" name="T60">
  ...
  <AccessPoint name="S1">
   <Server>
    <LDevice inst="Master" ldName="T60Master">
     <LN0 inst=""  lnType="LLN0_0" lnClass="LLN0"/>
     <LN lnType="LPHD_0" inst="1" lnClass="LPHD">
      <DOI name="Proxy">
       <DAI name="stVal" valImport="true" valkind="RO">     Indication of
               <Val>true</Val>                                Proxy
       </DAI>
      </DOI>
     </LN>
    </LDevice>
   </Server>
  </AccessPoint>
</IED>
```

Figure 8.24 SCL example of configuration of a proxy device.

8.2.2 Logical Nodes

Logical nodes represent functions that are used for automation, monitoring, and the creation of other distributed collections. As the application domains for IEC 61850 increase so do the number and types of logical node definitions. Currently, application domains include substations, wind power, hydroelectric, and distributed energy resources. The domains outside of substations and distribution functions are allocated a character in the logical node class (LNClass) name that represents the domain that is responsible for their definition. Additionally, there is a namespace URI defined for all logical nodes that also specifies the domain from which they are defined and administered.

The purpose of this book is not to replicate the detailed information from the various standards, but rather to provide insight into the standards. As an example, IEC 61850-7-4 is, at the time of this writing, 419 pages. It would be worthless to recreate all the information in this standard. However, it is worthwhile to provide a list of the LNClass character definitions and to show how logical nodes from various domains can be used to provide information for a distributed application. This list is provided in Table 8.4.

Prior to Edition 2.1 of IEC 61850-7-4, all logical node definitions inherited from logical node. However, with the introduction of UML modeling as part of Edition 2.1, there is a hierarchy of abstract logical node classes from which all the actual logical node definitions inherit.

Figure 8.25 shows a partial representation of the inheritance hierarchy of logical nodes. It shows that the ProtectionLN inherits from FunctionLN which inherits from StatisticalLN which inherits from DomainLN which inherits from LogicalNode. This inheritance structure (e.g., abstracts) are never instantiated directly. There are LNClass definitions that inherit from various levels of the abstract inheritance hierarchy and various examples of this are shown in the figure (e.g., LLN0, MMXN, and PIOC). These examples show how the attributes defined in the Abstract hierarchy appear in the LNClass definitions. One of the impacts of utilizing the inheritance hierarchy, is there is no mechanism to absolutely define the order of the data objects within the LNClass definitions.

However, SCL has a different perspective of logical nodes. It separates logical nodes via its use during the engineering process. The separation is based on ClientLNs and logical nodes contained by a server and logical device. Any logical node can be contained in a logical device and they provide information that can be exchanged between different IEDs. ClientLNs are used to configure subscriptions (e.g., reporting, GOOSE, and Sampled Values) but do not expose information that can be exchanged (e.g., read, reported, or published). ClientLNs are typically specializations from the NonProcessInterfaceLN abstract logical node. Typically, these have an Ixxx designation but can be any logical node that is not exposing information to be exchanged. ClientLNs are consumers of reporting, GOOSE, and Sampled Values. Additionally, ClientLNs can issue control commands and write information, set setpoints, select settings groups, and other functions. ClientLNs are defined not by the logical node hierarchy but by where they are positioned in the SCL file. This definition can be expressed as the following UML as shown in Figure 8.26.

Table 8.4 Substation and Distribution Related Functional Groups

Fuctional Group	*Designation*	*Most Typically Used*	*Usage*
System	Lxxx	LLN0, LPHD, LTIM	Provides information regarding the IED nameplate (LLN0), health (LPHD), and time (LTIM) as well as supervision for exchange mechanisms such as GOOSE, Sampled Values, reporting, and telecontrol communication channels. These are IED specific and typically would not be represented in a SLD.
Automatic Control	Axxx	ATCC, AVCO	Represent automatic control functions such as tap changer (ATCC) and voltage (AVCO). These functions are typically bound to primary or secondary equipment in an SLD.
Control	Cxxx	CILO, CSWI	Represents a set of functions that allow control of switches (CSWI) and interlocking (CILO). These functions typically interact with logical nodes that represent functionality of the primary or secondary equipment being controlled. As an example, a CSWI interacts with an XSWI which represents a switch which is bound to a position in a SLD. The "C" functions are typically bound to primary or secondary equipment in a SLD.
Generic Functions	Fxxx	FSCH, FSCC	Represents a set of functions that provide generic functionality such as schedules (FSCH) and the ability to control the selection of what schedule is being executed (FSCC).
Generic	Gxxx	GGIO	Represents a set of generic functions that expose internal information of the IED such as input/output information (GGIO), security, and others.
Interfacing and Archiving	Ixxx	IHMI	Represents a set of functions that typically represent client functions that are used for subscription purposes for reporting, GOOSE, and Sampled Values. The IHMI function is typically used to represent a SCADA system or substation operator interface (e.g., user interface). Many times, these functions do not require the construct of an IEC 61850 server. These functions are almost never bound to primary or secondary equipment in an SLD.
Non-electric	Kxxx		These functions were generalized from other domains such as hydroelectric where pumps, fans, tanks, filters, etc. are needed. However, other domains need similar functions such as industrial plants and distributed energy resources. Thus, they were generalized, as an example, to Kxxx from Hxxx. These functions are almost never bound to an SLD. The main exception would be KFAN as power transformers typically have fans.
Metering and Measurements	Mxxx	MMXU, MMTR, MMET	Represents a set of functions that provide data acquisition of various quantities such as power related information (e.g., amps, volts, frequency, etc.) of both AC and DC systems. The meteorological (MMET), environment (MENV), and hydrological (MHYD) functions are functions that were generalized from other domains and are useful in all identified domains. These power related functions (e.g. MMXU and MMTR) are almost always bound in a SLD.
Protection	Pxxx	PDIS, PTUV, PDIR, PIOC, and more	This set of functions is based on the ANSI/IEEE protection function specification[1]. There is not a one-to-one correlation between the IEEE numbers and the "P" nodes due to the IEC 61850 functions having configurable behavior. Currently distance (PDIS or IEEE 21), under voltage (PTUV or IEEE 27), directional (PDIR or IEEE 32), instantaneous overcurrent (PIOC or IEEE 50), as well as other protection mechanism are common. A "cheat sheet" of IEEE numbers versus IEC 61850 logical nodes is provided in this book. These functions are almost always bound to a SLD.

Table 8.4 (continued)

Fuctional Group	*Designation*	*Most Typically Used*	*Usage*
Power Quality Events	Qxxx		This set of functions allow events to be generated based on IED algorithms that detect frequency variations, transients, and other variations. These functions use information from other logical nodes such as Mxxx nodes for their analysis. These functions can be bound to an SLD.
Protection Related	Rxxx		This set of functions represent information that is generated based on protection or event detection actions in the IED. These events can create disturbance recordings known as COMTRADE files for retrieval. The logical nodes of RADR, RBDR, RDRE, RDRS all represent information related to COMTRADE. There are other functions in this group that expose information regarding fault location, direction, reclosing, and more. To be able to understand the power system topology of these functions, they should be bound to an SLD.
Supervision	Sxxx	SLTC, SSWI, SPTR	This set of functions provide information regarding tap changers (SLTC), switches (SSWI), power transformers (SPTR), and other primary or secondary equipment. If a "C" logical node exists for the same type of equipment the controller and supervision functions are typically both present in the SLD.
Instrument Transformers and Sensors	Txxx	TCTR, TVTR	This set of functions provide the representation of various types of sensors. Transmission and distribution systems are reliant on current transformers (TCTR) and potential (e.g., voltage) transformers (TVTR) for providing information to the actual devices in the substation. These functions are typically bound into the SLD and can be conveyed through devices performing Sampled Values called merging units. There are other types of sensors that provide angle, frequency, distance, pressure, and more.
Switchgear	Xxxx	XCBR, XSWI	This set of functions provide the representation of actual switchgear such as circuit breakers (XCBR) and switches (XSWI). These functions are bound to a SLD.
Power System Equipment	Zxxx	ZGEN, ZBAT, ZLIN	This set of functions provide information regarding equipment such as generators (ZGEN), batteries (ZBAT), and lines (ZLIN). These functions are typically bound to an SLD.

See http://www.ece.uidaho.edu/ee/power/ECE525/Lectures/L3/L3.pdf for the list of functions.

An example of the generated XML for the SCL representation of a ClientLN follows.

The XML, in Figure 8.27, shows that a single IED can contain a ClientLN (e.g., IARC) as well as a server. The server has two logical devices with other logical nodes including an IHMI. This example is provided to demonstrate that a logical node can be configured to expose information (e.g., via being included in a server) or to not expose information (e.g., not included in a server).

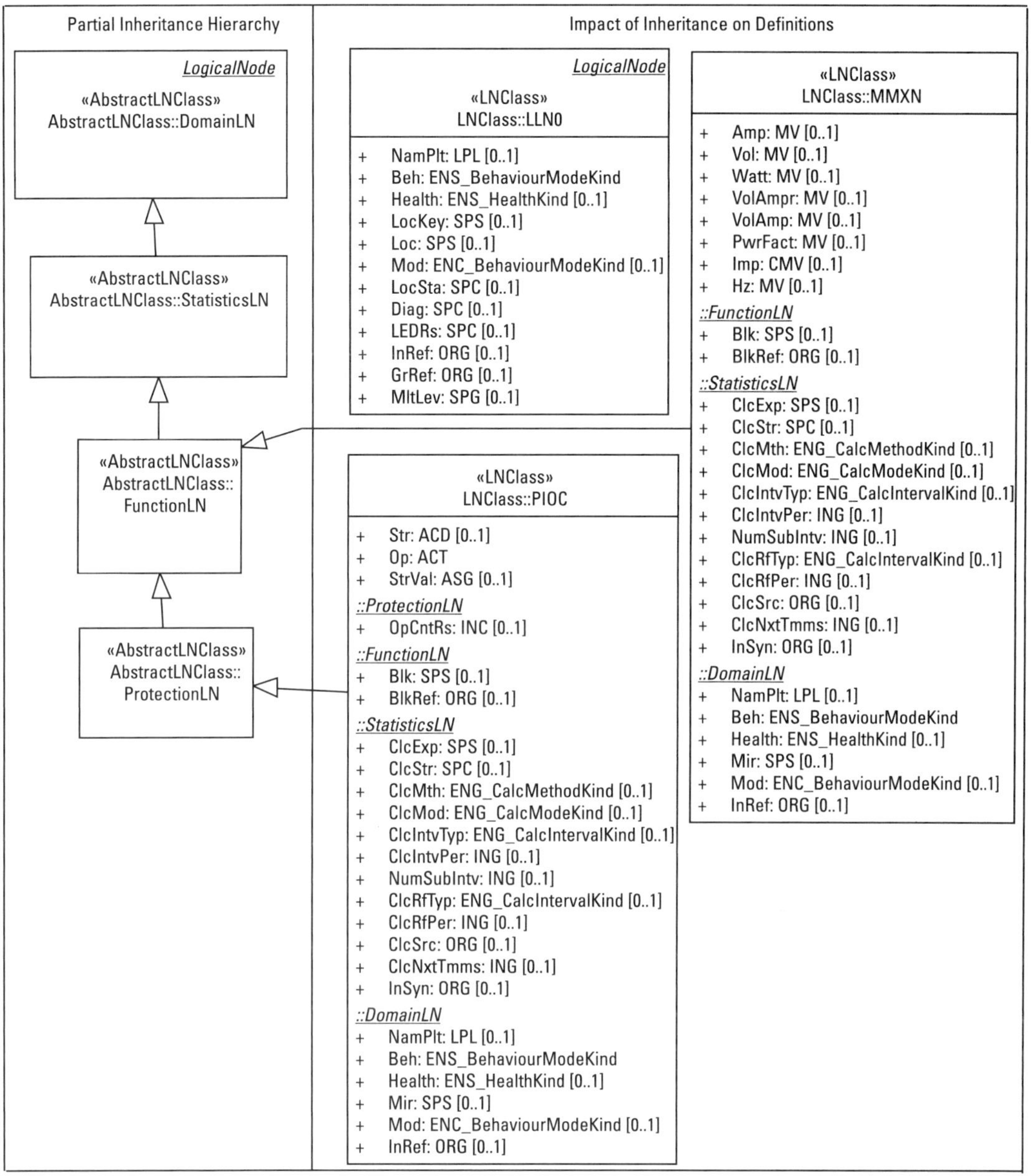

Figure 8.25 Example of logical node inheritance.

8.2.2.1 Structure

The definition of a logical node is different depending on the perspective (e.g., abstract, configuration, or instantiation). Figure 8.28 depicts the UML with the various perspectives.

The concept of a logical node is a reusable function. In the abstract, a logical node is composed of a set of data objects, which are composed of a sequence data attributes, which are in turn typically composed of a sequence of data. Logical nodes, data objects, and data attributes are collections of a combination of mandatory (e.g., must be present in the definition) and optional (may be present in the definition) information.

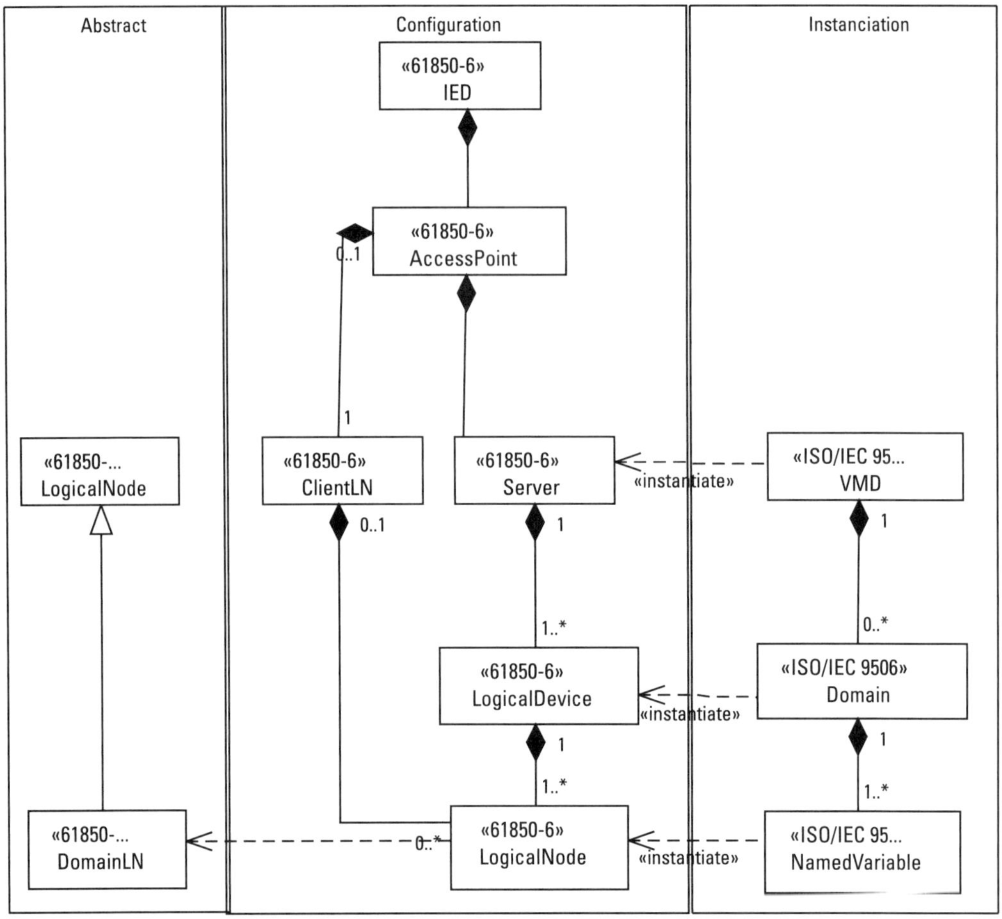

Figure 8.26 UML definition of ClientLN.

However, the configuration and engineering declaration of logical node instance definitions becomes much more complicated. A logical node instance references a logical node type (LNType) which contains the list of data objects that defines the set of data objects. At the LNType definition level, the order of the DataObject definitions is not required to match the actual definition in the standard (e.g., IEC 61850-7-4 or others). Figure 8.29 helps to demonstrate the concept.

The logical node instance is defined in the <LN> XML production. It contains the reference to the LNodeType. The reference is the value of the "id" attribute in the LNodeType production. There are three other attributes: prefix, lnClass, and inst. The lnClass value specifies the abstract logical node from which this instance inherits. However, the actual definition of the instance is defined by the referenced LNodeType. A single LNodeType may be referenced by multiple instances. The order of the <DO> definitions within a LNodeType is not defined and can be in any order.

The <DO> (e.g., data object) definition, within an LNodeType provide a "name" attribute whose value is the name of the data object, a reference to the DOType that defines the structure of the data object as shown in Figure 8.30.

```
<IED name="AA1D1Q01FN1">
        <AccessPoint name="S1">
            <LN lnType="IHMI" inst="1" lnClass="IHMI">
        </AccessPoint>
</IED>
```

Figure 8.27 SCL example of ClientLN.

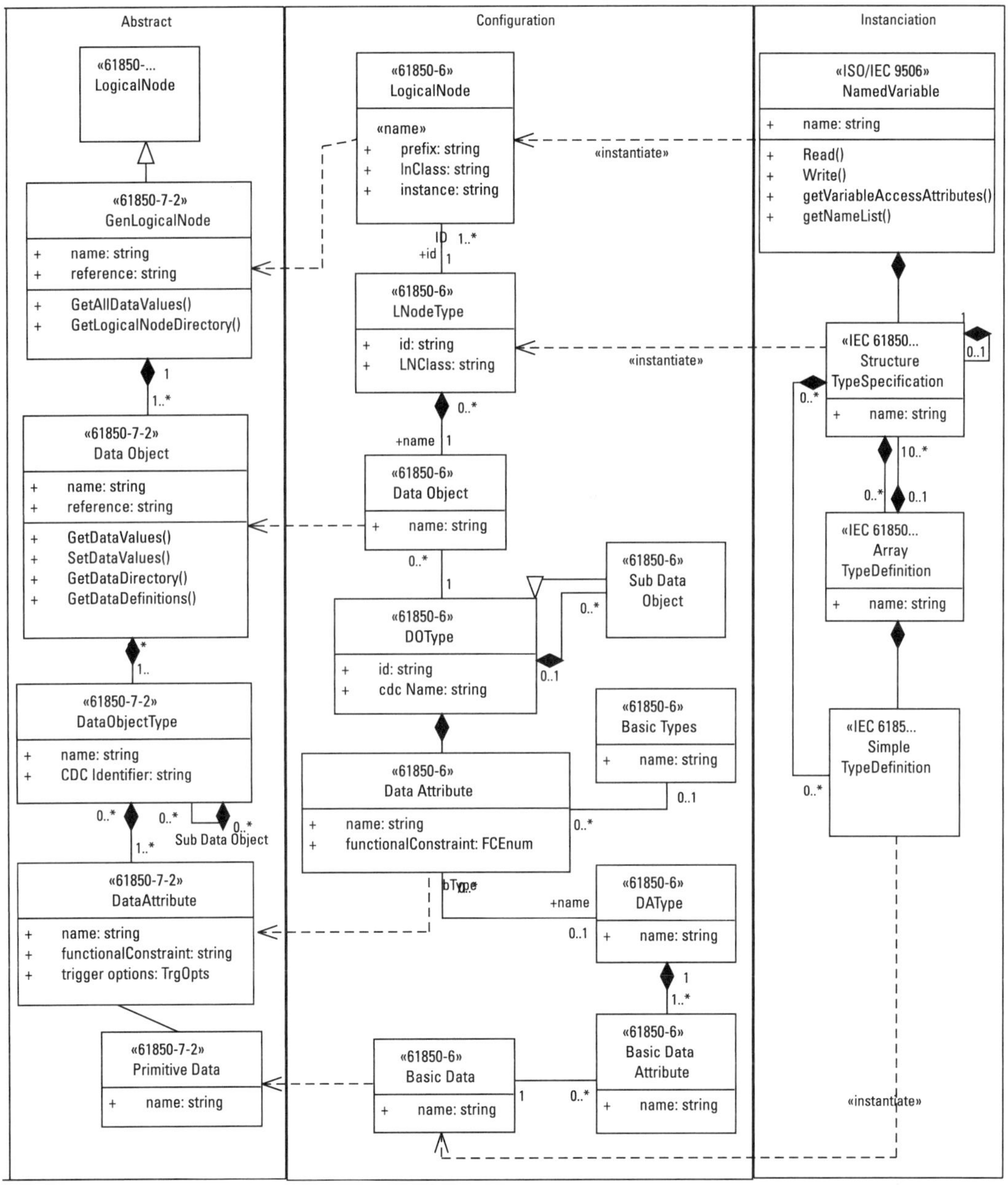

Figure 8.28 UML definition of logical nodes.

```
<LN lnType="PIOC_0" inst="1" lnClass="PIOC" prefix="PhsIoc">

  <LNodeType id="PIOC_0" lnClass="PIOC">
    <DO name="NamPlt" type="LPL_1"/>
    <DO name="Beh" type="ENS_0"/>
    <DO name="Str" type="ACD_1"/>
    <DO name="Op" type="ACT_2"/>
  </LNodeType>
```

Figure 8.29 SCL Example of logical node definition.

```
<LNodeType id="PIOC_0" lnClass="PIOC">
  <DO name="NamPlt" type="LPL_1"/>
  <DO name="Beh" type="ENS_0"/>
  ........
</LNodeType>

<DOType id="ENS_0" cdc="ENS">
  <DA bType="Enum" name="stVal" type="BehaviourModeKind" dchg="true" fc="ST"/>
  <DA bType="Quality" name="q" fc="ST" qchg="true"/>
  <DA bType="Timestamp" name="t" fc="ST"/>
  <DA bType="VisString255" name="d" fc="DC"/>
  <DA bType="VisString255" name="cdcNs" fc="EX"/>
  <DA bType="VisString255" name="cdcName" fc="EX"/>
  <DA bType="VisString255" name="dataNs" fc="EX"/>
</DOType>
```

Figure 8.30 SCL example of data object definition.

A single DOType may be referenced by multiple data object or sub-data object definitions. The DataAttribute <DA> definitions are defined in a similar manner but refer to a primitive type or structure definition. DOTypes are instantiations (e.g., may be a subset) of the common data classes defined in IEC 61850-7-3.

Enumerations (e.g., bType="Enum") are a special case for configuration. In the standard, all ordinals are defined, see Figure 8.31. User supported values are a subset of the standard definitions.

If an implementation needs a custom enumerated value, the ordinal value must be a negative value.

When the definitions are instantiated in IEC 61850-8-1, or IEC 61850-8-2, the result is a nested structure NamedVariable. The name of the variable is the combination of the attributes of the <LN> instance definition: <prefix><lnClass><inst>. This name of the instance is known as the logical node name (e.g., LNname). The value of "inst" may be no more than seven digits. The value of prefix + inst shall be no more than 12 characters. This means that the variable name is no more than 16 characters. The order of the nest structure is reordered slightly so that information of the same functional constraint can be retrieved. The first three levels of the variable is <LNname>.<Functional Constraint>.<Data Object Name>. The structure variable is contained in a domain whose name is the name of the logical device.

```
<EnumType id=" BehaviourModeKind">
    <EnumVal ord="1">on</EnumVal>
    <EnumVal ord="2">blocked</EnumVal>
    <EnumVal ord="3">test</EnumVal>
    <EnumVal ord="4">test/blocked</EnumVal>
</EnumType>
```

Definition in Standard

```
<EnumType id=" BehaviourModeKind">
    <EnumVal ord="1">on</EnumVal>
    <EnumVal ord="5">off</EnumVal>
</EnumType>
```

Subset Supported in Implementation

Figure 8.31 SCL example of enumeration definition.

Since the information in the named variable can be large, IEC 61850-8-1 and IEC 61850-8-2 provides a decomposition of the large variable into multiple smaller variables. The names of these variables are based on the main variable name and the nested structure names separated by "$".

Another way understand logical nodes is as a construction of Lego blocks; see Figure 8.32.

Figure 8.32 shows that one logical node might represent a measurement function. Within IEC 61850 the definition of the measurement unit function is defined as an abbreviation (e.g., LNClass) of MMXU. The definition (e.g., the LNodeType) consists of multiple data objects (DOs). The figure shows that two such data objects are measurements of frequency (e.g., Hz) and phase to ground voltage (e.g., PhV).

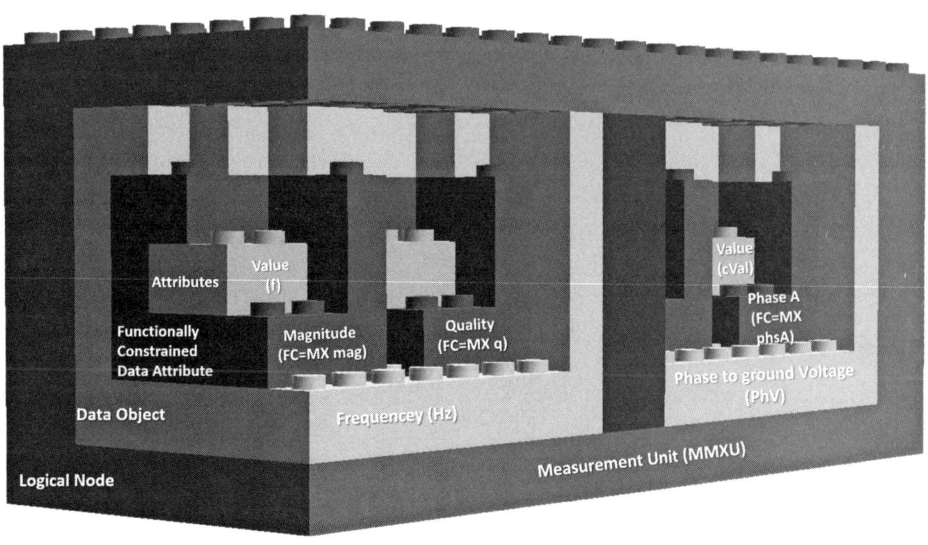

Figure 8.32 Lego representation of a logical node.

The frequency measurement has multiple data attributes whose functional constraint indicates a measurement (e.g., MX) and has the magnitude (e.g., mag) value (e.g., f). Another data attribute is the quality (e.g., q).

The functional constraints are a set of string values and are defined in Table 8.5.

The SCSMs (e.g., IEC 61850-8-1 and IEC 61850-8-2) add an additional FC of 'CO' so that the service parameters needed for the control model can be manipulated as data.

Logical nodes are grouped into functional groups as shown in Tables 8.4 and 8.6. The group to which a specific logical node type belongs is determined by the first character of the LNClass designation that has been assigned by the standards. The example shown in Figure 8.32 shows an example of an LNClass whose abbreviation is MMXU and therefore it belongs to the measurement functional group.

The logical node type (a.k.a. LNodeType) is defined as a combination of mandatory and optional data objects. The constraint that required a mandatory ordering of data objects was removed in Edition 2 of IEC 61850. Therefore, the list of data objects is a set of semantics whose names may appear in any order. Figure 8.33 depicts the relationships that create the logical node structure hierarchy for an MMXU.

Table 8.5 List of Functional Constraints

FC Abbreviation	*Indicates*	*Defining Standard*
MX	Measurement information	IEC 61850-7-2
ST	Status information	IEC 61850-7-2
SP	Setting (e.g., setpoint)	IEC 61850-7-2
SV	Substitution	IEC 61850-7-2
CF	Configuration	IEC 61850-7-2
DC	Description	IEC 61850-7-2
SG	Setting group	IEC 61850-7-2
SE	Setting group editable	IEC 61850-7-2
SR	Service response	IEC 61850-7-2
OR	Operate received	IEC 61850-7-2
BL	Blocking	IEC 61850-7-2
EX	Extended definition	IEC 61850-7-2
CO	Control	IEC 61850-8-1 and 8-2

Table 8.6 LNClasses for Non-Substation or Distribution Application Domains

Application Domain	*Administrative Working Group*	*Functional Group*	*Namespace*
Wind	IEC TC88	Wxxx	IEC 61400-25-2:<revision>
Substation & Distribution	IEC TC57 WG10	Many	IEC 61850-7-4: <revision>
Distributed Energy	IEC TC57 WG17	Dxxx	IEC 61850-7-420: <revision>
Hydroelectric	IEC TC57 WG18	Hxxx	IEC 61850-7-410: <revision>

Note: <revision> is a character that represents the revision of the standard. It is noteworthy that the value of this character implies the date of the release of the standard as well.

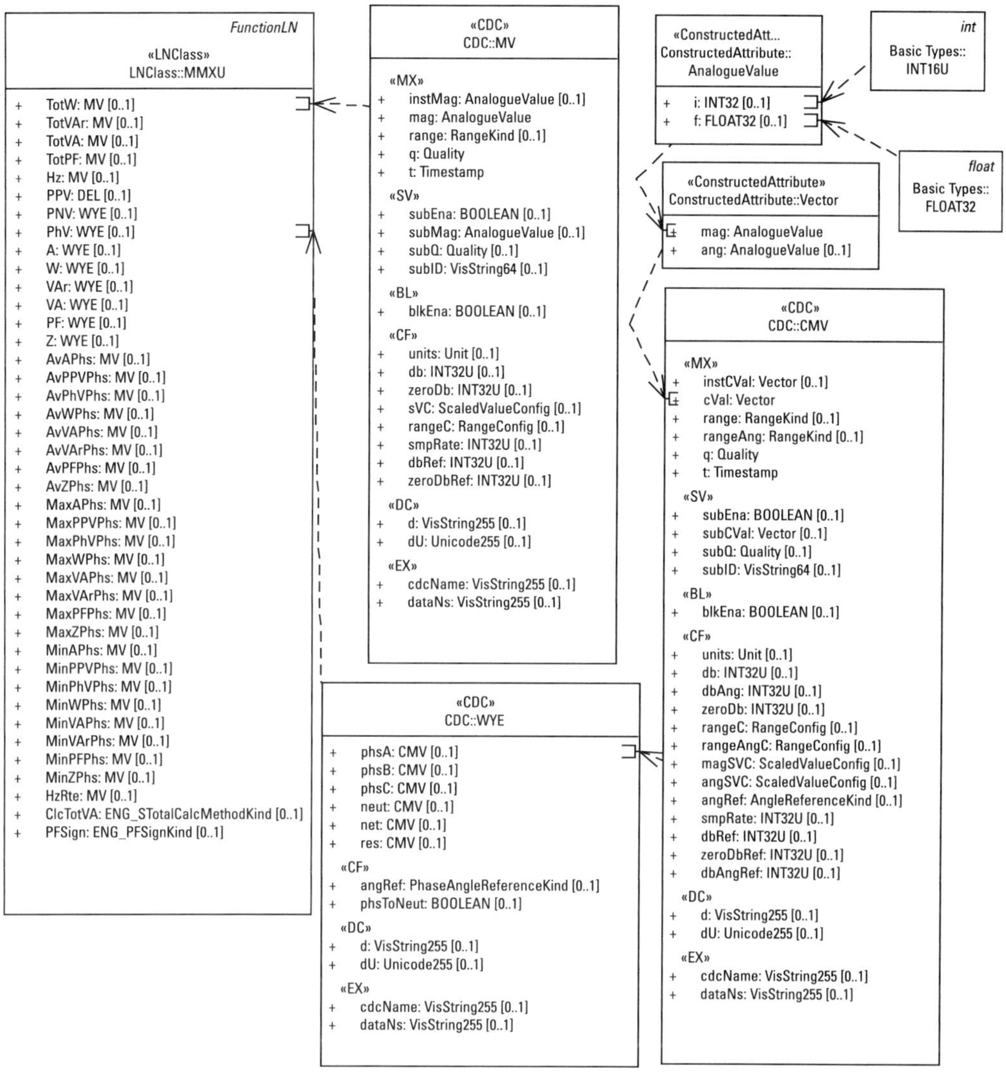

Figure 8.33 UML example of logical node structure.

Figure 8.33 shows that the LNodeType for the LNClass of MMXU is specified in IEC 61850-7-4. Each data object (e.g., Hz and PhV) are based on DOType definitions found in IEC 61850-7-3.

In the figure

- Hz is a DOType that references a common data class (CDC) defined in IEC 61850-7-3 named MV (measurement value). The MV CDC definition shows the groupings of CDC attributes based on functional constraints as well as named attributes such as 'mag.' The definition of 'mag' contains subattributes of a floating-point value (e.g., 'f')[7] or an integer value (e.g., 'i'). These attributes are defined to be basic data types defined in IEC 61850-7-2.

7. IEC 61850 has a stated preference for floating point values and many implementations no longer provide both floating point and integer values.

- PhV is similar in definition to Hz; however, not all of the data objects have a functional constraint (e.g., 'phsA'). It is these types of data objects that refer to other CDCs and that type of definition is known as a sub data object (SDO) in IEC 61850-6. The data object of phsA is defined to be a CDC defined as WYE. It is in the WYE definition where all the attributes are grouped to a functional constraint.

Figure 8.34 shows that logical nodes contain additional objects beside data objects, which were already discussed in Chapter 5. Besides the control blocks datasets are also contained in a logical node.

The abstract definition of a logical node (see Figure 8.34) can contain one or more different types of control blocks. To instantiate the control blocks concretely, the SCSMs have added the following additional functional constraints in order to indicate the different types of control blocks (see Table 8.7). Each of the control blocks has associated IEC 61850 service behavior that is controlled by the control block.

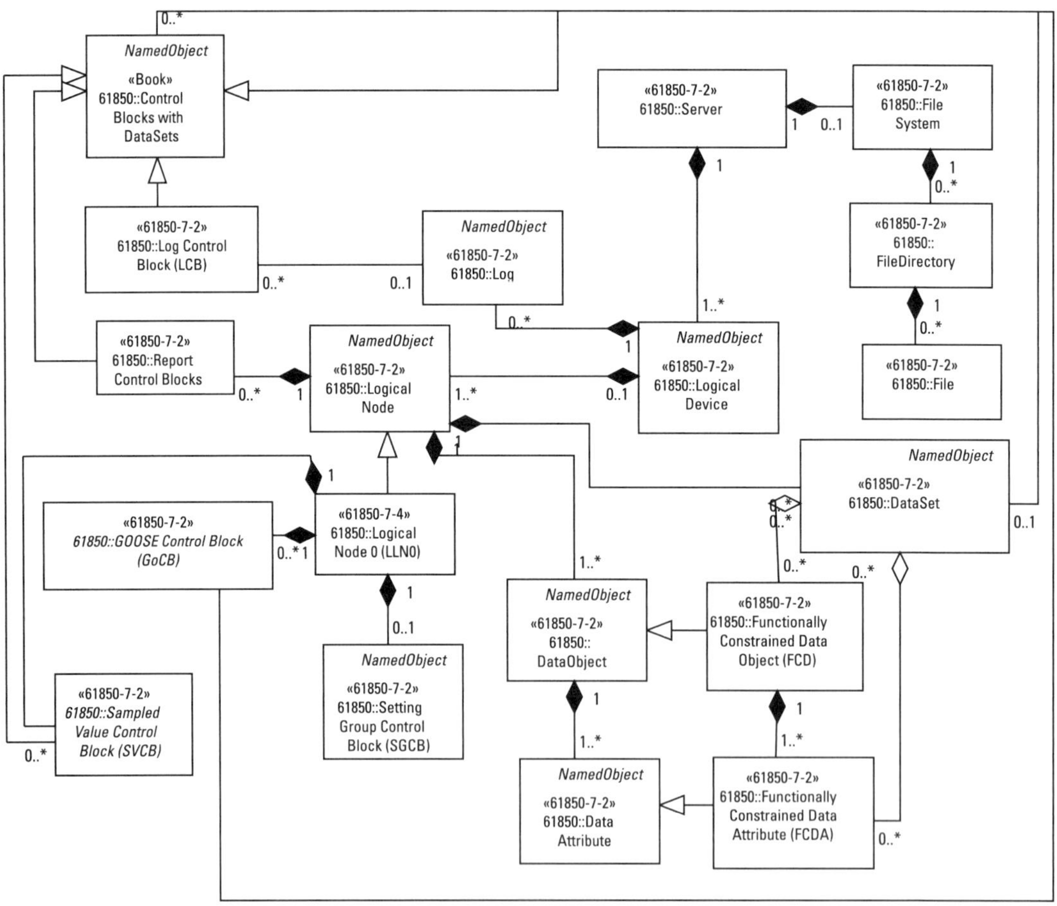

Figure 8.34 IEC 61850 server objects–duplicate.

Table 8.7 List of Control Block Functional Constraints

FC Abbreviation	*Control Block Indicated*	*Instantiated in*
RP	Unbuffered Report	IEC 61850-8-1 and 8-2
BR	Buffered Report	IEC 61850-8-1 and 8-2
GO	GOOSE (Layer 2)	IEC 61850-8-1 and 8-2
MS	Multicast Sampled Measured Value Control Block (Layer 2)	IEC 61850-8-1 and 8-2
RG	Routable GOOSE Control Block	IEC 61850-8-1 and 8-2
RS	Routable Sampled Measured Value Control Block	IEC 61850-8-1 and 8-2

8.2.2.2 Controls

Controls are typically used by substation or control center human beings to manually impact the state of the power grid. As an example, a control center is staffed with human beings that are tasked to maintain the operation and stability of the electrical grid. These individuals are typically known as operators. In some instances, operators must manually intervene to protect electrical grid assets from damage or to minimize the spread of power issues. Their actions require an interaction with the breakers/switches on the electrical grid to control the flow of power. This type of interaction is facilitated by what is known as control.

Controls allow a client to interact with a server and cause an action based on the issued command. In addition to controls being used to open/close breakers, they can be used to adjust automation processes and to synchronize multiple actions at the same time. IEC 61850 has a limited set of control interaction patterns:

- *Direct Operate:* This control basically commands a server to take an action immediately. The command action may be delayed due to checking for certain local constraints to be met (see the *Constraint Checking (Test and Check)* discussion in Section 8.2.2.2). If the local constraints are not fulfilled, the control may be refused. See the *Direct Operate Interaction Pattern* discussion in Section 8.2.2.2 for more information.
- *Select Before Operate:* Like Direct Operate with the exception that it provides a reservation of a control object so that only a single client can issue a control. This is the equivalent of placing a computer semaphore/lock on a resource or database entry so that only a single application can manipulate that resource. See the *Using the Select Operation: UEA SBQ* discussion in Section 8.2.2.2 for further information.
- *Time Activated Control:* There are certain situations where a command needs to be performed at a specific time of day. Typically, this type of control would be used to coordinate multiple controls across different IEDs. See the *Time-Activated Control* discussion in Section 8.2.2.2 for further information.

The interaction patterns consist of the use of a series of primitive services such as Operate, Cancel, Select, SelectWithValue, and TimeActivatedOperate. Figure

8.35 shows these services as UML operations.[8] These operations typically return either the success or failure of the operation. When the configuration parameter of ctlModel specifies an integration pattern with "enhanced-control", the return value of the operations is augmented by the ability to provide SupervisoryInformation that is delivered through the CommandTermimation service.

Typically, configuration can change the sequence of primitives that form an integration pattern. It is rare that configuration can create an additional pattern. However, the ctlModel value also allows the configuration of a control object to be status-only. If a control data object is configured for status-only the object will respond negatively to any of the service primitives. The configuration, instantiation, and use of this pattern will be discussed in the *Status Only Interaction Pattern* section of Section 8.2.2.2.

Controls are embedded data objects within logical nodes. Unlike control blocks, the object behavior (e.g., what is controlled) is defined by the data object and the interaction pattern can be controlled by the client that issues the control. There are specific common data classes defined to be controllable. The CDCs allow controlling of discrete, analog, and enumerated objects.

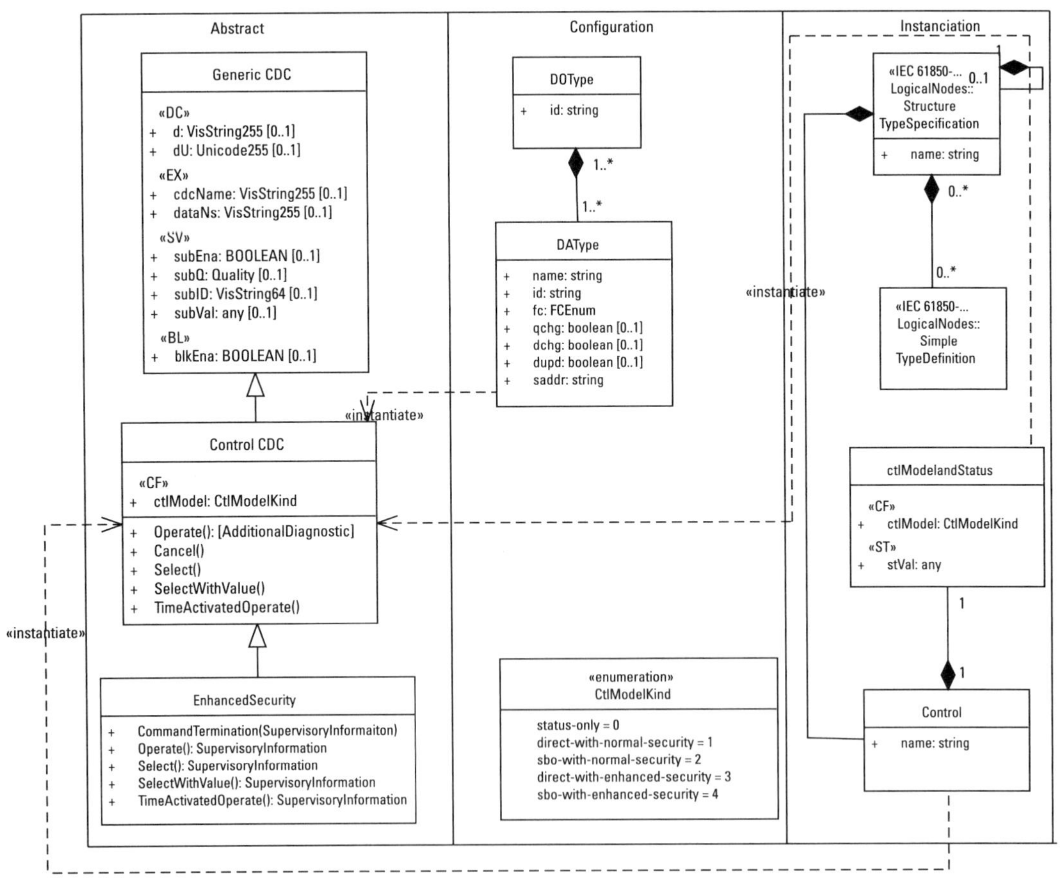

Figure 8.35 IEC 61850 control relationships.

8. The parameters that are conveyed by the operations are not shown.

Figure 8.35 shows the relationship of the abstract common data class (CDC) definition for control. The abstract portion of the UML is a partial representation of the actual CDC inheritance. The abstract control CDC has additional operations (e.g., Operate, Cancel, Select, SelectWithValue, and TimeActivatedControl) that are used to support the integration patterns. There is a mandatory configuration parameter that is used to specify the integration pattern that is to be supported. The operation/service definitions are defined in IEC 61850-7-2 and the abstract CDC definitions are found in IEC 61850-7-3.

Although IEC 61850-7-3 abstracts the control value as a service parameter (e.g., a FC=SR). However, actual object configuration and instantiation typically convert these to a FC=CO (e.g., control). The configuration and instantiation aspects of IEC 61850 instantiate the control value as part of the data (e.g., a data attribute). The IEC 61850-7-2 operations (e.g., operate, cancel) are also instantiated as DataAttributes. The control configuration types can be viewed in Figure 8.36.

Figure 8.36 shows, except for StatusOnly, that there three mandatory data attributes of any control object:

- *ctlModel*: This attribute is a configuration attribute that specifies the overall type of control (e.g. StatusOnly, DirectOperate, SelectBeforeOperate) as well as the interaction pattern to be used (e.g., Normal Security or Enhanced Security).*ctlVal:* This is a value that represents the actual control operation being commanded.
- *stVal:* This is the current value of the object that is to be controlled.

The interaction pattern of Normal Security, at its most rudimentary, is shown in Figure 8.37.

The interaction pattern for direct-with-normal-security starts in the abstract domain with a client issuing an operate service that contains the specification of which server (e.g., association) over which to issue the control, the data object to control, and the control value (e.g., open/close). The abstract operate service is instantiated by IEC 61850-8-1 and IEC 61850-8-2 as being mapped to a ISO 9506 (e.g., MMS) write request. The Write_req is issued over the specified association and contains the data object as a named variable and the control value. If the constraint checks succeed, the command begins to be executed and a successful Write_response is returned. In turn this becomes a successful Operate_response. When the commanded position is reached and the status of the object changes, it is also possible to return a report of the status change.

The difference between normal and enhanced security is the requirement of an additional message known as CommandTermination. This message is used to return that the actual control has completed. In the case of a successful control, the message indicates that the commanded movement (e.g., open or close) has completed. In the instance of a failed control, the message returns AdditionalDiagnosticInformation.

Figure 8.38 shows that intermediate information is returned by an MMS InformationReport that contains a NamedVariable representing the DataAttribute Oper written with the Operate command. The MMS Write_response+ returns Success (e.g., that the Oper structure had been successfully written). Regardless of

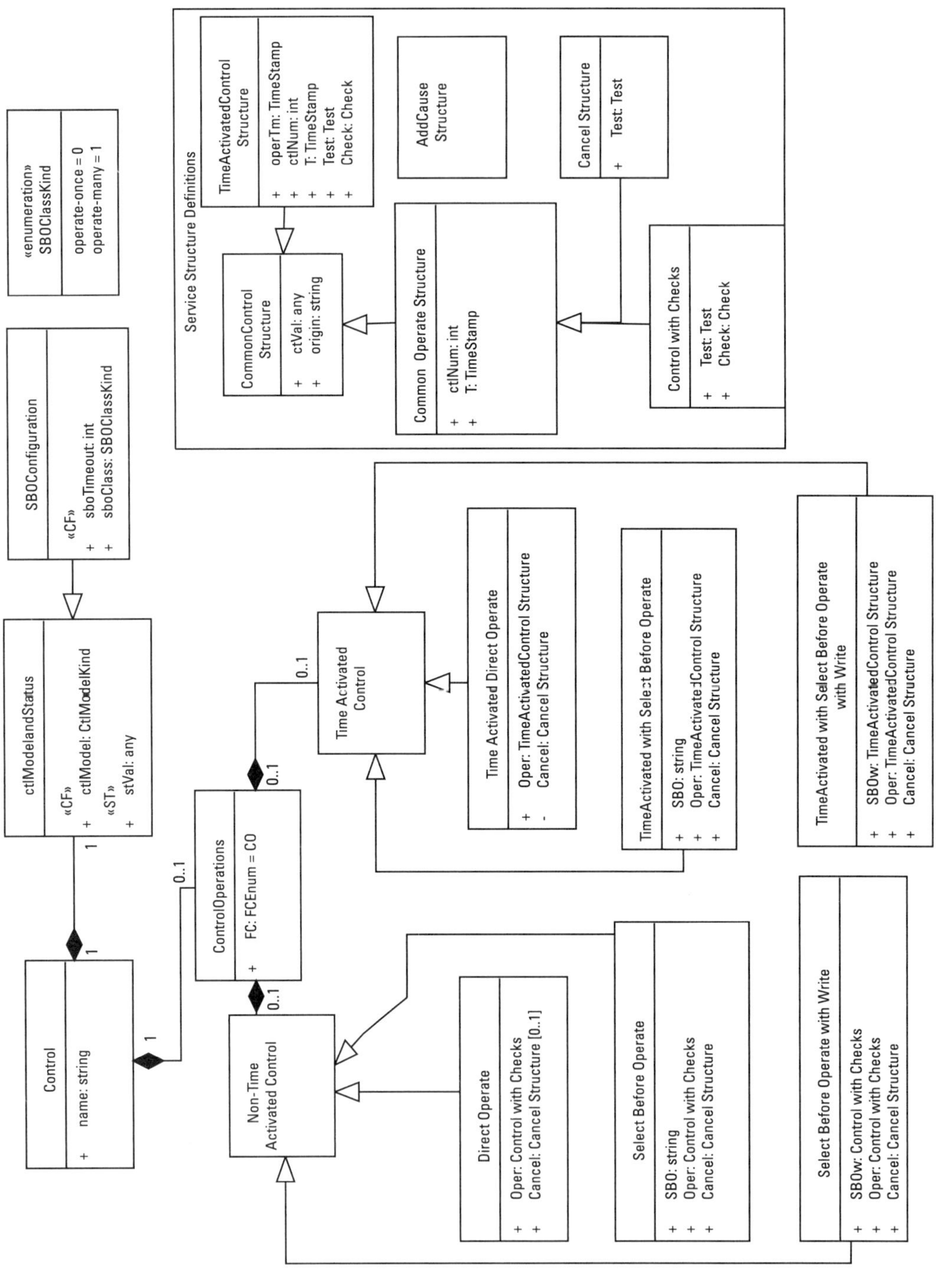

Figure 8.36 Control types and data structures.

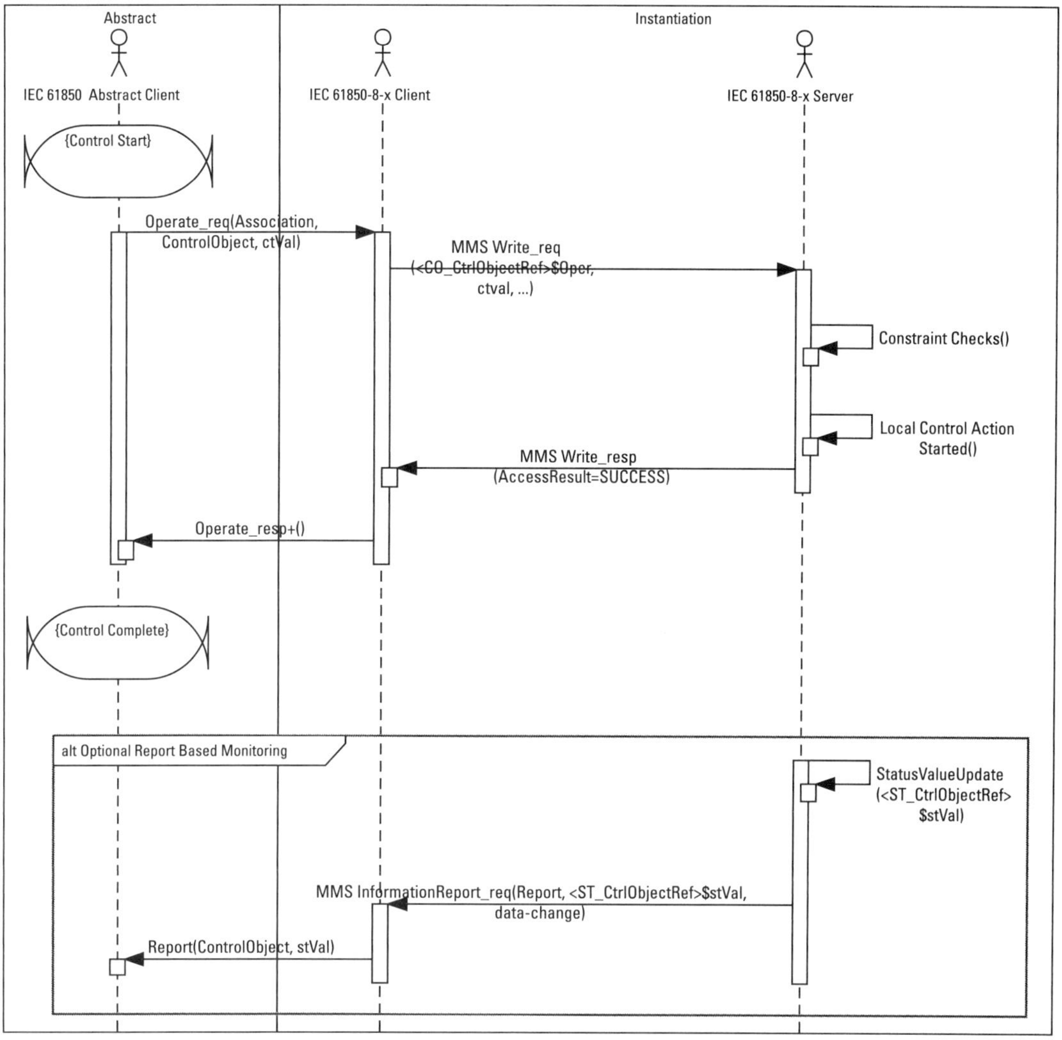

Figure 8.37 Direct with normal security interaction pattern.

the state diagrams in IEC 61850-7-2, both the intermediate information and the Operate_response need to be received to declare that the Operate has completed.

The pattern for a failed Operate, using enhanced-security, is shown in Figure 8.39. There are two major differences between Success and Failure. The first is that the InformationReport includes a NamedVariable that represents the last detected application error regarding the control operation (e.g., LastApplError) and the Oper structure that was utilized to initiate the control. Additionally, the Write_response is returned with a AccessResult whose value indicates a Failure. It is the combination of the reception of the LastApplError and Write_response(Failure) that provides the indication that the control is complete.

As a general rule, if the client receives an InformationReport that contains LastApplError, the control is in the process of terminating and no further control related indications will occur. This applies to all enhanced control related operations (e.g., Operate, Cancel, SelectwithValue, and TimeActivated) with the exception of Select.

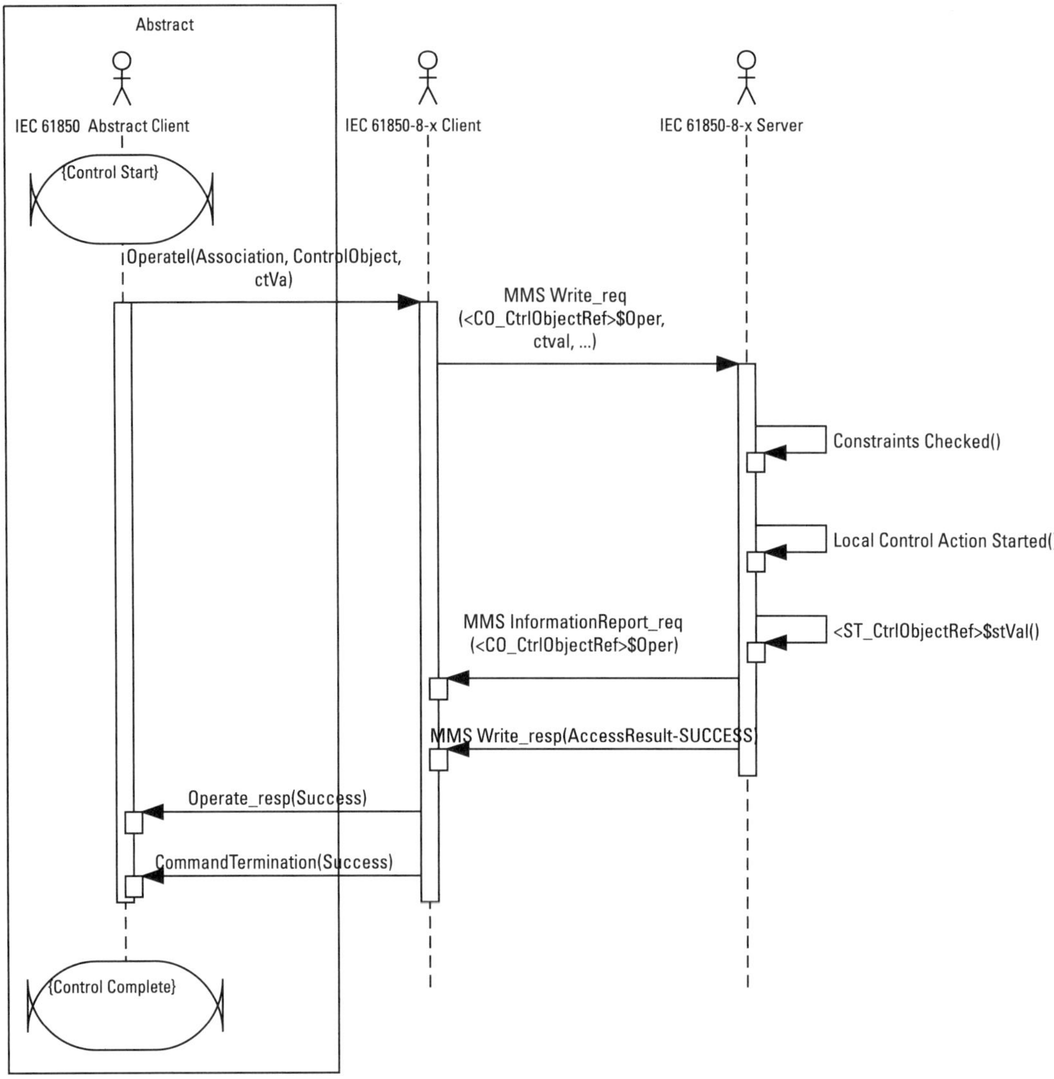

Figure 8.38 Direct with enhanced security indicating success.

The actual structure definitions are driven by the integration patterns that are supported as well as the value of the ctlModel configuration parameter. In many situations, servers will provide control object instantiations that appear to provide all the services required to support all the integration patterns. This may not be true in all cases as not all services or integration patterns may be supported.

Inquiring minds might ask how this could be. The answer lies in a couple of rules of the IEC 61850 configuration language. The advent of IEC 61850 Edition 2 concluded that enumerations should only include the values that are appropriate for the use of that enumerated value. The following example shows how a subset of ctlModel enumerated values can restrict a Control CDC to be status-only.

Figure 8.40 shows that the only ctlModel allowed is status-only and thus the presence of SBO, SBOw, Oper, and Cancel structures become invalid to utilize.

The next option to restrict the ctlModel is to initialize the value of ctlModel within the type definition. In Figure 8.41, there are multiple values in the

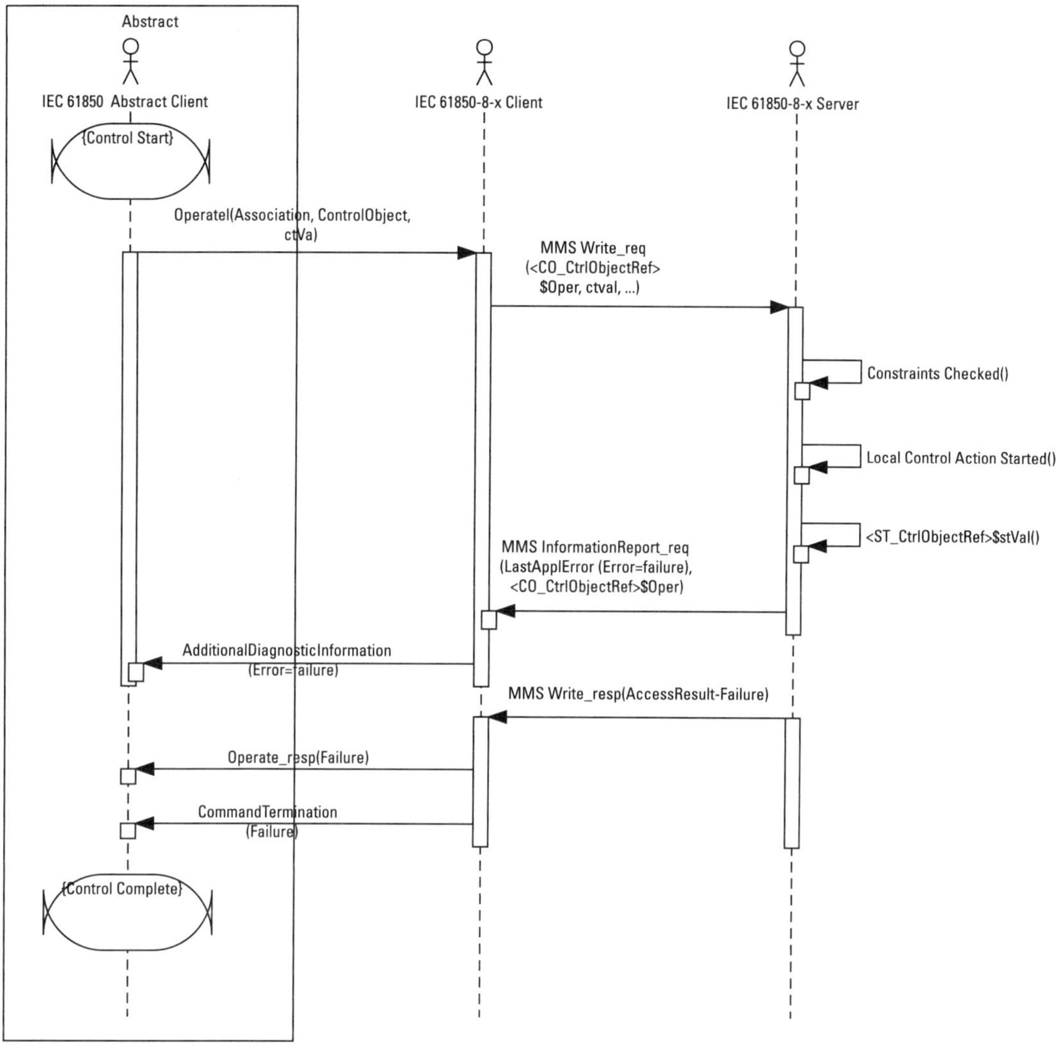

Figure 8.39 Enhanced security with failure.

```
<DOType id="DPC_1" cdc="DPC">
   <DA bType="Dbpos" name="stVal" dchg="true" fc="ST"/>
   <DA bType="Quality" name="q" fc="ST" qchg="true"/>
   <DA bType="Timestamp" name="t" fc="ST"/>
   <DA bType="Enum" name="ctlModel" type="StatusOnly" fc="CF"/>
   <DA bType="VisString255" name="d" fc="DC"/>
 </DOType>

 <EnumType id="StatusOnly">
   <EnumVal ord="0">status-only</EnumVal>
 </EnumType>
```

Figure 8.40 Control object restriction via enumeration restriction—Option 1.

```
<DOType id="DPC_1" cdc="DPC">
  <DA bType="Dbpos" name="stVal" dchg="true" fc="ST"/>
  <DA bType="Quality" name="q" fc="ST" qchg="true"/>
  <DA bType="Timestamp" name="t" fc="ST"/>
  <DA bType="Enum" name="ctlModel" type="restricted_ CtlModel " fc="CF"/>
  <DA bType="VisString255" name="d" fc="DC"/>
</DOType>

<EnumType id=" restrict_CtlModel ">
  <EnumVal ord="0">status-only</EnumVal>
  <EnumVal ord="4">sbo-with-enhanced-security</EnumVal>
</EnumType>
```

Figure 8.41 Control object restriction via DOType ctlModel Value Initialization—Option 2.

enumeration. However, the ctlModel value, within the DOType is initialized to sbo-with-enhanced-security through the use of the <Val> element.

Since the value is initialized in the type definition and the value is defined as read-only (e.g., valKind="RO"), this is the only value that is usable for any DataObject that utilizes this specific type definition. The configuration rule that prevents other tools from modifying the value is that the data types defined by the device, or device configuration tool (IED configuration tool) are not allowed to be changed by other tools. There is a benefit to this initialization approach in that a client can read the ctlModel and immediately know the model.

There is another initialization method that provides the same benefits as the type initialization method. This method (see Figure 8.42) initializes the value at an individual logical node instance basis.

The instance initialization method has a benefit over the type initialization method in that different data objects (DO) can be initialized to different ctlModels even though the same type definition is being used. Data object initialization (DOI) specifies the name of the data object whose values are to be initialized (e.g., Pos). Data attribute initialization (e.g., DAI) specifies the name of the data attribute whose value is to be initialized (e.g., ctlModel). Like the type restriction method, it includes the directive that indicates it is read-only, indicating it may not be changed during run-time. The valImport="false" directive provides information to other

```
<LN lnType="XCBR1" inst="1" lnClass="XCBR">
      ....
    </DOI>
    <DOI name="Pos">
      <DAI name="ctlModel" valKind="RO" valImport="false">
        <Val>sbo-with-enhanced-security</Val>
      </DAI>
    </DOI>
      ....
</LN>
```

Figure 8.42 ctlModel Value Initialization using instance initialization—Option 3.

tools that they may not change the specified value during the SCL engineering process.

IEC 61850 clients are expected to process SCL system configuration description (SCD) files and to understand the initialized values of ctlModel and behave accordingly. As such, Option 2 and Option 3 are typically used to initialize the ctlModel value. This type of configuration removes the need of clients to set the ctlModel.

In the instances where the value is not read-only, IEC 61850 clients can change the ctlModel during interaction with a aerver. In this case (e.g., Option 1),clients must either set the ctlModel value or behave based on acquiring the value that is present and could have been set by another client. If there are two clients attempting to control the same data object, with different ctlModels needed, this can create issues for the first client that set the ctlModel value and that client could attempt to perform a control that is refused since the ctlModel is incorrect. As an example, the first client could set the ctlModel value to direct-operate and the second (e.g., last one to set the value) sets the value to sbo-with enhanced-security. The first client would attempt to set the values in the Oper structure (see Figure 8.36). Since the interaction pattern did not include a "select" before the operate, the control will fail. This is the reason why Option 1 initialization is to be avoided.

Not only does enhanced control provide a message indicating the actual completion of the control, it provides additional diagnostic information in the case where the control fails. The LastApplicationError structure contains the information shown in Figure 8.43.

The value of CntrlObj specifies the instantiated control object to which the operate service was issued. Therefore, it contains the hierarchy of the names logical device, logical node, and a functionally constrained data attribute (FCDA) that specifies the actual object. For the SCSMs of IEC 61850-8-1 and IEC 61850-8-2, the representation of the hierarchy is a single VisibleString (no Unicode allowed in this value) and has a maximum size of 129 characters. The following example shows the structure of the value which includes the FCDA:

<LogicalDevice>/<LogicalNode>$<FC>$<DataObject>$DataAttribute

which might have a value such as

<AirportHV_1>/CSWI1COPos$Oper

which would indicate a control issued for the position (e.g., Pos) of a switch (e.g., CSWI1).

Since the ErrorKind allows the value of error to indicate "No Error," LastApplError can be used to indicate error information or AdditionalDiagnostic information. Thus, the reception of LastApplError really doesn't necessarily indicate an error. The value of error must be checked. The standard doesn't provide a matrix indicating which AdditionalCauseKind values can be used to indicate a status instead of an error. Table 8.8 provides some guidance.

The values of Org (e.g., Origin) and ctlNum mimic the values in the actual service utilized.

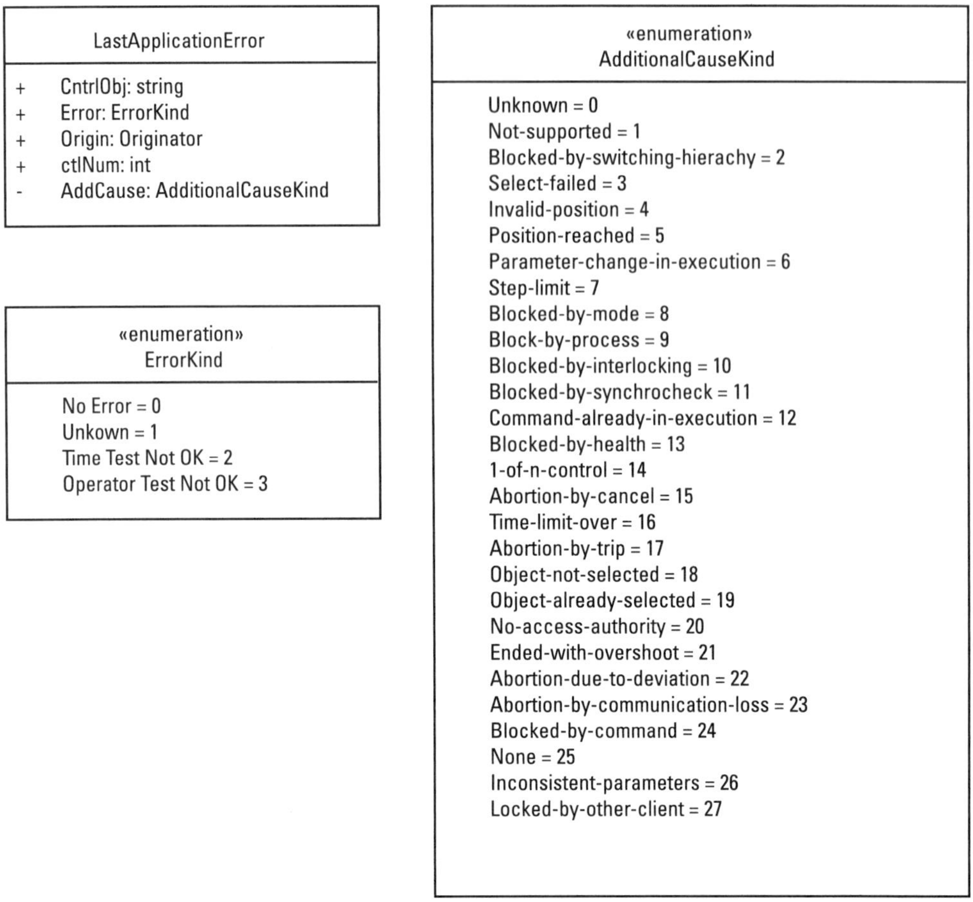

Figure 8.43 LastApplicationError definition.

There is a complication in the fact that IEC 61850 does not define the meanings of the values in AdditionalCauseKind. Some of the values are self-evident, others leave the definition to the server implementer's choice and can in some cases inhibit interoperability. Table 8.9 is this author's definition and rationale for the definitions.

The TimeActivatedOperate operation and the values of ctlModel in Figure 8.35 provides the rationale for the UML in Figure 8.36 to show specialization for TimeActivated patterns. Between the combinations of ctlModel and operations, there are several major interaction patterns. There are several interaction patterns that have fallen out of favor of the industry or have not been widely implemented. These are sbo-with-normal-security and any pattern that deals with TimeActivation. A discussion of these and the other patterns occur in the following chapters.

Status Only Interaction Pattern

This could have been called the NO-CONTROL-POSSIBLE interaction pattern. This interaction pattern is used when the ctlModel value is set to "status-only." Figure 8.40 discusses the different mechanisms through which status-only can be persisted and not-changed (e.g., Option 2 and Option 3). If the status-only

Table 8.8 Last Application Error Reason Matrix

Error Value	*AddCause Value*	*Status meaning*
No Error	Unknown	Provides no additional information but could be used to indicate a positive command termination. This is not recommended since "None" would be clearer as to the status being indicated.
No Error	Position-reached	The position (e.g., open or closed) has been reached as requested by the control. It can be used to indicate a positive command termination.
No Error	None	Provides no additional information but can be used to indicate a positive command termination.
No Error	Step-limit	Can indicate that the maximum step value (e.g., for a tap changer) has been reached.
No Error	1-of-n-control	When select-before-operate is used for control, there are parameters that allow the configuration of operate-once or operate-many (e.g., within the same selection). If the parameter is set to "operate-many," this combination would be returned.
No Error	Object-already-selected	Indicates that the same client that has already selected a control has tried to select it again.
No Error	Other values	This author believes that this combination should not be allowed.
Unknown	Unknown	This is a horrible combination as it represents a ternary logic state.* It doesn't provide any information that would allow the client to determine if there was success or an error. This author recommends that this combination never be used except to indicate that the server can't determine success or failure.
Unknown	Position-reached	Indicates that the commanded position was reached, but an unidentified issue (e.g., not able to be conveyed) was detected but the control succeeded. This author believes that this combination should be used to indicate the need for additional investigation (e.g., maybe maintenance).
Unknown	None	Another horrible combination that creates an uninterpretable ternary logic state.* It doesn't provide any information that would allow the client to determine if there was success or an error. This author recommends that this combination never be used and recommends the combination of unknown/unknown be used instead.
Error	Unknown	Indicates that the cause of the error is not one of the allowed values. In some cases, implementations cheat and do not implement the logic required to determine the other discrete values of AdditionalCauseKind and therefore return unknown.
Error	Object-already-selected	Would typically indicate that a different client has already selected (e.g., placed a semaphore lock) on the control object. This author would recommend the return of locked-by-other-client for this condition.
Error	None	This author believes that this combination should be disallowed, and that error/unknown is the more appropriate semantic combination.

* Russia actually implemented a ternary computer (–1,0,1) where the –1 value was used to indicate "maybe".

configuration is configured as permanent, the control object is not required to contain any data attributes of the CO functional constraint since it is never controllable. Therefore, such a control object would have no Oper, SBO, SBOw, or Cancel.

Status-only is typically used to allow the status value (e.g., stVal) to be monitored but not controlled. An example of this might be the position of a monitored, but not controllable, switch. In this case it is desirable to know the position of the switch even though it cannot be commanded to open or close but could change state based on other entities/logic in the automation process besides control.

Table 8.9 Interpretations of the Value of AddCause in LastApplError

Value	*Interpertation*
Unknown	Is used to indicate that the server has detected a diagnostic issue but either doesn't have the process knowledge to match a different value or the value is not in the set of values.
Not-supported	Is typically used to indicate that the control service is not-supported. This would typically occur when a control service is attempted that does not match the value of ctlModel. As an example, an operate service is attempted on a control object that is configured to be status-only.
Blocked-by-switch-ing-hierarchy	IEC 61850 and electric grids can implement a sequence of switching that must be followed. If a switch operation is attempted out-of-sequence, this error might be returned.
Select-failed	This reason would be returned is either select-before-operate (SBO) or SBOw operation fails. It would be preferable, in the case where the control object is already selected by a different client to return Locked-by-other-client.
Invalid-position	Indicates that the commanded position was not obtained (e.g., open or closed).
Position-reached	The position (e.g., open or closed) has been reached as requested by the control. It can be used to indicate a positive command termination.
Parameter-change-in-execution	This value would typically be used if the original parameters for a control were changed after the sequence for control was started but not completed.
Step-limit	The position of tap changers, and some other types of motorized controls, are controlled via steps. This value, in combination with error returns either that the step was reached or that the step value commanded was not available or reached.
Blocked-by-mode	This value would be returned if the logical node, in which the control object is contained, has value of the behavior (e.g., Beh DataObject) of "blocked" or "test/blocked." This value is determined through the logical device/logical node hierarchy of Mode (e.g., Mod) DataObject values.
Block-by-process	If the server has an understanding of the actual automation process, there may be instances where the state of the process may inhibit the control command. This value would be returned in that instance.
Blocked-by-inter-locking	Interlock checking is the process of one system checking the state of another system (e.g., breakers) to determine if the other entity is in the correct state to allow the local entity to change state. If the client selects that the interlocking constraint to be checked (explained in *Constraint Checking (Test and Check)*) and the check fails, this value would be returned.
Blocked-by-syn-chrocheck	Synchrocheck checking is the process of one system checking the sinusoidal phase, or frequency, of the electrical system. Certain operations can only be performed if systems are synchronized properly and the phases of the systems match. If the client selects that the synchrocheck constraint to be checked (explained in *Constraint Checking (Test and Check)* and the check fails, this value would be returned.
Command-already-in-execution	Have you ever been impatient about an elevator and kept pressing the button? If the elevator was IEC 61850, it would return this value to indicate that the elevator is already executing that request.
Blocked-by-health	There are three potential health data objects whose values might cause the return of this value. The external health of the equipment being controlled (EEHealth), the logical node health of the logical node in which the control object is found, or the physical health of the server (e.g., PhyHealth). If any of these values inhibit operation, this value might be returned.
1-of-n-control	When select-before-operate is used for control, there are parameters that allow the configuration of operate-once or operate-many (e.g., within the same selection). If the parameter is set to "operate-many," this combination would be returned.
Abortion-by-cancel	A control that is underway, but not complete, may be canceled. If the cancel was successful, this is the value that should be returned.
Time-limit-over	In many situations, the process monitors the time it takes for certain movements (e.g., open or close) to occur. This value might be returned if the movement, or completion of the control, exceeds the expected time.

Table 8.9 (continued)

Value	*Interpertation*
Abortion-by-trip	Indicates that a protection event has occurred that caused an interruption of the control.
Object-not-selected	This value would be returned if the ctlModel indicated the use of select-before-operate and the client did not appropriately select the control object.
Object-already-selected	See Table 8.8.
No-access-authority	This value would be returned if the client does not have the appropriate rights (e.g., role, rights, privileges) to access or control the control object.
Ended-with-overshoot	This value might be returned if the controlled caused a motion that transitioned through the desired position.
Abortion-due-to-deviation	This value indicates that a control started but was terminated due to a process issue that was detected.
Abortion-by-communication-loss	IEC 61850 facilitates distributed automation that requires communication. This value might be returned if a control has started had the communication link required to check the state of the process, interlock, or synchrocheck was not available.
Blocked-by-command	This value indicates that the action did not occur due to the block attribute being True.
None	This value should be reserved to be used with no error and would indicate that there is no additional diagnostic information.
Inconsistent-parameters	This value might be returned if the parameters used for the control are not appropriate and don't match the expected pattern.
Locked-by-other-client	This value should be used to indicate that another client has previously placed a semaphore lock (e.g., selection) on the control object.

Direct Operate Interaction Pattern

Note that for status-only, this is the simplest interaction pattern as it was originally designed to only utilize the Operate operation (see Figure 8.37). The application of enhanced security caused a more complicated implementation to support both direct-with-normal-security and direct-with-enhanced-security. A compromise was made and the equivalent of a CommandTermination message is now allowed even for direct-with-normal-security.

Most implementations return the "CommandTermination" (e.g., LastApplError) message regardless of the use of enhanced-security. Therefore, there is no real difference today between the two direct operates. However, to explain the separation of normal and enhanced-security in IEC 61850-7-2, one "CommandTermination" is discussed as AdditionalDiagnosticInformation (e.g., normal-security) and the other is a true CommandTermination (e.g., enhanced-security). This hybrid pattern is shown in Figure 8.44.

The figure shows that the differentiation between a true error and AdditionalDiagnosticInformation is determined by the value of the error attribute of the LastApplError structure. If the value is of Error=none, then it represents the delivery of addition diagnostic information which is contained in the other members of the structure.

During the evolution of IEC 61850, a requirement arose to be able to monitor and audit the control interaction of all clients and a server. The direct operate sequence was extended to provide this capability through tracking and interacting with a supervisory logical node instance of LTRK.

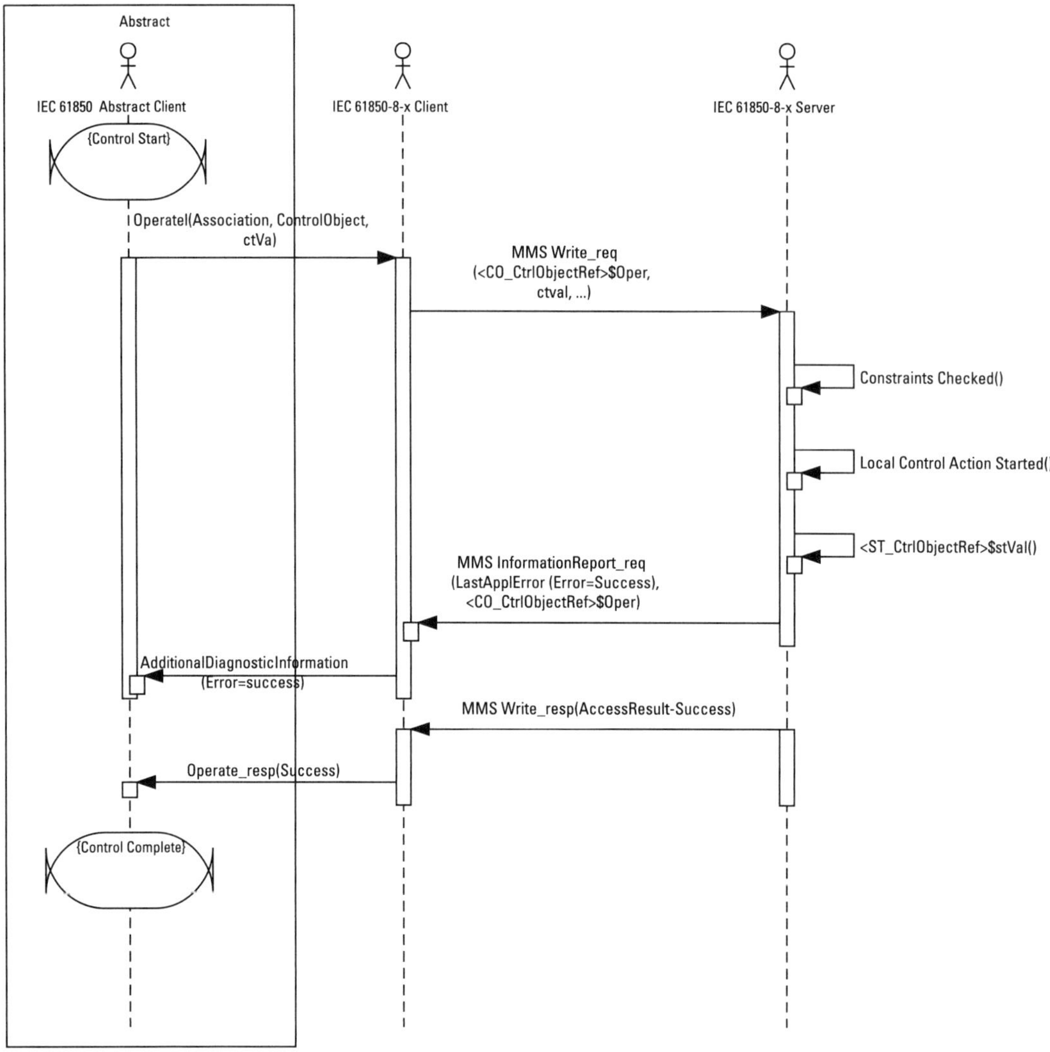

Figure 8.44 Direct operate hybrid pattern.

Figure 8.45 shows a simplified interaction providing tracking. It is worthwhile to note the errors detected in the sequence update tracking.

Select Before Operate Interaction Pattern

Note: If you ask most utility grid operators if they use select-before-operate, they will respond "yes." If you ask them "how do you know?" they will tell you they select the control that they are going to issue, and the display asks "are-you-sure"; thus they have selected and verified the selection prior to the control being issued. This is not server/device enforced select-before-operate and not what is discussed in this chapter.

The inception of the select-before-operate construct was driven by the necessity to avoid misapplication of the electrical grid due to poor communication technology. In the early days of bit-synchronous communications noise could cause a change of a transmitted bit the changed/flipped bit undetected by the receiving entity. This

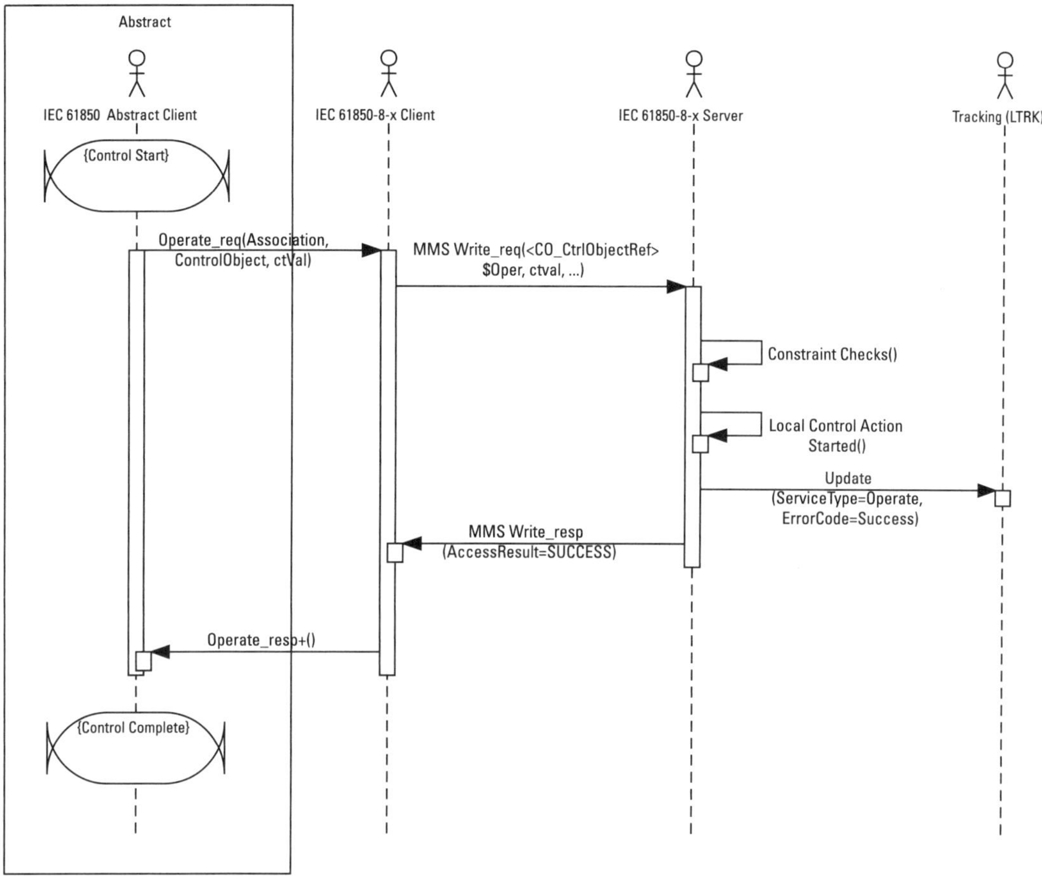

Figure 8.45 Direct operate with tracking.

inability to detect the change could cause misoperation when the bit flip changed the entity to be controlled and the wrong switch was opened or closed. In some situations, this misoperation could be catastrophic, causing fires and explosions.

To ensure that the correct switch was operated, the concept of select-before-operate was developed. A select is sent for an object (e.g., switch). Only a selected object could be controlled and the control itself ended the selection.

Figure 8.46 demonstrates a normal Operate of Switch 1 when the network/ noise did not cause a "bit" change. The operate command was issued by the SCADA system and opened Switch 1 since the address was a value of "1." The next sequence shows how a bit change would cause misoperation. The SCADA system issues the same operate command; however, the address is changed from "1" to "0" due to undetected message corruption on the noisy network/transmission medium (such as older radio or modem based systems.) The device receives the operate command and opens switch 0, which is the incorrect switch.

Select before operate was designed to minimize the probability of a misoperation through issuing a select to the device (e.g., switch 1). If the operate address changed, the device would refuse the operate since it had not previously received a selection for the switch that was attempted to be operated.

As communication technologies shifted from bit-synchronous to serial technology, the technologies attempted to address the issue of undetectable bitchanges.

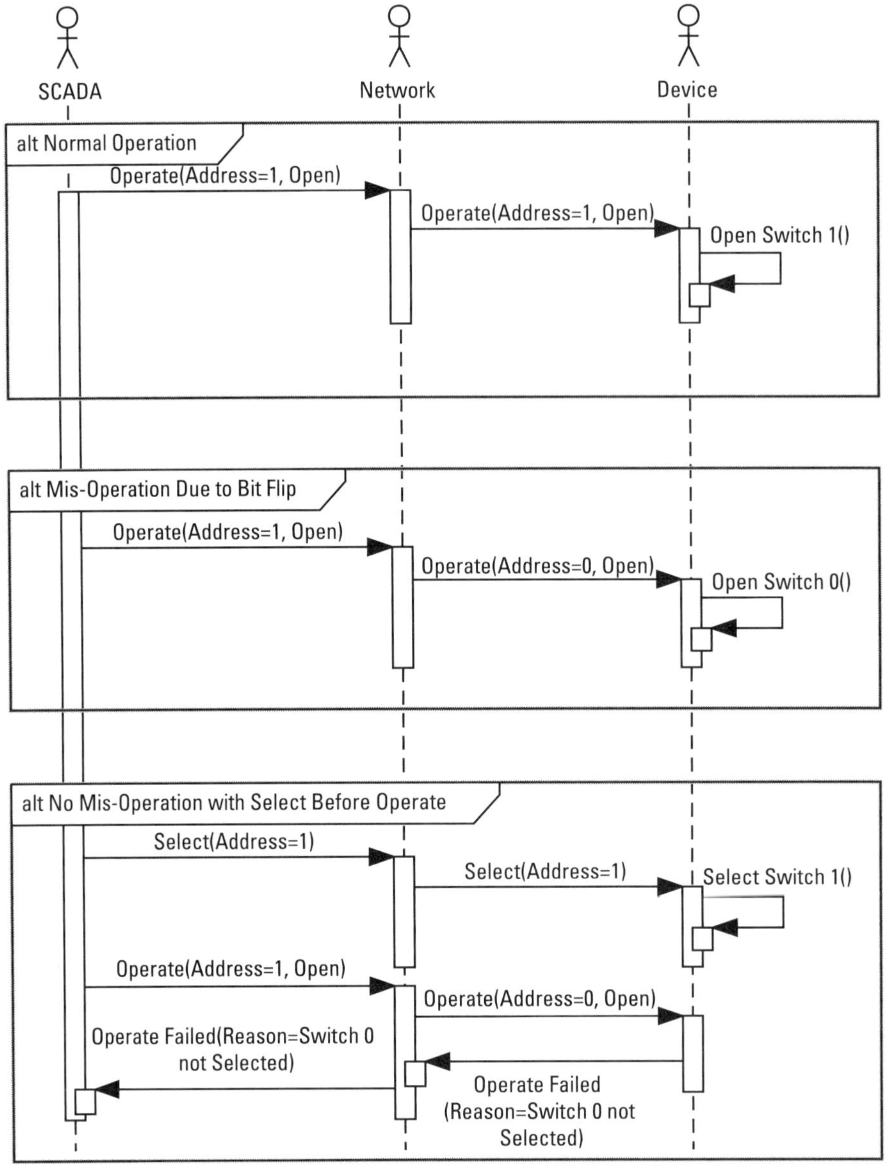

Figure 8.46 Original reason for SBO: network/noise caused message corruption.

RS-232/RS-432/RS-485 was first set of widely deployed serial communication technology. These shared a basic format for each byte of data sent over the transmission medium. Figure 8.47 shows the basic parts of RS-232.

The number of start bits, data bits, and the use of parity were all configurable parameters. The number of start bits was typically 1 or 2 and was used to address the capacitance of the transmission media. Originally ASCII was the data being sent and that required only 7 bits. As binary information was required for transmission, the number of data bits was expanded to 8 bits. The parity bit was added to provide a limited bit change protection.

The parity bit adds limited protection since it is used to determine if the summation of TRUE bits is odd. If so, the parity bit is set true. It can be used to detect

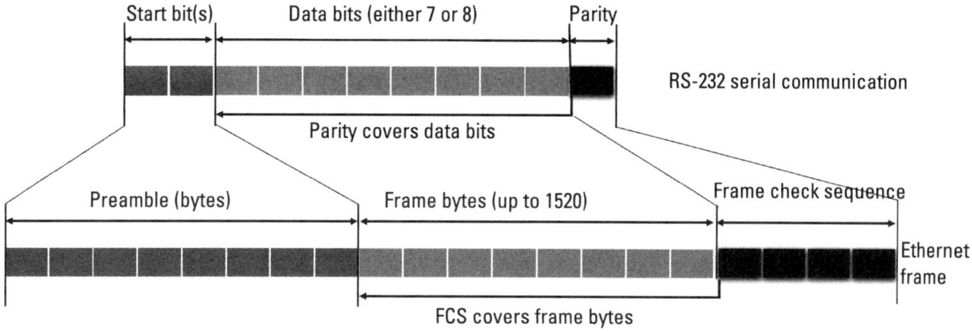

Figure 8.47 Bit change detection in serial protocols.

a single bit change from true to false or false to true. The ability to detect these changes decreases as the number of bits change. As SCADA protocols evolved, the metric of Hamming distance[9] was developed. The higher the Hamming distance (e.g., the greater the number) the more bit changes could be detected.

The concepts from RS-232 were leveraged in Ethernet, but for different reasons. Start bits became preamble bytes that were used to provide carrier sensing to detect if there is already a device transmitting on the media. Data bits became data bytes. The parity bit became a cyclic redundancy calculation (CRC) known as the frame check sequence (FCS). The Hamming distance for the maximum length Ethernet Frame is four. However, a Hamming distance of six is typical. The Hamming distance of Ethernet is in alignment with the Hamming distances of IEC 60870-5 and DNP 3.0 whose Hamming distances are four and six, respectively.[10]

The need to use select-before-operate in order to prevent misoperation has decreased substantially as the communication media noise immunity has increased. However, the utilities still require it for critical operations. The advent of devices (e.g., servers) allowing multiple clients simultaneous access created another use for select-before-operate (SBO): resource locking.

Locking of grid related resources requires that deadlocks (e.g., locked forever) be avoided. A design that allowed a deadlock would inhibit the overall operation of the grid and provide a security related denial of service (DOS) attack vector. In many computer technologies a resource lock/semaphores require additional programming to prevent deadlocks. The typical deadlock protection mechanism is to provide a parameter or configuration that reserves the resource for a limited period.[11].

In IEC 61850, additional configuration parameters are provided in the control data object. The additional parameters are exposed optionally but must exist in the server control logic. The primary parameter that prevents deadlock is the sboTimeout value (see Figure 8.36). The value is specified in msec and if exposed in the controllable DataObject may be written by a client. The other optional SBO parameter is sboClasses which allows specification of how many controls (e.g., operates) are allowed prior to the semaphore reservation being removed. The default value for

9. See https://en.wikipedia.org/wiki/Hamming_distance for more information about Hamming distance.
10. The use of IP and TCP Checksums further increase the Hamming distance.
11. A source code example can be found at https://stackoverflow.com/questions/19647344/golang-how-to-timeout-a-semaphore.

this is a single operation. Depending on the relative values of sboTimeout and sboClasses, either the timeout or number of operations can terminate the resource lock. The other mechanism to release the resource is to issue a cancel operation.

Figure 8.36 shows that there are two operations used to implement SBO: Select and SelectWithValue. Either of these operations can be used with sbo-with-normal-security or sbo-with-enhanced-security. These will be detailed in the following sections.

Using the Select Operation: UCA SBO The Select operation was provided in IEC 61850 to provide a modicum of backward compatibility with UCA 2.0 and IEEE TR 1550. However, the SelectWithValue operation is allowed to be used with enhanced security and Select is not allowed. This makes SelectWithValue the preferred operation today. In the future, the option of Select may become deprecated in the standard as its use decreases further.

Figure 8.48 shows that the control sequence starts with the client issuing a select operation. The abstract operation specifies the association and control data object that is desired to be selected. The abstract operation is instantiated as an MMS Read_request containing the object reference of the SBO structure of the specified data object. If there is no previous lock on the control object, the MMS Read_response contains the object reference of the Oper structure that has been locked for use by the client that issued the Select. The client can then issue an Operate operation. The sequence shows that the reception of the Operate removes the lock (e.g., operate-once) and returns success. When the successful Write_response is received, that is translated to a successful Operate and the control sequence is complete.

If an error occurs on the Select (see Figure 8.49), the following sequence occurs. The figure shows that a different client had previously locked the control data object. A client that does not own the lock attempts to select the control data object.

Since there is a previous lock, the select fails. Tracking is updated with the fact that the Select failed and the MMS Read_resp returns a NULL value thereby indicating a Select failure to the IEC 61850 client. This pattern of SBO is disallowed for sbo-with-enhanced-security.

Using the Select Before Operate with Value The general sequence of SBO, using SelectWithValue, is like that of Select. SelectWithValue can be utilized as sbo-with-enhanced-security and it is typically used this way (see Figure 8.50).

The SelectWithValue operation is mapped to a MMS Write_request that sets the contents of the SBOw data attribute in the control data object. If the control object is not locked by a different client, the control is locked and a MMS Write_response is returned indicating SUCCESS. This message is then translated into a successful SelectWithValue response. Once the lock is obtained, and the response is received, the client can issue an operate service.

If a lock has been placed on the control data object by a different client, the SelectWithValue operation will fail (see Figure 8.51). When used with sbo-with-enhanced-security, the LastApplicationError is returned with diagnostic information. This example shows that the AddCause is indicating that the resource is locked by a different client. The reception of this diagnostic information and the response to the SelectWithValue completes the sequence. Since the LastApplicationError was sent with an error indication, there is no opportunity to operate successfully.

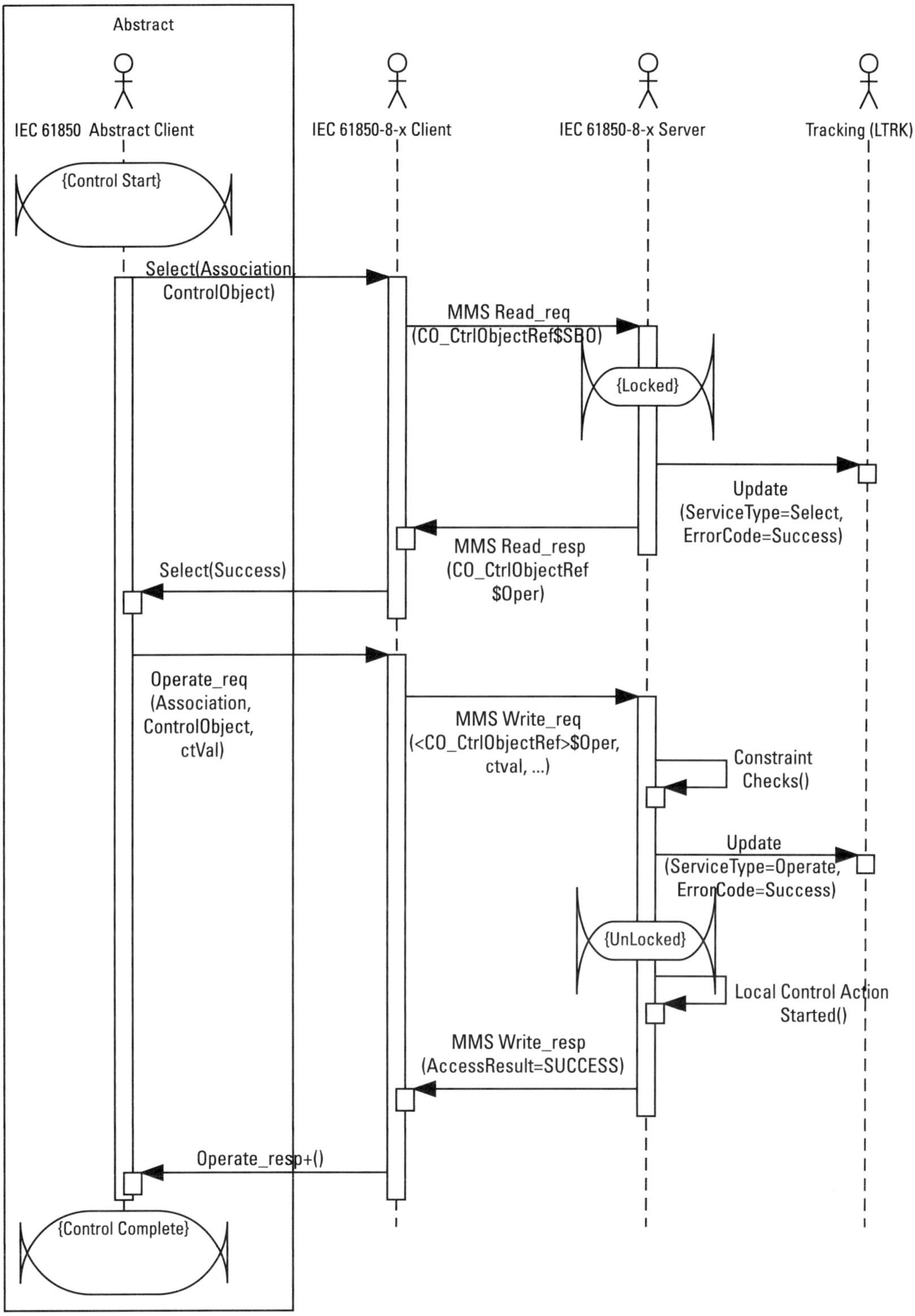

Figure 8.48 Select before operate sequence.

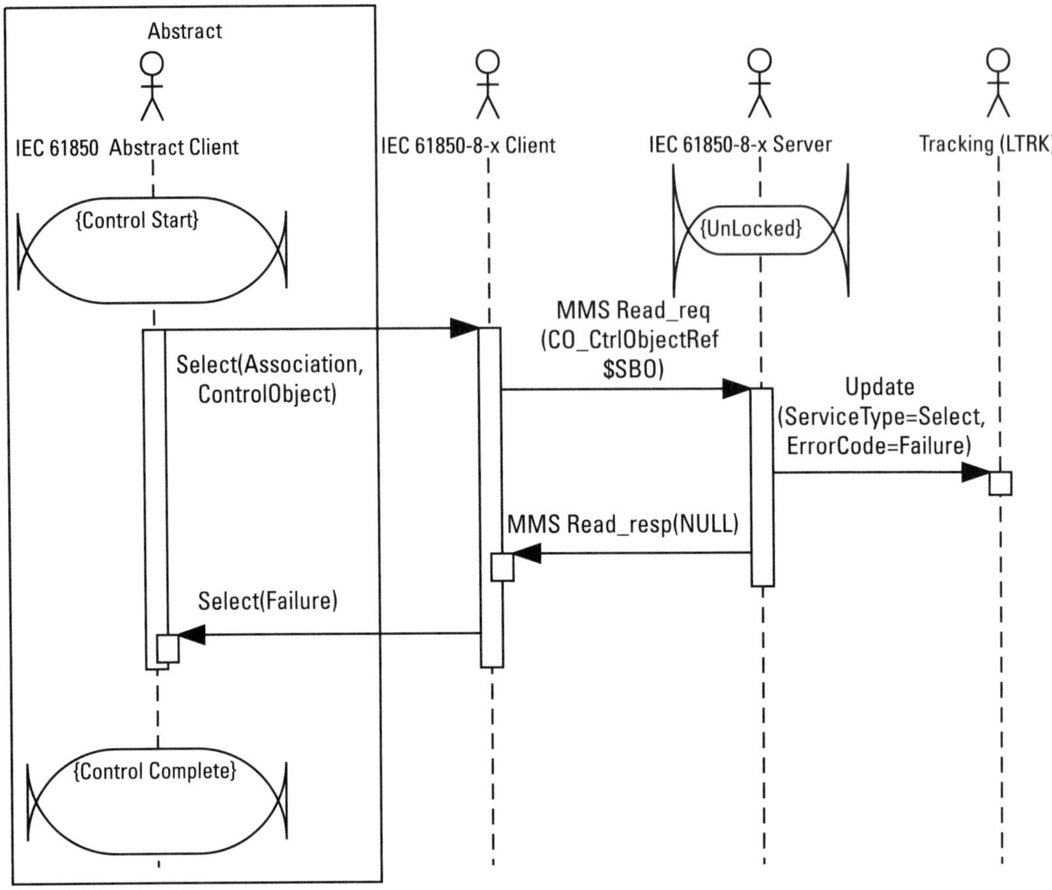

Figure 8.49 Select before operate sequence with error.

Time-Activated Control

Examples of this control are found in our houses, apartments, or condominiums in the control of lights. In the case of lights, a timer can be used to automatically set the time of day at which the lights are to be turned on. The time which the lights are controlled to turn on is a time-activated control. Besides deferring turning on a strand of lights, multiple timers can be used to synchronize one strand of lights with another.

To perform an IEC 61850 time-activated control, the client utilizes the abstract operation of TimeActivatedOperate. Unlike, other abstract control services, this operation utilizes the constructs of other concrete services (e.g., MMS Write to Oper).

Figure 8.52 shows the sequence of the TimeActivatedOperate. The abstract operation has an additional parameter which represents the time at which the control action is desired. The MMS Write_request is used in a similar manner to the Operate operation mapping, but also conveys the desired operation time (e.g., operTm). The value of operTm is the IEC 61850 timestamp at which the control (e.g., ctVal) is to be executed. Since the value of operTm is not NULL, the server knows that this is a time-activated control and may update LTRK.

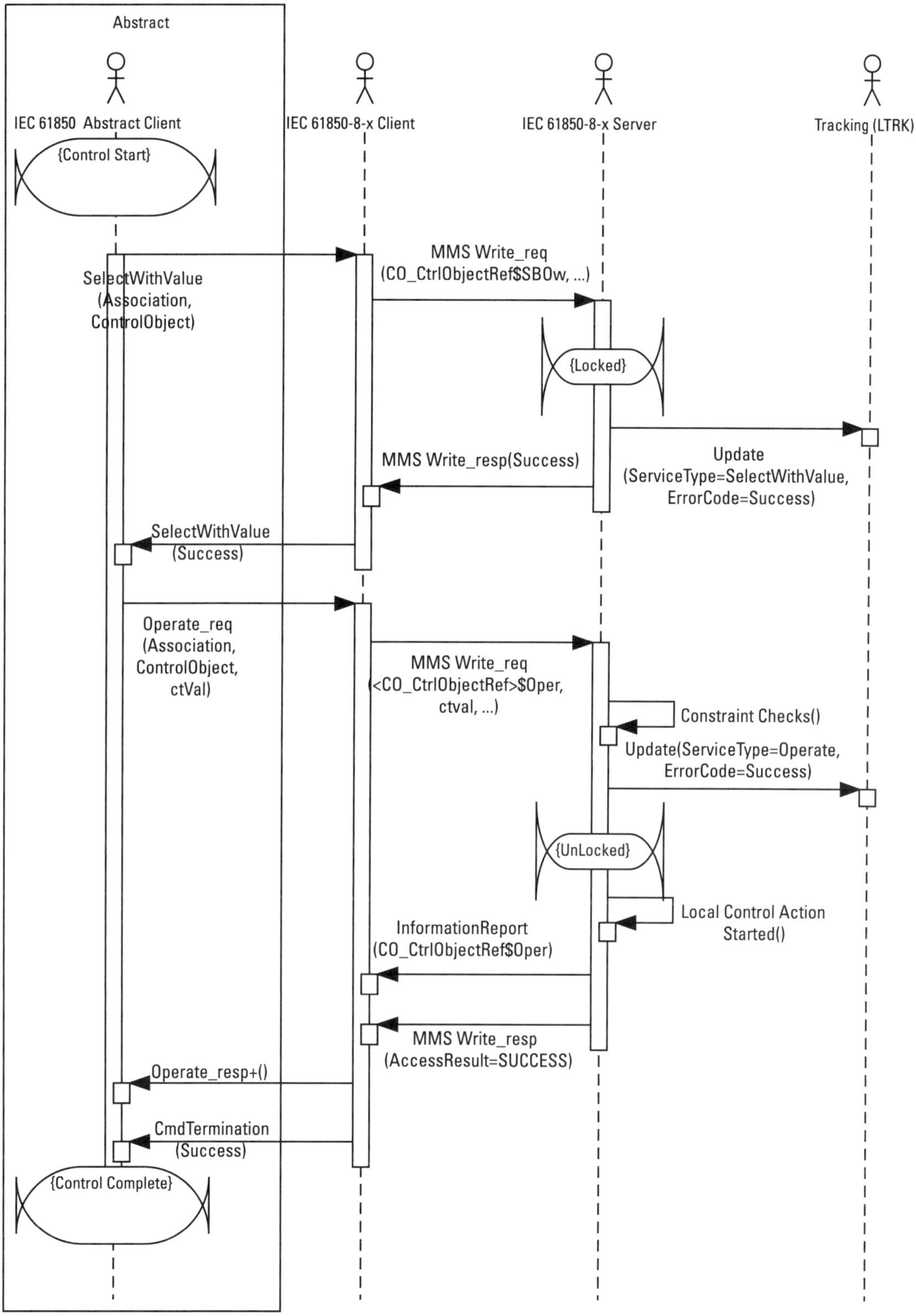

Figure 8.50 Select-before-operate with enhanced security—success.

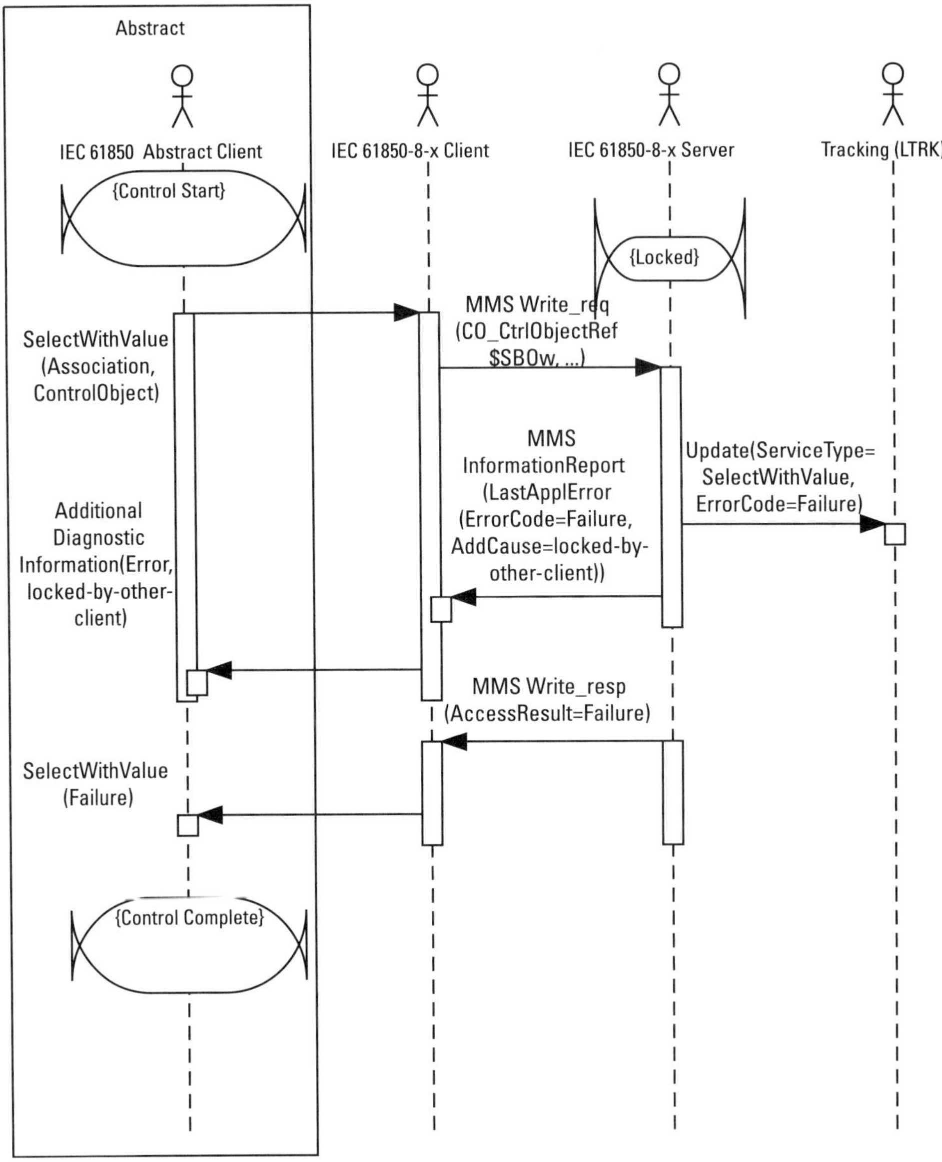

Figure 8.51 Select-before-operate with enhanced security—failure.

The representation of a NULL value may be different between different SCSMs. IEC 61850-8-1 utilizes ASN.1 with basic encoding rules (BER). ASN.1 represents a value through a TLV paradigm. A NULL value has a length of 0 and no value. IEC 61850-8-2 utilizes ASN.1 but with XER and represents a NULL value as "<operTm/>" or "<operTm></operTm>". More detailed information on ASN.1, see Section 11.3.

After the control occurs or is canceled, it is strongly suggested that the server set the value to operTm to NULL. It is also strongly suggested that if the client sets operTm to a time not in the future, that the value be refused, the Select or Control should fail with LastApplError indicating inconsistent-parameters.

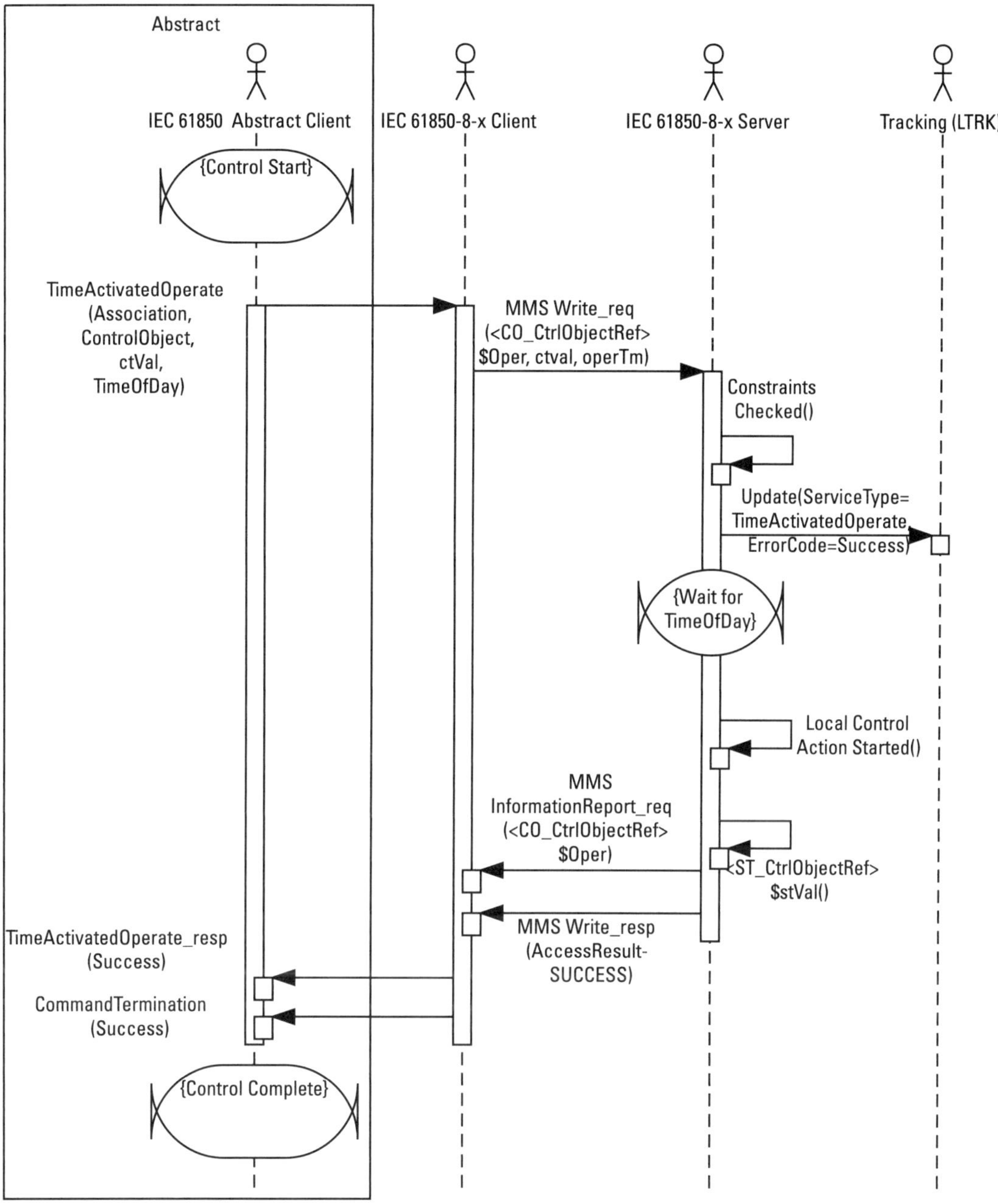

Figure 8.52 Example of time-activated control with enhanced security.

When the desired time is reached, the control is executed. The precision of the time activation is not only dependent on the implementation, but also the accuracy of the time synchronization. Since this control is delayed, it is suggested that it be utilized with a ctlModel of sbo-with-enhanced-security.

Constraint Checking (Test and Check)

There are two attributes in the structures that add additional processing to a control: test and check.

Test is a Boolean value that when true indicates to the server that the control is being executed for test purposes. If the logical node in which the control data object is contained does not indicate that it is in test mode, the constraint check shall fail and therefore the control operation shall fail.

The values of check are used to indicate if synchrocheck and/or interlocking checks are required. If the control operations places either of these to true and the server check fails, the constraint check fails and so does the control.

8.2.2.3 Datasets

The concept for the IEC 61850 dataset was borrowed from ISO 9506 where the name of the object was NamedVariableList. An IEC 61850 dataset is not a set[12] of data, rather it is an ordered list of object references that provide the name of the data object(s) that are the members of the dataset (see Figure 8.53). It is through this object that values from the different referenced member objects can be acquired in a single request, thereby improving communication efficiency. the dataset construct also allows the member object values to be set (e.g., written) in a single transaction. Although these communication efficiencies are significant, they do not constitute the primary use of datasets within an IEC 61850 context. The main use is for the definition and communication of information through GOOSE, Log, and Reporting in the respective control blocks.

The definition of a dataset can be done via SCL configuration or at runtime by a client using the CreateDataSet service/operation. There are a couple of constraints placed on the dataset, and its members, that are not constraints in the ISO 9506 DefineNamedVariableList service.

A dataset is defined and contained within a specific logical node. The members are references to data, which is defined as either functionally constrained data (FCD) or a functionally constrained data attribute. An FCD constrains a member of a subset of the data object of a particular functional constraint (FC). Therefore, the entire information in a data object or logical node is not allowed to be a single member.

The rationale of restricting members to be a functionally constrained subset of information is steeped in history and a desire to be able to obtain the information needed to be exchanged in the most efficient mechanism possible, the ability to separate write-only information, and the ability to group information based on access privileges (see Figure 8.54).

If a client accessed the entire logical node instance (e.g., data) of a logical node, it would access status (ST), configuration (CF), description (DC), and extension (EX) information for each of its data objects (e.g., TotVAh, TotW). The values returned in such an information exchange would create a large response and in some instances might fail since the data might include write-only information. The issue is further exacerbated when a dataset is created that could contain information from multiple logical nodes,

The dataset abstract was defined to allow automation systems to exchange process information efficiently. This would typically mean the exchange of status (ST) and measurement (MX) data attributes. Therefore, the mechanism of functionally

12. A set implies that order does not matter.

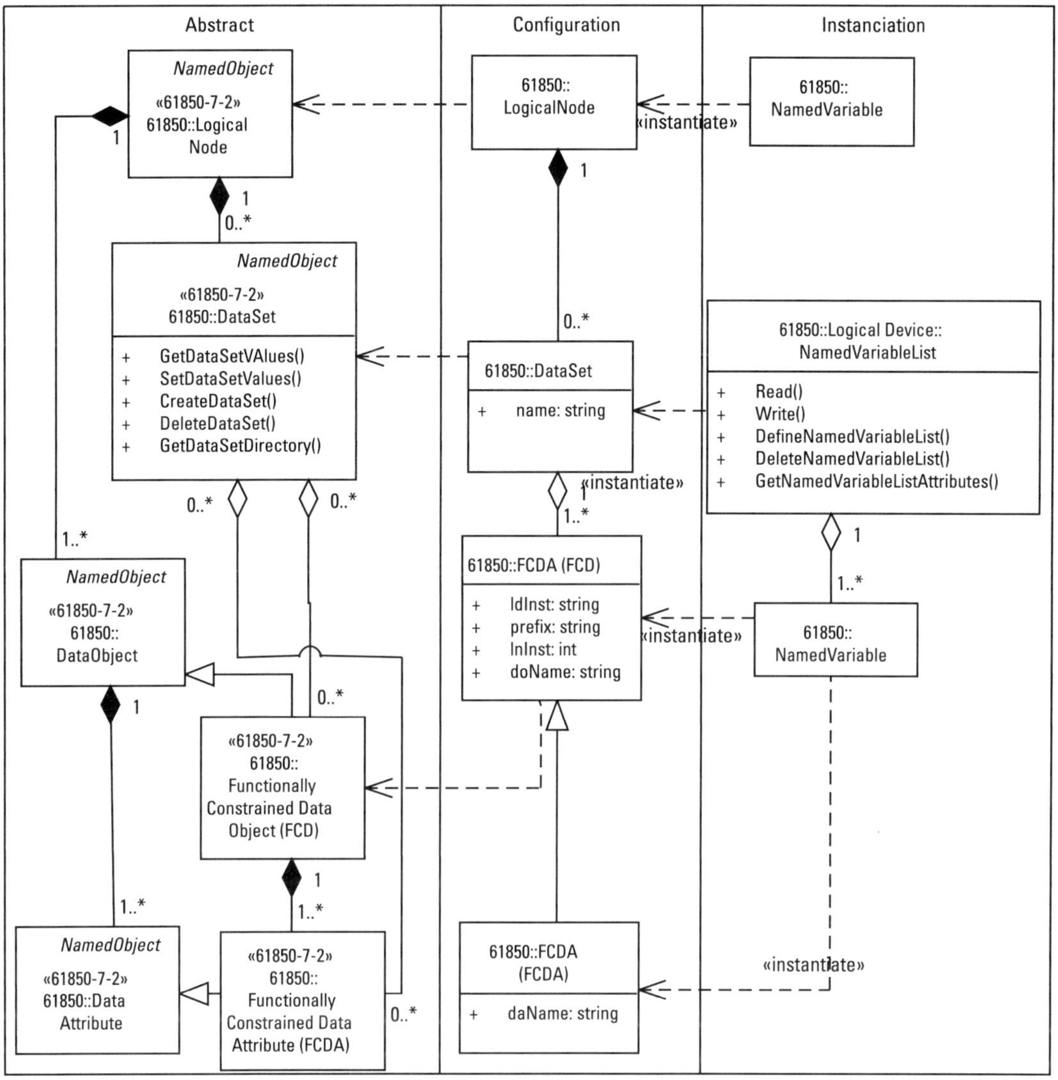

Figure 8.53 UML for IEC 61850 DataSet.

constrained data (e.g., data objects) was developed so that a dataset member could represent all of the status values as a single dataset member.

Figure 8.55 shows a snippet of a SCL configuration of a dataset whose name is "statusInfo." It is defined as being contained in LN0 of a logical device named "FCD01" within an IED named "IED1." The IEC 61850 abstract object reference for the dataset would be as defined in IEC 61850-7-2 <LDName>/<LogicalNode Name>.<DataSet Name>, which would result in

IED1FCD01/LN0.statusInfo

In the IEC 61850-8-1 and IEC 61850-8-2 SCSMs, there is a mapping defined that maps a dataset to an MMS NamedVariableList (NVL). The abstract dataset is defined as part of a logical node, which is always part of a logical device. A Logical Device is mapped to an MMS domain. Therefore, configured datasets are mapped

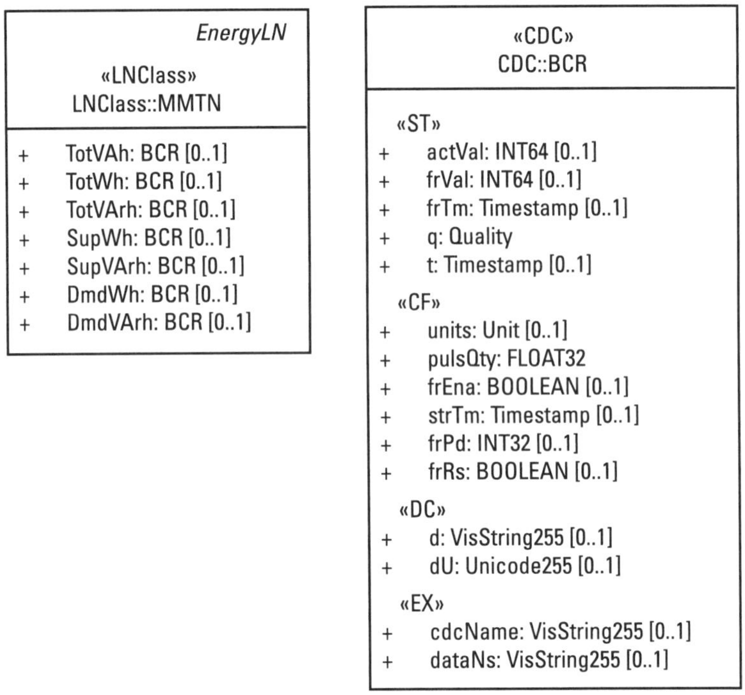

Figure 8.54 Need for FCD example.

```
<IED name="IED1">
    <Server>
        <LDevice inst="FCD01">
            <LN0 lnType="LN0" lnClass="LLN0">
                <DataSet name="statusInfo" desc="">
                    <FCDA doName="TotVAh" fc="ST" ldInst="FCD01" lnClass="MMTN" lnInst="1"/>
                    <FCDA doName="TotWh" fc="ST" ldInst="FCD01" lnClass="MMTN" lnInst="1"/>
                    <FCDA doName="TotVArh" fc="ST" ldInst="FCD01" lnClass="MMTN" lnInst="1"/>
                </DataSet>
            </LN0>
        </LDevice>
    </Server>
</IED>
```

Figure 8.55 Example of SCL configuration of a DataSet with FCDs.

to domain-specific NVLs. ISO 9506 specifies that NVLs are their own discrete objects and are not contained by named variables (e.g., the mapping for logical node). Thus the SCSM must provide a name mapping for the NVL name that provides this abstract information. The SCSM mapping specifies that the NVL name shall be: <LogicalNode Name>$<DataSet Name>. The SCSM specification of the example would be a NVL definition of a domain-specific NVL whose domain is IED1FCD01 and the NVL would have the name of LN0$ statusInfo.

The SCSMs map the abstract data objects into functionally constrained data object subsets by design. The dataset member configuration for TotVAh would result in a NVL reference of a domain-specific named variable where the domain name would be "IED1FCD01" and the named variable would be

"MMTN1STTotVAh." This specifies a reference to all of the status (ST) values of the TotVAh DataObject in the logical node (e.g., named variable) "MMTN1."

An analysis of the configuration construct utilizes the XML tag of <FCDA>, which would lead many to conclude that this is defining a functionally constrained data attribute (FCDA). This tag can be utilized to define either an FCD or FCDA. The FCDA definition includes an additional XML attribute of daName.

Figure 8.56 shows the members that are FCDs and three members which are FCDAs that are the members with the doName="SupWh" which have the daName attribute present. An FCDA is any data attribute, or nested data attribute, as defined by the CDCs.

The requirement of having a daName is driven by the level at which the functional constraint (FC) is specified. The MMXU PhV is a WYE class consists of ComplexMeasuredValues (CMVs). It is at the CMV level with the FC of MX is defined. Therefore, the FCDA definition is split between the value of doName and daName. The doName attribute value is specified to also contain the array number for those objects that contain arrays. An example of including an array index is shown for MHAN. If an array index is specified, the SCSMs typically map this type of member to an MMS VariableAccessSpecification containig an AlternateAccessSpecification.

The abstract operations for datasets are mapped to MMS services as shown in Table 8.10.

The DefineNamedVariableList, DeleteNamedVariableList, and GetNamedVariableListAttributes either succeed or fail (e.g., they are atomic). However, the read and write services treat the NVL objects references atomically. Therefore, a read or write of a dataset may be partially successful (e.g., some member access may succeed and others may fail) and the service itself return a positive response but may contain a mix of access results indicating success or failure. As data sets

```
<IED name="IED1>
  ....
   <Server>
    <LDevice id="FCD01">
      <LN0 lnType="LLN0 " inst="" lnClass="LLN0">
       <DataSet name="statusInfo" desc"">
        <FCDA doName="TotVah" fc="ST" ldInst="FCD01" lnClass="MMTN" lnInst="1"/>
        <FCDA doName="TotWh" fc="ST" ldInst="FCD01" lnClass="MMTN" lnInst="1"/>
        <FCDA doName="TotVarh" fc="ST" ldInst="FCD01" lnClass="MMTN" lnInst="1"/>
        <FCDA doName="SupWh" daName="actVal" fc="ST" ldInst="FCD01" lnClass="MMTN" lnInst="1"/>
        <FCDA doName="SupWh" daName="q" fc="ST" ldInst="FCD01" lnClass="MMTN" lnInst="1"/>
        <FCDA doName="SupWh" daName="t" fc="ST" ldInst="FCD01" lnClass="MMTN" lnInst="1"/>
        <FCDA doName="PhV.phsA" daName="ctVal.mag.f" fc="MX" ldInst="FCD01" lnClass="MMXU" lnInst="1"/>
        <FCDA doName="HzVol.har.1" daName="ctVal.mag.f" fc="MX" ldInst="FCD01" lnClass="MHAN" lnInst="1"/>
       </DataSet>
      </LN0>
          .....
    </LDevice>
   <Server>
</IED>
```

Figure 8.56 Example of SCL configuration of a data set with FCDs and FCDAs.

Table 8.10 Mapping of IEC 61850 Dataset Operations

IEC 61850	*MMS Service*
GetDataSetValues	Read
SetDataSetValues	Write
CreateDataSet	DefineNamedVariableList
DeleteDataSet	DeleteNamedVariableList
GetDataSetDirectory	GetNamedVariableListAttributes

are referenced in control blocks, if a dataset is referenced by a control block, the dataset is not allowed to be deleted via the DeleteDataSet service.

8.2.2.4 Control Blocks

Figure 8.57 depicts the generic abstract model for control blocks.

The configuration of the control block, and runtime interaction with the control block, provides control of the following services: settings groups, GOOSE, Sampled Values, reporting, and logging. Except for the setting group control block, all other control blocks are configured including a dataset reference. It is this reference, and the specific type of control block, that allows an IEC 61850 Server to be configured with knowledge of which data objects on which IEC 61850 service is to be applied.

The abstract definition is for a specific class of a control block, and the services it controls, which are abstractly specified in IEC 61850-7-2. The abstract control block's relationship to other parts of IEC 61850 is shown in Figure 8.58. All classes of control blocks can be contained by LLN0. Figure 8.75 has an example showing the mappings between the abstract, configuration, and instantiation domains for the GOOSE control block. The actual configuration of control blocks might look as shown in Figure 8.59.

Figure 8.57 shows the concept of SCL configuration of a logical node containing control blocks. The example utilizes LLN0 since it can contain all the different classes of control blocks. All of the control blocks except for setting control are configured, including the specification of the name of the control block via the name attribute value of the control block. This is required since there may be multiple instances of each type of control blocks configured per logical node. Since there can only be one setting group control block, its name is preassigned by the standard.

The instantiation of IEC 61850 control blocks are contained as a MMS named component that is contained within a named component representing a functional constraint whose name is specified in the SCSM and relates to IEC 61850-7-2. The functional constraint is a named component within the named variable that represents a specific logical node.

Each type of control block, when used, impacts the server either through enabling or disabling information acquisition and generation or modification of the automation and protection settings. Table 8.11 details the allowed placement (e.g., containership), the purpose of a specific control block, and the IEC 61850 abstract services that are controlled by the block.

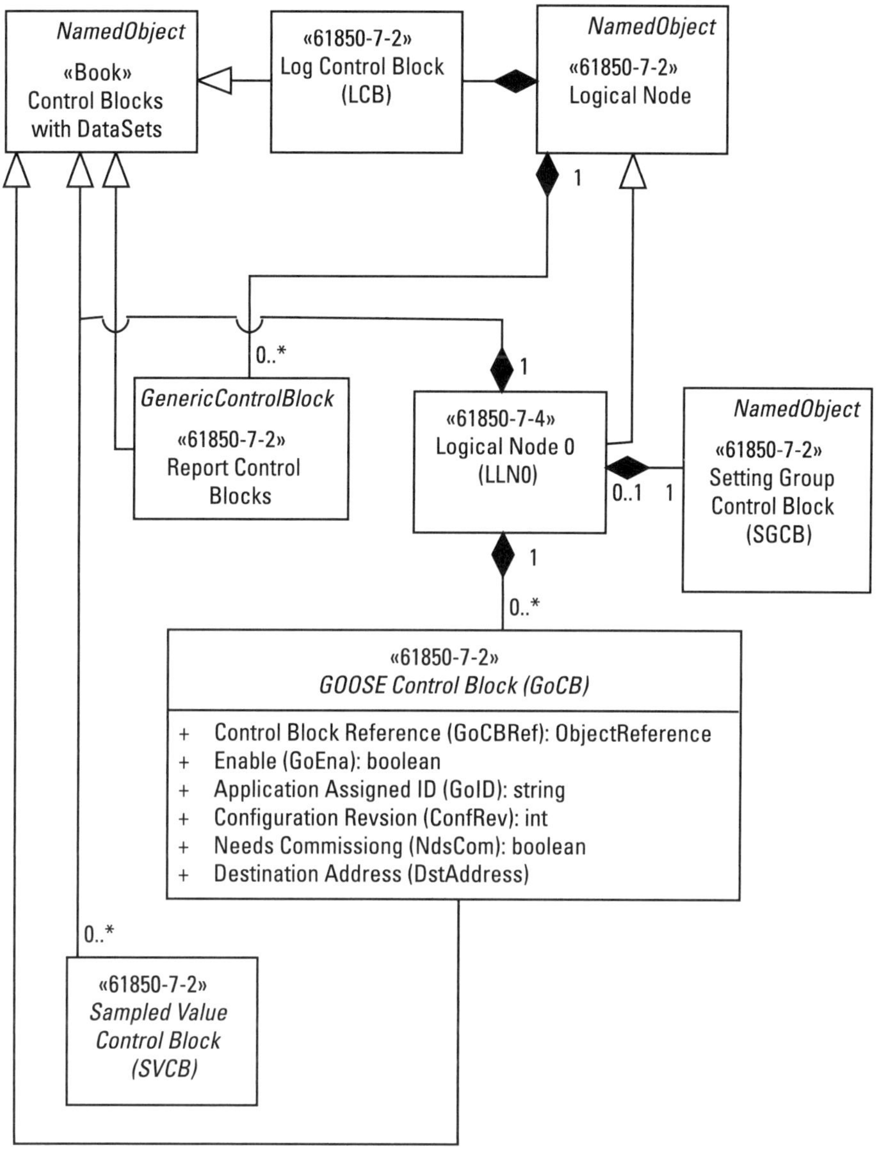

Figure 8.57 UML for abstract control blocks.

Figure 8.60 depicts the generic process flow that is controlled by either a report or log control block. The server contains process data whose values, qualities, and timestamps change based on the actual real-time process. This information is filtered by two different filters. One filter restricts the information to be buffered through the use of a dataset reference. If process data changes, and the data is not a member of the dataset specified by the control block, the data is ignored as part of that log or report process (e.g., per the control block).

The other filter further constrains the data, to a configured set of reasons, for the data to be placed into a first-in-first-out buffer. The base reasons are

- *Data Change (dchg):* Include the in the buffer if the data value of the data has changed.

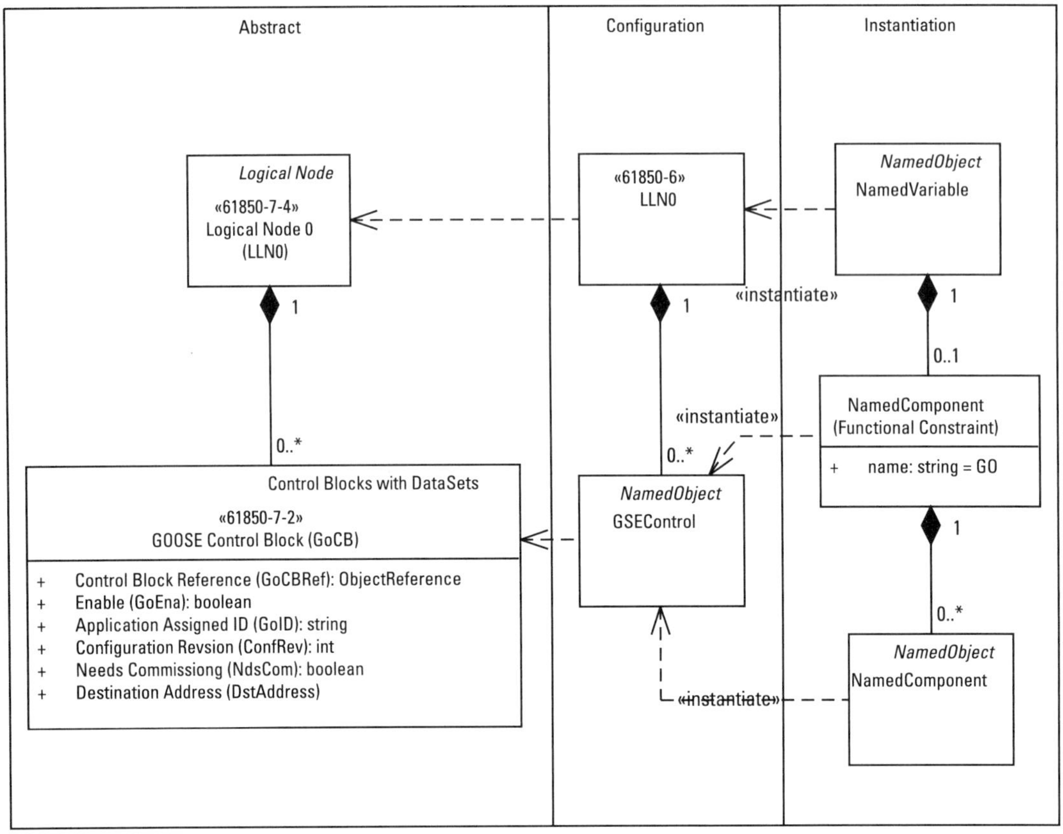

Figure 8.58 Control block mapping example.

```
<IED name="IED1>
  ....
   <Server>
     <LDevice id="FCD01">
       <LN0 lnType="LLN0 " inst="" lnClass="LLN0">
           <GSEControl name="gocb" .... >
           <ReportControl " name="A_URCB" ...">
           <SampledValueControl name="MSVCB01" .... >
           <LogControl name="lcblog" .... >
           <SettingControl numOfSGs="1" actSG="1" >
       </LN0>
           .....
     </LDevice>
   <Server>
</IED>
```

Figure 8.59 Example of LN0 configuration with control blocks. (The SCL/XML in the example does not show all of the information required for validation.)

- *Quality Change (qchg):* Include the data in the buffer if the data quality of the data has changed (e.g., from good to bad).
- *Data Update (dupd):* Include the data in the buffer if the timestamp of the data has changed.

Table 8.11 Types of Control Blocks

Control Block	*Service Control*	*LogicalNode Containership*	*Purpose*
Setting Group	Settings Groups	LN0	Used to activate and edit automation and protection settings groups.
GOOSE	GOOSE	LN0	Used to enable, disable, and configure the transmission of Layer 2 or Routable GOOSE.
Sampled Value	Sample Values	LN0	Used to enable, disable, and configure the transmission of Layer 2 or Routable Sample Value.
Reporting	Reporting	Allowed in all LogicalNodes	Used to enable, disable, and configure unsolicited information delivery to a client.
Log	Logging	Typically, LN0 but allowed in other LogicalNodes	Used to enable, disable, and configure the archiving of information that can be queried via multiple clients.

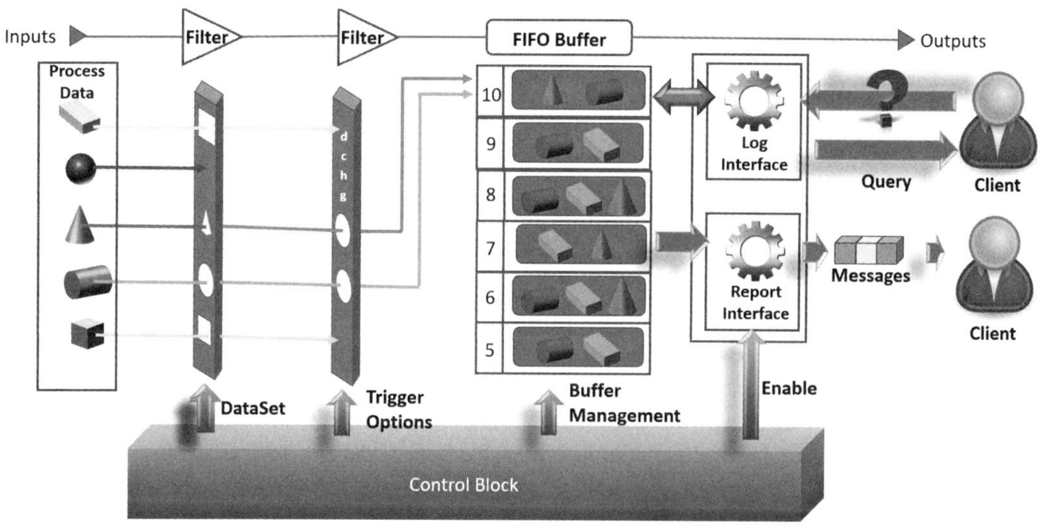

Figure 8.60 Generic reporting and logging flow.

If the data passes the dataset filter and trigger option filter criteria, then the data has met the conditions to be placed into the buffer.

Consider the following example where a dataset member is an amp measurement that has a value, quality, and timestamp and the trigger options are dchg and dupd. If the amperage value changes or the timestamp of the measurement changes, the measurement will pass the filter and be placed into the buffer. If the quality changes from good to bad, and neither the value or timestamp changes (this is rare), the new information would not pass the filter and therefore not be placed into the filter. An oddity is that if the measurement dataset member contains only a value, then a timestamp change in the process data would not ever pass the trigger options filter of dupd. Therefore, it is important that dataset members typically include value, quality, and timestamps.

The difference in how the information which passes the filter criteria are placed into a first-in-first-out (FIFO) buffer will be detailed in the report and log control sections. The log buffer, containing qualifying event information, is queried by a client through an interface that interacts with the buffer to return the requested information. Reporting (if enabled) sends information in an unsolicited fashion to the enrolled client. However, both log and report interfaces allow a client to recover historical events if they are still in the FIFO.

GOOSE, Sampled Value, and setting control blocks operate in a different manner than reporting and logging. the specifics of each class of control block can be found in the next sections.

Object References Most of the instantiated control blocks contain references to other IEC 61850 abstract classes (e.g., typically a dataset) that are abstractly shown as an association. References, or associations, to other classes are exposed in the instantiated control blocks as an object reference. The value of the object reference has the format of

<Logical Device Name>/<Logical Node Name>.<name of instantiated object>

As an example, a dataset named "foobar" that is contained within "LLN0" in logical device "Example" would have a value of Example/LLN0.foobar.

Setting Groups

A setting group is a set of operational configuration parameters that impact the operation of a device. A common example of the concept are the parameters which control the operation of a computer based on whether it is plugged in or using battery power as represented in Figure 8.61.

The configured values for power and sleep settings represent two settings groups. The first group is the configured parameters for operation if the computer is plugged in. The second is for operation if the computer is on battery power. The

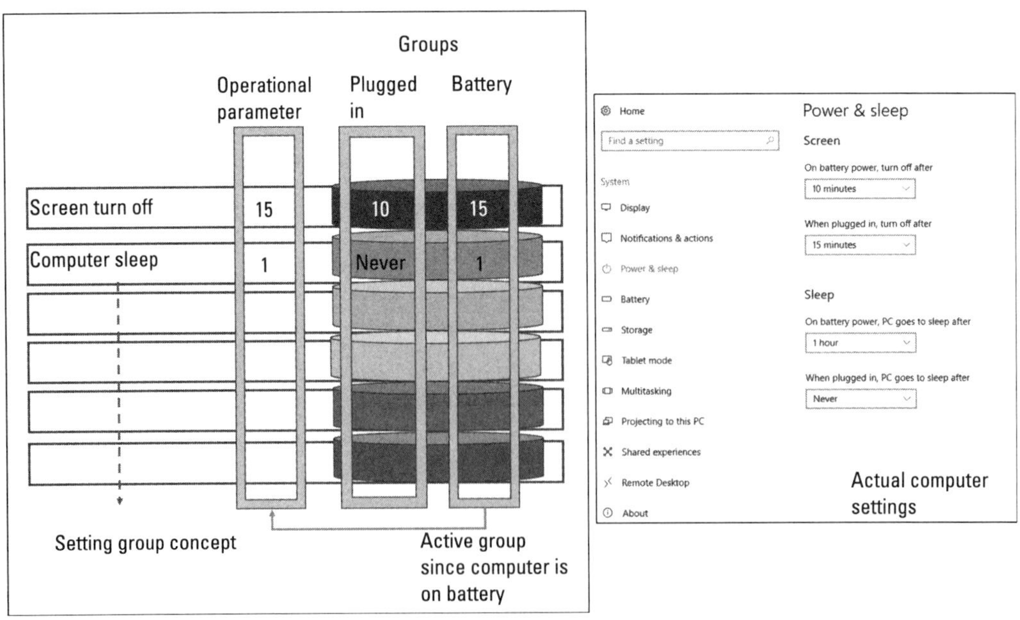

Figure 8.61 Example of computer power settings shown as setting groups.

actual operational parameters (e.g., the parameters in actual use) are determined by the operational condition of the type of power being supplied to computer (e.g., plugged in or battery). The example shows that the computer is operating on battery power and therefore the battery group of configured parameters is being utilized as the operational parameters of the computer. Besides the computer selecting the operational parameters from a configured group, there are some other characteristics that translate to the actual IEC 61850 setting groups as the configuration parameters are persistent, editable, and apply to the entire computer operation.

The configured parameters of IEC 61850 parameter groups typically deal with automation and protection functions. These types of parameters are distributed among different logical nodes. Thus, IEC 61850 setting groups allow the grouping of parameters from different logical nodes. The information that can be used within setting groups are provided through data objects that utilize the following CDCs: single point setting (SPG), integer status setting (ING), analog setting (ASG), enumerated status setting (ENG), and time setting group (TSG). The use of these CDCs provides the option to be in a setting group and exposes setting group participation via exposing functional constraints of SG and SE. Otherwise, the FC will be SP (e.g., not participating in setting groups).

Figure 8.62 shows that the server is configured with three groups (1, 2, and 3). The configured parameters are shown as abstract values. The groups provide the nominal voltage (VNom) setting for CYSN1 and CYSN2 (analog settings). The directional mode (DirMod) settings are enumerated values that allow the selection of nondirection, forward, or reverse. The true or false value is to indicate if the environmental function (e.g., MENV) is participating in carbon trading. It shows that the active group (the source of the operational parameters) is Group 2. Unlike the power settings of the computer, the active group is typically set to be a client, but can be set due to local logic. The active parameters are exposed as FC=SG. Since

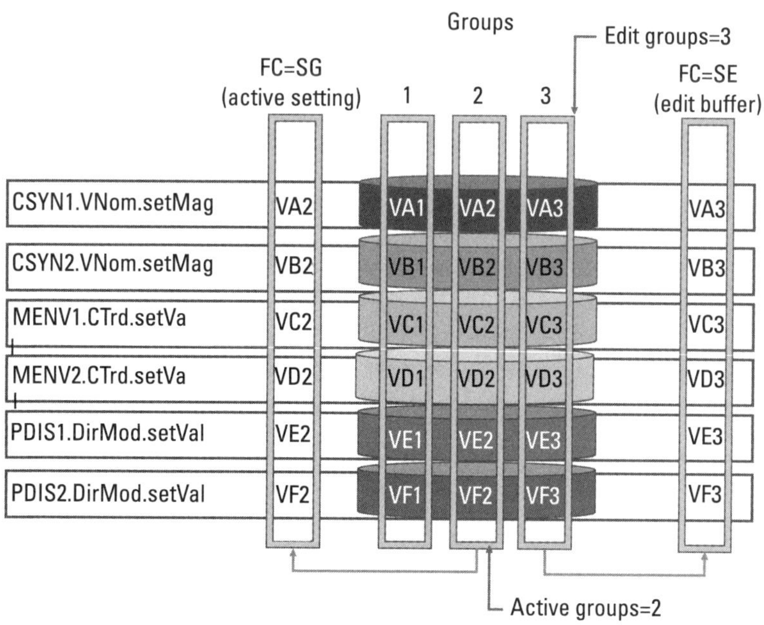

Figure 8.62 Example of IEC 61850 setting groups.

automation and protection is a critical function and parameters may need to be evaluated for consistency, IEC 61850 introduced the concept of an edit buffer. The values in the FC=SE must be committed into a specific group.

Figure 8.63 shows the abstract attributes and the number of setting groups; how the default active setting group can be configured along with whetherthe resvTms is present or not. The instantiation of a setting group control block (SGCB) is a substructure of the LLN0 named variable and is an MMS named component of the name SGCB. Because of mapping rules in IEC 61850-8-1, it can also be its own named variable having the relative name of LLN0$SGCB.

The mappings of the operations to an instantiated service is as follows:

Abstract Service	*MMS Service*	*Component Accessed*
SelectActiveSG	Write	LLN0.SGCB.ActSG
SelectEditSG	Write	LLN0.SGCB.EditSG
ConfirmEditSGValues	Write	LLN0.SGCB.CnfEdit
SetEditSGValue	Write	AnyLN.SE.xxx
GetEditSGValue	Read	AnyLN.SE.xxx AnyLN.SG.xxx
GetSGCBValues	Read	LLN0.SGCB

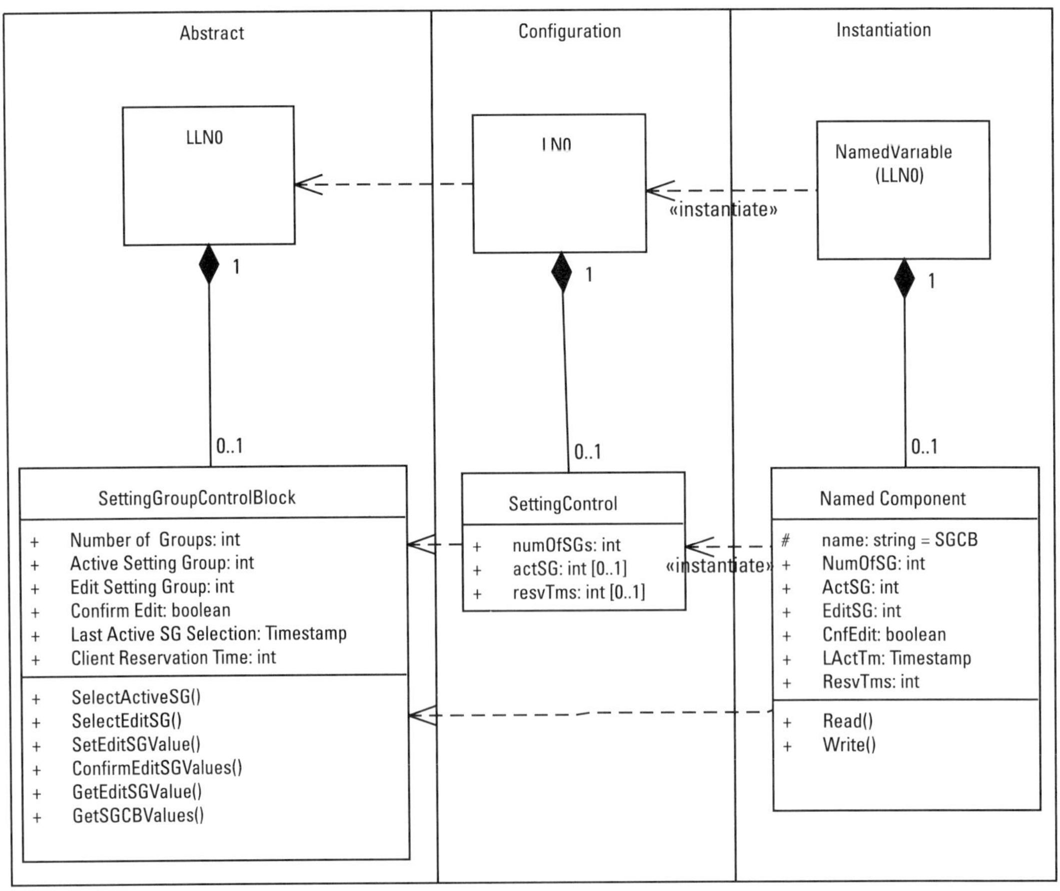

Figure 8.63 UML for setting group.

The abstract services map to either an MMS read or write service. It is the object on which the MMS service is issued that completes the mapping.

SelectActiveSG and SelectEditSG is translated to an MMS write. The value of the desired group to activate, or edit, is written. The value may not be greater than the maximum number of setting groups declared in the IED capability description (ICD) file of the device. The numbering of groups starts at 1.

The sequence for both SelectActiveSG, see Figure 8.64, and SelectEditSG begins in the abstract with the 61850 client issuing the appropriate abstract service request (e.g., in the example SelectActiveSG). The abstract service request is mapped to an MMS write request containing the object reference of the setting group control block attribute to be written and the value of the group to be selected. The IEC 61850 server receives the request and updates the appropriate value in the internal control block. Once there is an update, the tracking structure (e.g., LTRK) is updated. A successful MMS write response is returned indicating the success of the write. This is translated to the abstract response and the select service is completed.

- SetEditSG is translated to an MMS write to the logical node of any FCD or FCDA, whose functional constraint is SE (e.g., indicating editable).
- GetEditSGValue is mapped to an MMS read. There are two uses of this service in that it is used to obtain the values of either the editable values (e.g., SE functional constraint) or the active value (e.g., the SG functional constraint).

The SCL configuration of the actual setting group information may appear similar to the following:

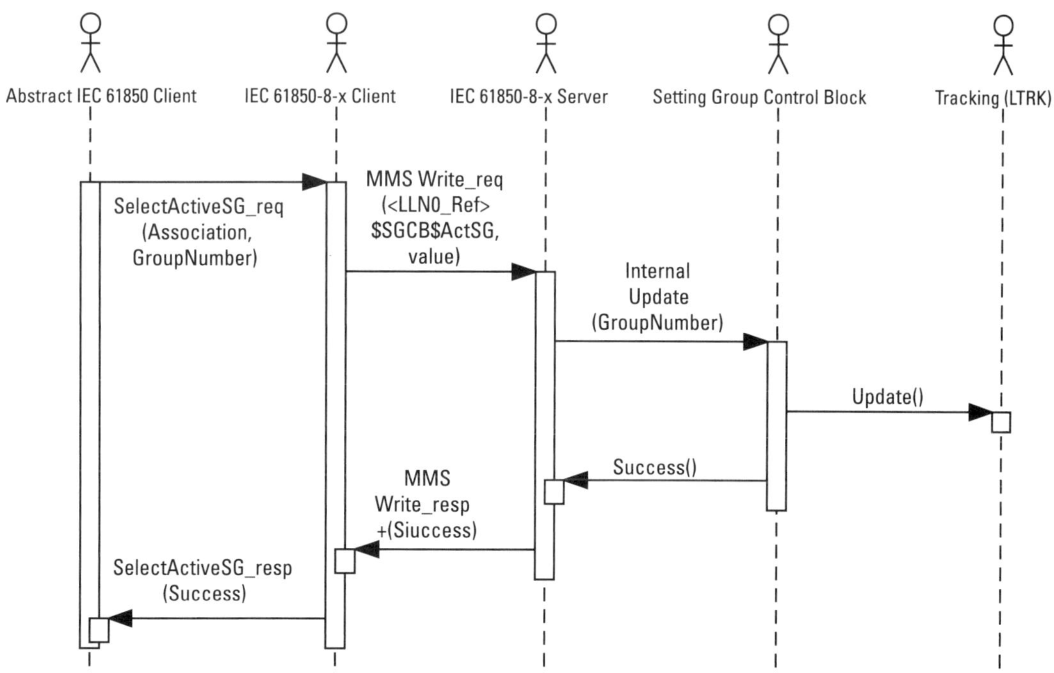

Figure 8.64 Sequence of SelectActiveSG.

Figure 8.65 shows that the total number of setting groups (e.g., numOfSGs) is three and that the initial group is group one. What is not obvious is therequirement that if the active group is changed, and a power cycle occurs, the last written value of actSG is restored and not the value in the SCL file.

There is another configuration requirement regarding setting groups regarding the declaration of the ability to provide setting group support.

GOOSE

Users typically utilize GOOSE for unsolicited multicast message delivery in automation systems. The delivery of these messages needs to be high performance and reliable. The reliability of GOOSE message delivery was previously explained. GOOSE control blocks control the configuration and control of GOOSE multicast messages (see Section 3.2.2).

Conceptually a GOOSE control block (GCB) (see Figure 8.66) is used to specify which process data of interest is to be sent via the control of a specific GCB. This filtering is performed through the control block containing a reference to a dataset (see Section 8.2.2.3). If the reference to the dataset is null (e.g., not defined), the control block should not begin transmitting a GOOSE message. In Edition 1 of IEC 61850, it could begin transmitting but with an indication that the control block was not properly configured (e.g., NdsCom=true). This is still an allowed behavior in Edition 2 but provides little information exchange enhancement as the failure to receive a GOOSE by the enabling publisher/client conveys similar information as if a GOOSE was received indicating NdsCom. The members of the dataset represent the process information of interest to be sent via GOOSE. The last known values of the process data are buffered and the GOOSE Interface (i.e., the software that is responsible for sending GOOSE messages) detects value changes in the buffer. If the GCB is enabled, GOOSE messages will be sent to the destination address that is provided in the configuration information. Enabling a GCB starts a state machine that controls the messaging of GOOSE and it is the GOOSE interface software that manages the retransmission, and state machine as shown in Figure 8.67, of GOOSE messages.

```
<IED name="IED1>
  ....
   <Server>
     <LDevice id="FCD01">
          <LN0 lnClass="LLN0" inst="" lnType="LNN0_Type" desc="General">



            <SettingControl numOfSGs="3" actSG="1" >
       </LN0>
            .....
     </LDevice>
   <Server>
</IED>
```

Figure 8.65 Example declaration of the setting group control block.

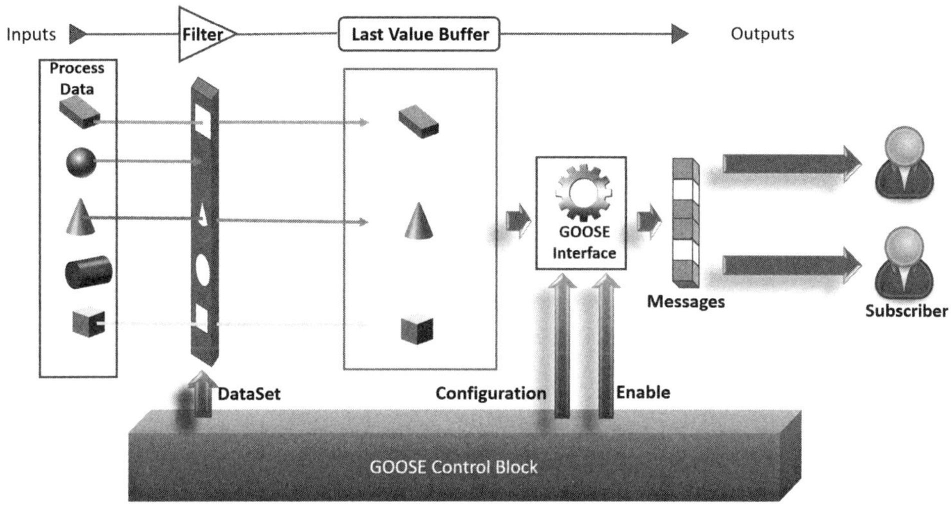

Figure 8.66 GOOSE control block and event flow concept.

To detect undelivered messages, the GOOSE messages contain information related to the detection of a value change in the buffer (e.g., state) and an identification of the sequence of retransmission (e.g., sequence). The subscriber state machine, shown in Figure 8.68, also shows that the retransmission timer is based on a locally defined retransmission curve that provides information of when to expect the delivery of the next message to a subscriber (e.g., time allowed to live (TAL)).

Prior to IEC 61850 Edition 2 Amendment 1, the relationship between TAL and the retransmit timer (t) was a local issue. Sometimes, implementations decided that t=TAL and this inevitably caused TAL timeouts on the subscriber due to network latency. Not only that, but if t=TAL, in many situations the guaranteed. Edition 2.1 Amendment 1 specifies that the retransmit time must be less than one half of TAL.

Table 8.12 shows where the various elements of the GOOSE message are derived from which provide the contents of the general GOOSE message shown in Figure 8.69.

The control block also exposes configuration information related to addressing and quality of service parameters that are used typically in the header or communication profiles. There are two such profiles that result in two different GOOSE control blocks. The first GOOSE (e.g., present in Edition 1) is typically referred to as Layer 2 GOOSE. This profile takes the GOOSE message and transmits it directly embedding the GOOSE within Ethertype 0x88b8 (e.g., the information is not routable and does not utilize IP, TCP, or UDP). The newest version of GOOSE is transmitted using UDP/IP multicast and is referred to as R-GOOSE. The communication profiles for GOOSE communication is shown in Figure 8.70.

Since both GOOSE methodologies utilize multicast, a mechanism to control the paths/distribution of the published packets is also specified within IEC 61850. L2 GOOSE utilizes virtual LANs (VLANs) as specified by IEEE 802.1Q to provide this control. Only Ethernet ports configured to receive a VLAN value (referred to as a VLAN tag) will allow egress of a message with that VLAN value. It is typical that switch ports can be configured to support a limited number of VLAN tags. Since the number of VLAN values/port is limited, it is important to take this

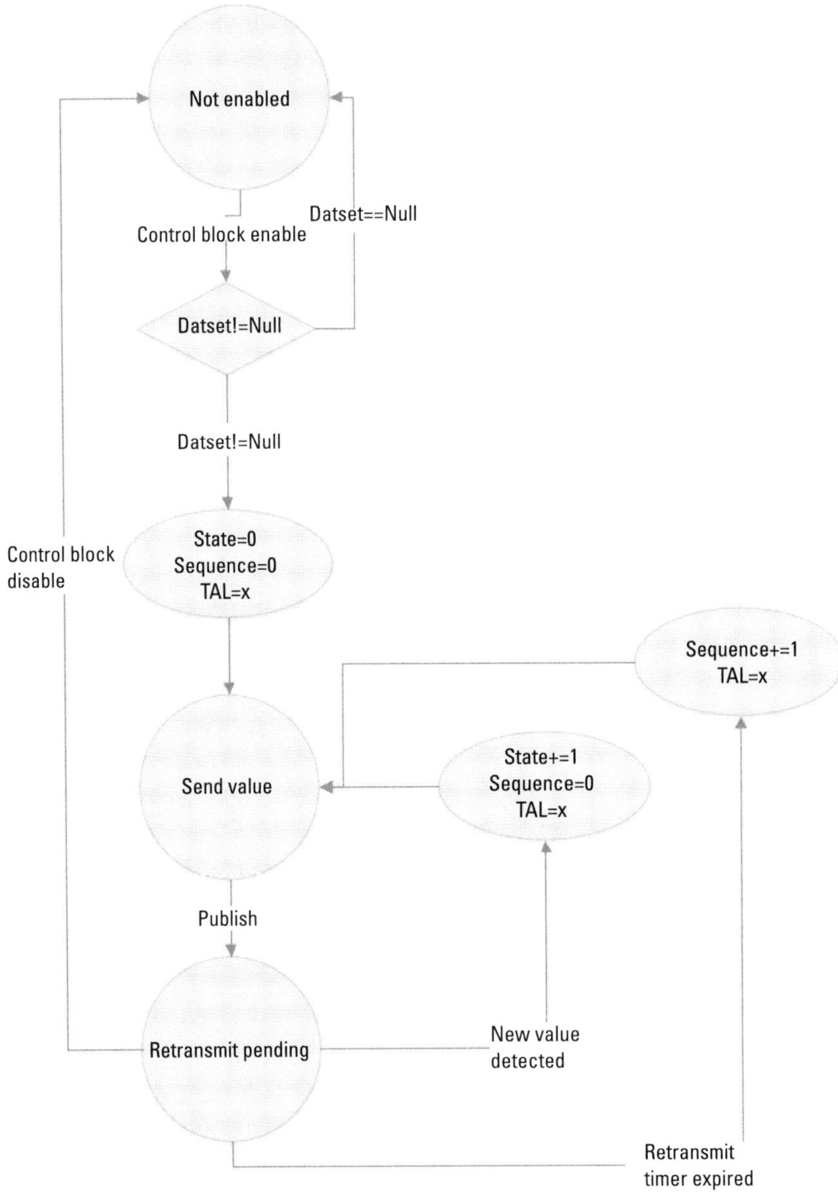

Figure 8.67 GOOSE publisher state machine.

constraint into account when engineering a system and in selecting the appropriate Ethernet switches.

Although R-GOOSE also provides the ability to utilize VLANs, the use of the Internet Group Management Protocol (IGMP) allows routers to determine the appropriate paths based on subscribers posting requests for group (e.g., destination multicast address) membership. There are two different versions of IGMP: Any Source Membership (ASM) and Source Specific Membership (SSM). R-GOOSE specifies the use of SSM which is known as IGMPv3. The use of SSM allows only messages from a specific publisher to be routed to the destination. The use of SSM improves security and decreases bandwidth over the routed network. There is a

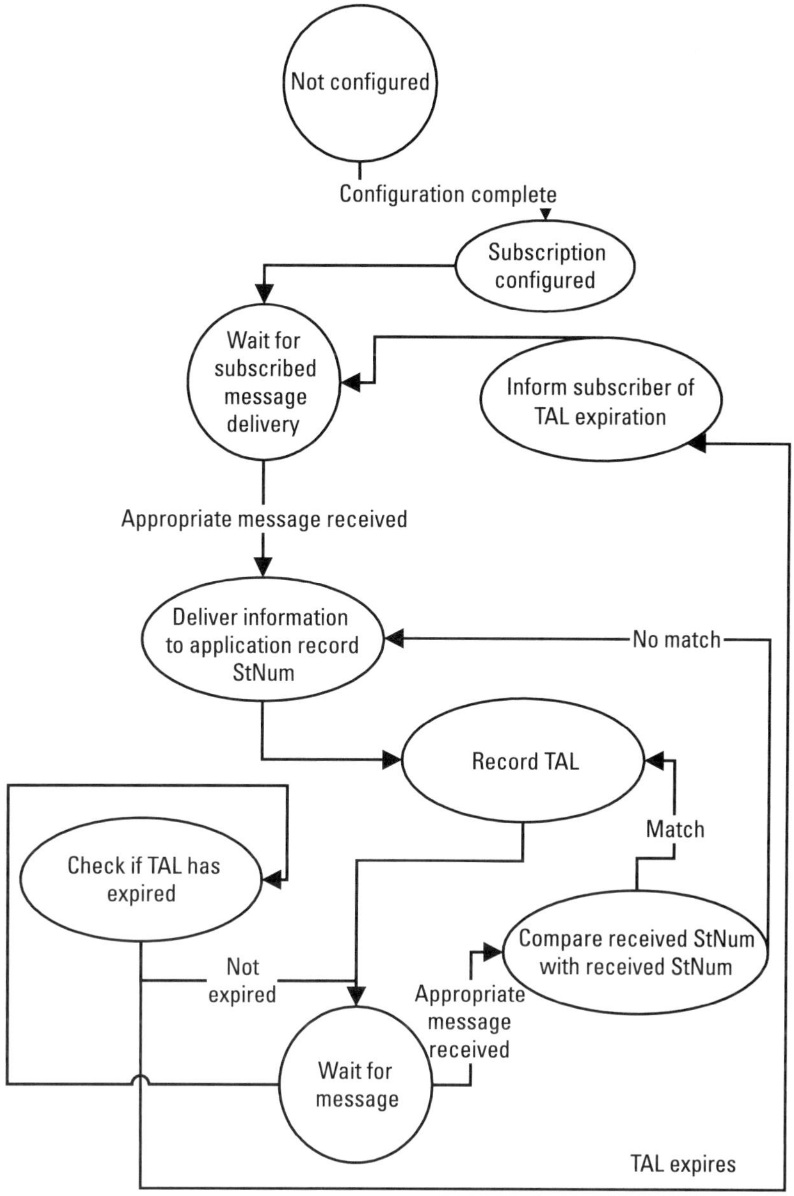

Figure 8.68 GOOSE subscriber state machine.

side benefit to utilizing IGMP in that Ethernet switches capable of IGMP snooping can provide a filtering (like VLANs) that does not require advance configuration for a specific port. The use of IGMP snooping switches does increase the complexity of network message diagnosis. Since ports that do not post subscriptions do not egress all packets, network design should include the configuration of a mirror port.

Table 8.12 GOOSE Payload Contents

IEC 61850-7-2 Attribute	*Values Typically Come from*	*Description*
DatSet	SCL	The value is an object reference that refers to a dataset whose member values are to be transmitted.
GoID	SCL	Is a user-configured string value that allows subscriber and diagnostic tools to easily filter GOOSE messages. If there is not one configured, the value of gocbRef is the default value.
GoCBRef	Object model and SCL	This is a visible string (also known as an object reference) reference to the actual control block that is contained in LN0 within a logical device.
T	State machine	This is the timestamp at which a value change (e.g., state change) was detected. The value does not change for retransmissions.
StNum	State machine	This value is provided by the state machine and increments when a value change (e.g., state change) was detected. The value of 0 is reserved for when the control block is enabled. If the value rolls over, it rolls to a value of 1.
SqNum	State machine	This value is provided by the state machine and is initially 0 on a state change (e.g. when stNum changes value).
ConfRev	SCL	A number that specifies the configuration revision of the dataset. As members are added, deleted, or changed this number must be changed in order that the subscriber can be guaranteed that it is receiving the appropriate information.
Simulation or test* **	Local	This is a Boolean value that indicates if the GOOSE message is being issued with simulated data (e.g., not real process data). IEC 61850 reserves the right to set the value to True for test equipment although other publishers may also be able to set this value.
NdsCom	Local	In Edition 1, the Boolean value of True was reserved to indicate that the publisher had detected a major problem (e.g., either configuration or other) and is requesting assistance. In Edition 2 implementations, it would be rare to see a value of True being transmitted as the data in the payload is invalid in either instance. Many implementations now will not transmit a GOOSE instead of setting the value to True.
GOOSE data	DataSet Process	The transmission of GOOSE Data results in two payload fields being transmitted. The first is the number of data that is to be included in the payload. This value is the number of dataset members of the dataset referenced by the control block. The actual values of the members are also provided by the Process and GOOSE Interface.

*The name and processing on the subscriber side was changed from Test in Edition 1 to Simulation in Edition 2.

**See Section 8.2.2.7 for a description of using Simulation.

Figure 8.69 General construction of GOOSE message.

Since GOOSE messages are used for automation and real-time[13] control, ensuring that these critical messages are delivered egressed at a higher priority than a VOIP as an example. To understand the concept of egress priority, see Figure 8.71.

If a new message is to be added into the Ethernet switch egress buffer, and it has a similar or lower priority to other messages already in the buffer, it will be added to the end of the buffer.

13. Everyone has a different definition of what real time or fast is!

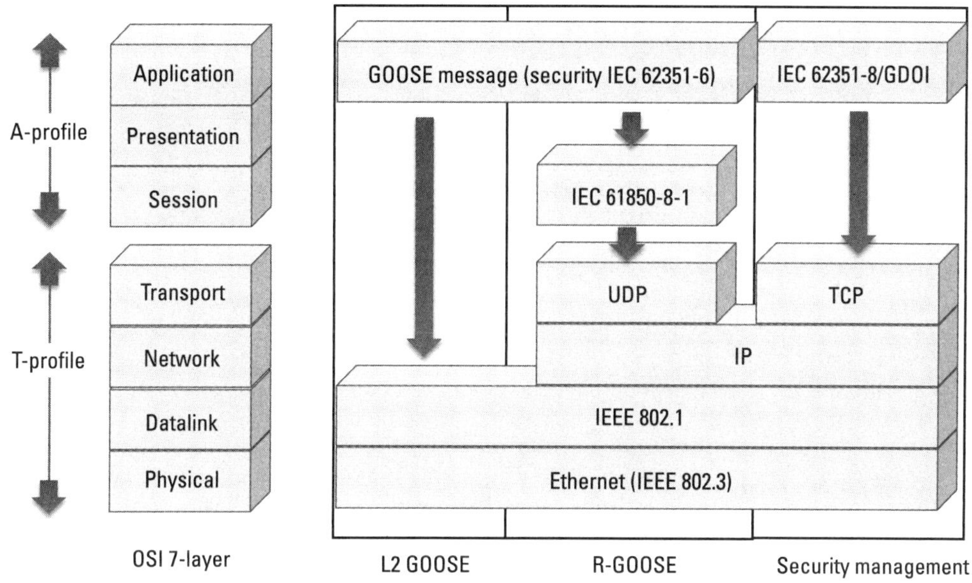

Figure 8.70 High level representation of GOOSE communication profiles.

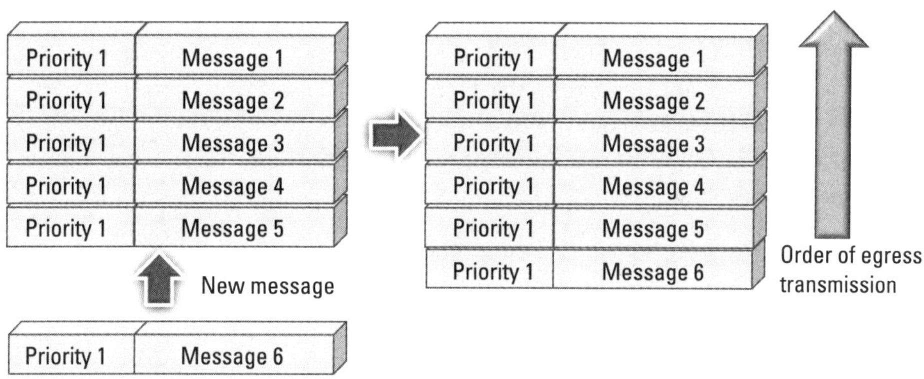

Figure 8.71 Egress example of new message with low or similar priority.

If a new message is to be added into the Ethernet switch egress buffer, and it has a higher priority (see Figure 8.72) than other messages already in the buffer, it will be moved to be the first message to be transmitted. However, this will not disrupt the transmission of a message already in the buffer.

Ethernet switch priority is based on IEEE 802.1Q and has values from 0-7. Seven is the highest egress priority with zero being the lowest. It is typical that Ethernet switches may not support all values of IEEE 802.1Q regarding re-ordering of the egress buffer. In certain implementations, the switches have individual buffers for each priority and egress and the buffers are serviced at different rates to avoid deadlock and failure to service the lowest priority messages. In these implementations, multiple priority values could be assigned to the same buffer (e.g., a buffer might service VLAN priorities 0, 1, and 2). In this type of implementation, a priority 2 message might not be egressed prior to a priority 0 message. In most of these

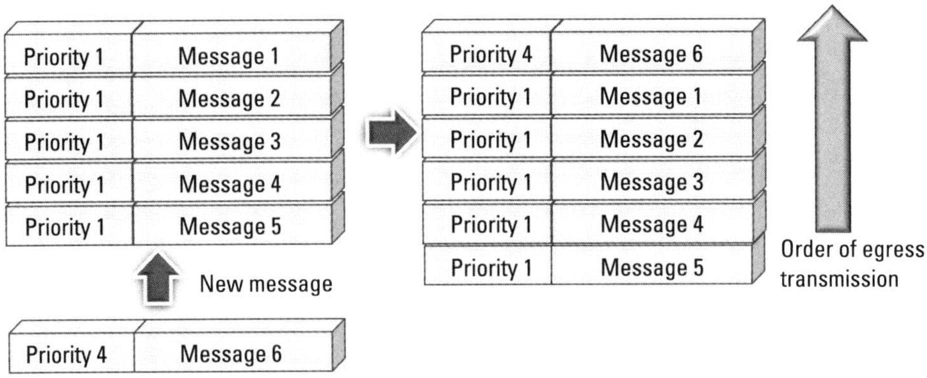

Figure 8.72 Egress example of new message with higher or similar priority.

types of implementation, there are typically three egress buffers. It is important to understand this implementation issue to design a high-performance system.

R-GOOSE also needs a mechanism to control the egress of routed messages (e.g., implemented by routers and not switches). Since this is a routing function, the field utilized is used within IP and is specified as the differentiated service field per RFC 2474. The RFC reserved 6 bits known as the differentiated service codepoint (DSCP). The other 2 bits allocated via the RFC are used to indicate the use of explicit congestion notification (ECN), which is also recommended for use in R-GOOSE systems. The use of ECN (RFC 3168) is used for diagnostic purposes as it can only notify that there was congestion in the communication path.

Once the publisher transmits a GOOSE and the network delivers the message to a subscriber, there is a state machine that the subscriber implements. There is a simple state machine and a complex state machine that is less prone to spoof and replay attacks. An efficient and simple state machine implemented in the subscriber, might look like Figure 8.73.[14]

The reception of GOOSE starts when the subscription configuration is complete in the subscriber. This includes not only application-level configuration, but the appropriate communication subscription (e.g., posting multicast and IGMP subscriptions). The state machine waits for the reception of the first message delivery for a specific subscribed message. If the message is appropriate, the appropriate information is passed to the application, the state number (StNum) and time allowed to live (TAL) are recorded for use at other stages in the state machine. The machine then enters a loop where it is waiting for the reception of the next message and checking if the TAL has expired. If the next message is not delivered prior to TAL expiring, it is legal to declare a message delivery failure although many implementations provide a little delay before declaring this to account for network latency packet delay variation (PDV) as defined in RFC 3393. If there is a message delivery failure declared (e.g., TAL expired), the state machine transitions to a wait state where the TAL is not being checked and informs the application of the message delivery failure. IEC 61850 does not specify how an application should behave on being informed of a message delivery failure although it is typical that

14. The state machine shown is enhanced and has more detail than the one in IEC 61850-8-1.

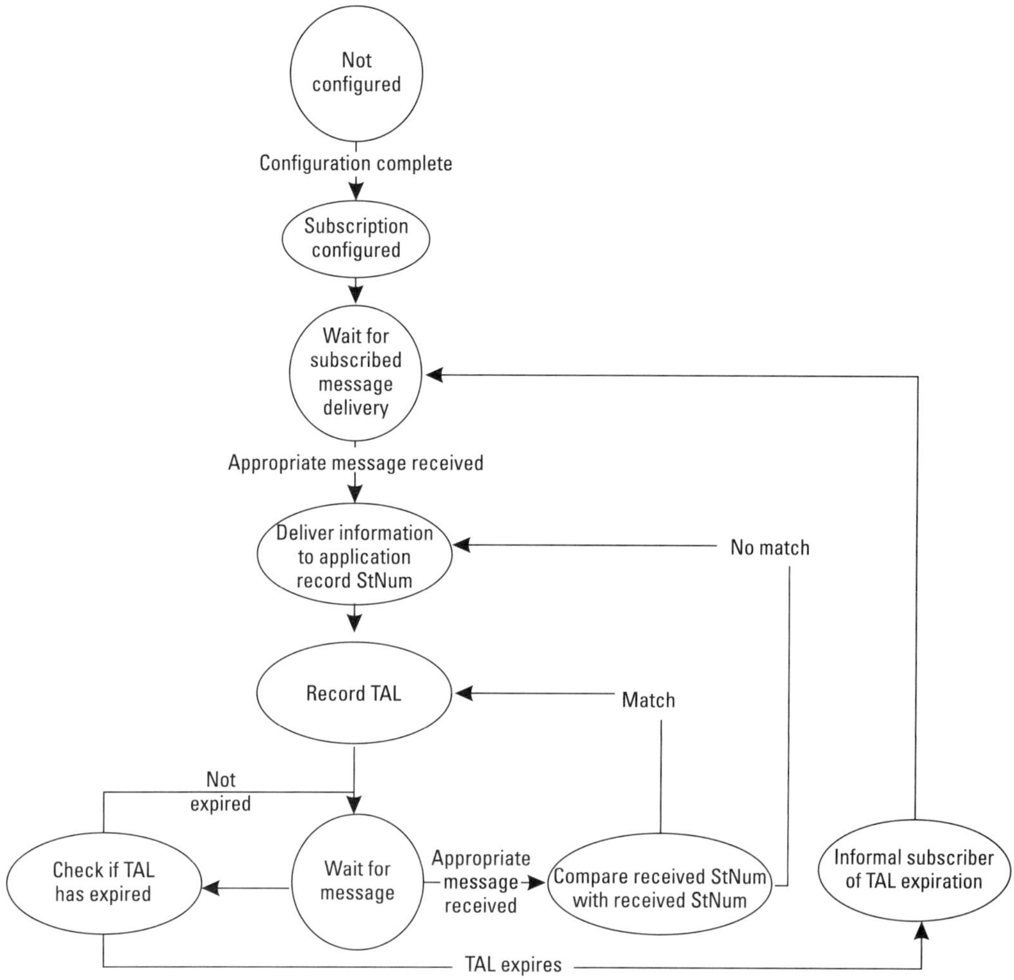

Figure 8.73 Simple GOOSE subscriber state machine.

the subscriber's GOOSE interface provides Bad or LastKnownValue qualities for the data contained in the message.

If a message is received prior to TAL expiration, the received StNum and the recorded StNum are compared. If the values match, then only the TAL needs to be recorded since there have been no data value changes. If the StNum values are different, this indicates that there have been data value changes that need to be considered for delivery to the application. Therefore, the appropriate information is delivered to the application and that information should be delivered. Not all information conveyed in a GOOSE message is normally required by a single application. Since GOOSE is a multicast, information that is needed by multiple subscribers can be published in a single message. Therefore, subscribers need to be able to be configured to understand which data from a specific GOOSE message is being utilized by the application.

One might want to know if the TAL mechanism works in instances where the network latency is greater than the TAL value as may be the case over wide area networks. It is important to understand that the value of TAL is the elapsed time at which to expect the delivery of another message. Therefore, if there is a $latency_{nom}$

that represents the nominal delay over the communication path (e.g. constant), the $latency_{delay}$ (e.g., actual delay latency) could be represented by (8.1):

$$latencydelay = latencynom + latencypdv \tag{8.1}$$

at a state change (e.g., S1)

$$latencydelay\ [S1] = latencynom + latencypdv[S1]$$

for a retransmit (e.g., R1):

$$latencydelay\ [R1] = latencynom + latencypdv[R1]$$

In order to avoid a TAL expiration (for $latency_{delay}\ [R1] \geq latency_{delay}\ [S1]$):

$$\begin{aligned} TAL &\geq latencydelay\ [R1] - latencydelay\ [S1] \\ TAL &\geq (latencynom + latencypdv\ [R1]) - (latencynom + latencypdv[S1]) \\ TAL &\geq latencydpv\ [R1] - latencypdv[S1] = \text{Packet Delay Variation (PDV)} \end{aligned} \tag{8.2}$$

Thus, it is the variance in network latency (e.g., PDV), see (8.2), that has an impact on if a TAL expiration will be detected and is not dependent on the nominal latency. However, if the $latency_{delay}$ for a state change exceeds the needs of an application, then there is an issue with the network design supporting the application.

The retransmission curve, and therefore the curve for TAL, is defined locally or configured outside the IEC 61850 standard. IEC 61850 places constraint on the maximum value of TAL which is 1 minute. Therefore, the maximum allowed retransmission value is 30 seconds. It is worthwhile to note that to achieve guaranteed[15] message delivery within 3 msec (per IEC 61850-5), there must be at least two transmissions within that 3-msec period. It is also required that when enabled, a GOOSE continuously transmits. This requirement is so that a new subscriber can be updated with the current process information without having to request it (e.g., within 1 minute). Thus, many implementations' retransmission curve is exponential in nature. Typically, it starts with a 2-msec retransmission and ends with a 30-second retransmission interval.

Figure 8.74 shows two typical strategies for implementing retransmission curves. One follows a doubling of the retransmission interval until the value approaches 30 seconds (e.g., 2, 4, 8, etc.). The other starts with a fast retransmission but rapidly achieves its end retransmission interval (e.g., 2, 4, 2000). IEC 61850 does not mandate that the slowest retransmission interval is 30 seconds, only that it is less than 30. The fastest retransmission interval is referred to the minimum retransmission interval (MinTime). The slowest retransmission interval is referred to as the maximum retransmission interval (MaxTime). These values, and the actual retransmission curves must be evaluated in order to determine if an implementation provides the performance required for a specific application. It is rare that

15. "Guarantee" is not 100% but 99.8% probabilistically. See Section 3.2.2 for further details.

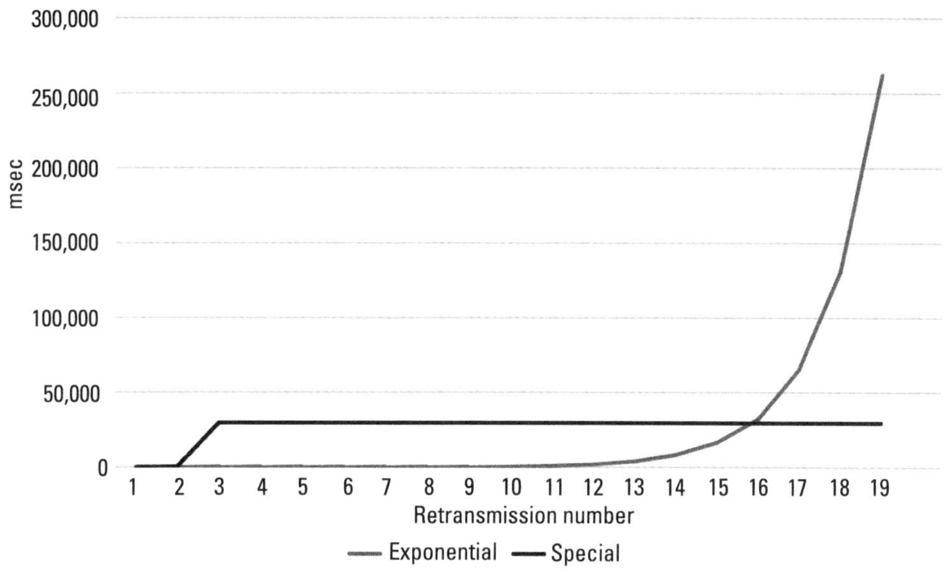

Figure 8.74 Example of GOOSE retransmission curves.

these values can be modified by IEC 61850 configuration, however some implementations do allow this.

The UML, in Figure 8.75, shows the abstract attributes that are exposed in all currently defined GOOSE control blocks. There are currently two different GOOSE control blocks: one for Routable GOOSE and the other for Layer 2 (L2) GOOSE. Each of these control blocks has different physical address/destination address structures. The instantiation of a GOOSE control block is a substructure of the LLN0 named variable and is an MMS named component of the name assigned by configuration (e.g., SCL) and will be contained by a named component of 'GO' or 'RG' for L2 GOOSE or R-GOOSE, respectively. Because of mapping rules in IEC 61850-8-1, it can also be its own named variable having the relative name of LLN0GO<name>. It is typical for most implementations to name L2 GOOSE control blocks as gocb1 through gocbx although this is not required.

The mappings of the operations to an instantiated service is as follows:

Abstract Service	*MMS Service*	*Component Accessed*
GetGoCBValue	Read	LLN0.GO.<name>.GoEna LLN0.RG.<name>.GoEna
SetGoCBValue	Write	LLN0.GO.<name>.<x> LLN0.RG.<name>.<x>

The abstract services map to either an MMS read or write service. It is the object on which the MMS service is issued that completes the mapping.

- SetGoCBValue is translated to an MMS write. This service might as well be named EnableGoCB as the only writeable value in the instantiated control block is the enable attributed. It is used to enable and disable the transmission of the GOOSE message.

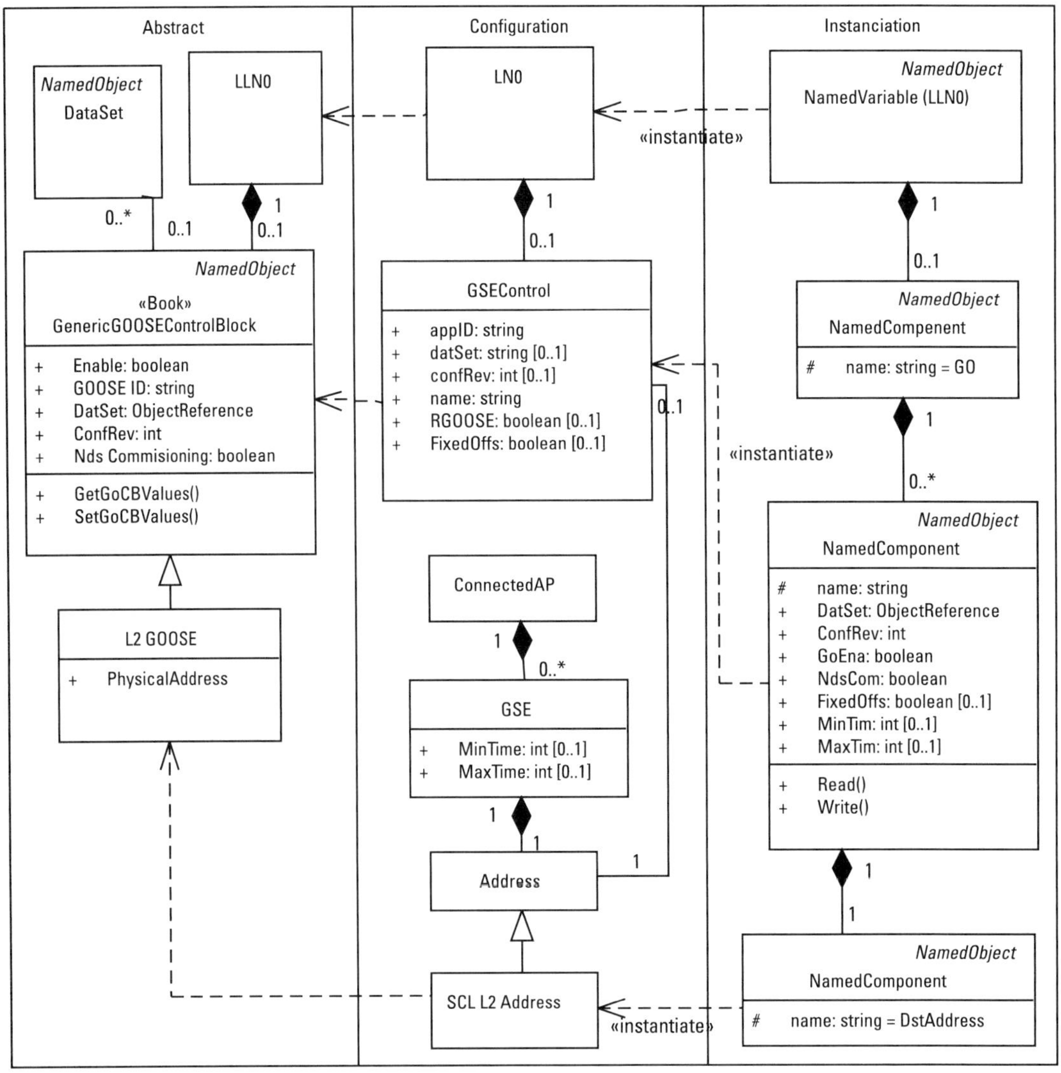

Figure 8.75 Generic GOOSE control block UML.

The sequence for SetGoCBValue, shown in Figure 8.76, begins in the abstract with the 61850 client issuing the appropriate abstract service request (e.g., in the example SetGoCBValue_req). The abstract service request is mapped to an MMS write request that contains the object reference of the enable attribute to be written. The IEC 61850 server receives the request and updates the appropriate value in the internal control block. Once there is an update, the tracking structure (LTRK) is updated. A successful MMS write response is returned indicating the success of the write. This is translated to the abstract response and the select service is completed.

- GetGoCBValue is mapped to an MMS read. This service is utilized to retrieve the current values of the control block.

The SCL configuration of both an L2 GOOSE and R-GOOSE control block may appear in Figure 8.77.

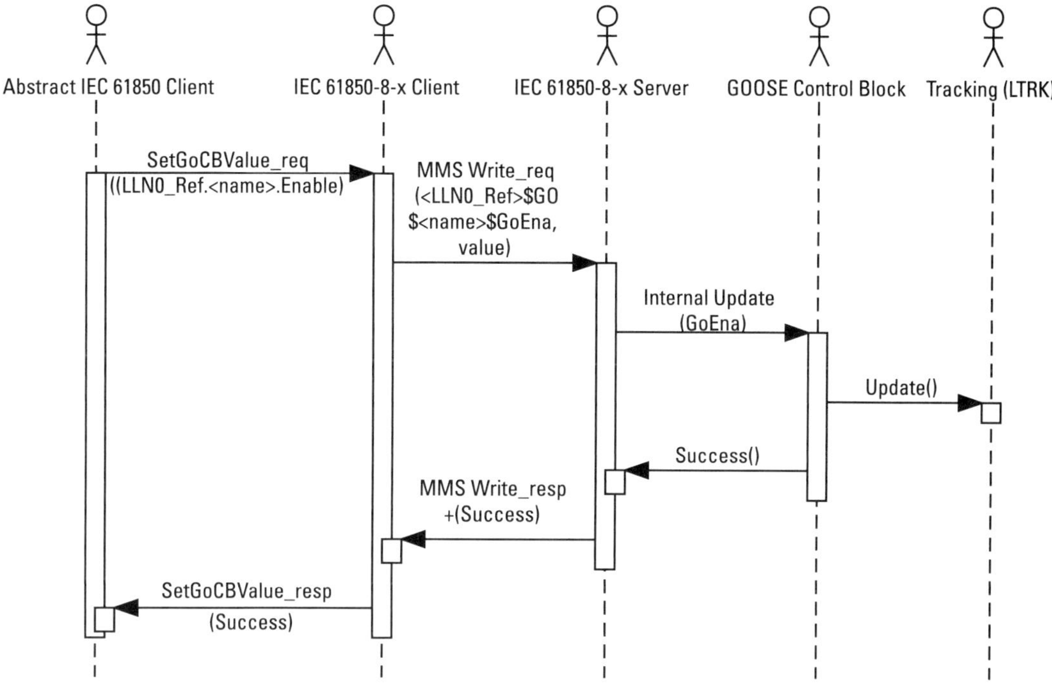

Figure 8.76 Sequence of SetGoCBValue.

Figure 8.77 shows the relationships of the different L2 and L3 addressing structures for GOOSE (e.g., the abstract physical address) and the two types of GOOSE Control blocks. The example also shows the configuration of the GOOSE control block name (e.g., GSE1 or GSE2), the reference to the dataset to be transmitted (dataset1 or dataSet2), the configuration revision (confRev) and the application ID (appID). It also shows that in order to differentiate between a L2 GOOSE and R-GOOSE, the <Protocol> element must be specified, and it is required to be understood by an SCL tooling (e.g., mustUnderstand="true").

The MinTime and MaxTime communication attributes are related to the TAL and retransmission rate. The value of MinTime indicates the minimum TAL value specified in msec. The value of MaxTime has been a topic of intense debate in 2018. The final interpretation is that the value of MaxTime represents the maximum time between retransmission of GOOSE messages. There is also a correlation between and the rate of retransmission. TAL must be at least two times greater than the retransmission interval in order to prevent inadvertent expirations.

The <IEDName> element within <GSEControl> is the mechanism that allows the SCL engineering process to configure which GOOSE messages should be subscribed to by one or more subscribers. Since GOOSE is a multicast message, there may be multiple subscribers listed and individual <IED> elements. The apRef attribute specifies the access point through which the subscribing IED should receive the GOOSE message. If the IED has only a single access point, this attribute is not required to be present.

This example does not show the use of fixedoffs. This configuration capability was not present in Edition 1 but was added to provide an encode efficiency that allows GOOSE processing to be committed to programmable logic arrays (PLAs)

```
<Communication>....
 <ConnectedAP apName="E" iedName="IED1">
    <Address>....
     <GSE cbName="GSE1" ldInst="CB2">
      <Address>
       <P type="MAC-Address">01-0C-CD-01-35-12</P>
       <P type="APPID">0001</P>
       <P type="VLAN-PRIORITY">4</P>
       <P type="VLAN-ID">2</P>
      </Address>
     <MinTime multiplier="m" unit="s">5</MinTime>
     <MaxTime multiplier="m" unit="s">10000</MaxTime>
    </GSE>
   ....
   </ConnectedAP>
</Communication>
<IED name="IED1>
  ....
    <GSEControl name="GSE1" datSet="DS1" confRev="2" appID="GSE1">
       <IEDName apRef="AP">AA1D1Q01KF2</IEDName>
    </GSEControl>
    <GSEControl name="GSE2" datSet="DS2" confRev="15" appID="GSE2">
       <Protocol mustUnderstand="true">R-GOOSE</Protocol>
       <IEDName apRef="AP">AA1D1Q01KF2</IEDName>
     </GSEControl>
  ...
</IED>
```

Figure 8.77 Example Declaration of the GOOSE control blocks.

instead of software. If the attribute is present, and has a value of "true," the encoding of the GOOSE message is not encoded by the specific rules of Abstract Syntax Notation 1 (ASN.1) Basic Encoding Rules (BER). ASN.1 BER specifies a tag, length, value encoding where the lengths of fields can vary. If fixedOffs="true", the length of particular data types is not variable. Table 8.13 shows the difference of encoding between BER and the fixed offset version.

The table compares the encoding (e.g., bytes that will appear on the wire) between the two methods. The length of the BER values varies based on the actual value. The maximum length is always encoded in fixed offset mode. It is worthwhile to note that even with fixed offset encoding there is processing involved to determine the actual offset of the primitive fixed length encoded fields since the MMS encodings (e.g., ASN.1) has variable structures and therefore the offset of those fields need to be precomputed or determined as needed. The advantage of the fixed offset encoding/decoding is that the offset/precomputation of location can be performed in advance, and provided to the PLA instead of the PLA having to implement the additional logic to determine the offsets as a message is being processed.

Figure 8.78 shows an example of the capture of an L2 GOOSE trace. The GOOSE Ethertype is 88B8 hex and accounts for the first goose after the 802.1Q fields (e.g., priority and VLAN). The AppID in that Ethertype is shown as being defined in the communication section defining the destination addressing information. The example in Figure 8.77 shows the AppID as being 0000 hexadecimal. The example also shows an AppID defined as part of the GSEControl configuration. This value is typically referred to as the GOOSE ID (GOID).

Table 8.13 Example of Encoding ASN.1 BER and Fixed Offset

	Tag Value of A0	
Integer Value 32-bit	*BER* Encoded*	*Fixed Offset*
1	A0 01 01	A0 05 00 00 00 00 01
127	A0 01 7F	A0 05 00 00 00 00 7F
128	A0 02 00 80	A0 05 00 00 00 00 80
–128	A0 01 80	A0 05 FF FF FF FF 80

*The actual BER encoding rules are discussed in Section 11.3.

```
⊞ Ethernet II, Src: IaLink_01:89:e3 (00:03:b3:01:89:e3), Dst: 09:00:00:00:00:01 (09:00:00:00:00:01)
⊞ 802.1Q Virtual LAN, PRI: 4, CFI: 0, ID: 0
⊟ goose
    AppID*: 257
    PDU Length*: 110
    Reserved1*: 0x0000
    Reserved2*: 0x0000
  ⊟ PDU
      IEC GOOSE
      {
        Control Block Reference*:    mydom/mygcRef1
        Time Allowed to Live (msec):  8
        DataSetReference*:    mydom/mydataset1
        GOOSEID*:    testAppID1
        Event Timestamp:  2004-01-06 22:10.20.000000  Timequality: 00
        StateNumber*:    1
        SequenceNumber*:    3
        Simulation Bit TRUE
        Config Revision*:    32
        Needs Commissioning*:    FALSE
        Number Dataset Entries:  3
        Data
        {
          FLOAT:  4.234000
          FLOAT:  6.234000
          FLOAT:  8.235000
        }
```

Figure 8.78 Capture of L2 GOOSE.

Figure 8.79 shows that R-GOOSE encapsulates the IEC GOOSE PDU, as defined by L2 GOOSE, within a different communication profile and is protected by security.

The structure shows that the session encapsulating protocol can differentiate between the contents being GOOSE, Sampled Values, tunneled L2 packets, and GSE management. It also shows that there is security information carried in the header as well as a security message authentication code (MAC) that signs the entire payload. The security information provides a warning as to the next time that the ID of the key is expected to change as well as the time at which the current ID took effect. The KeyID is an identification of the actual symmetric key provided by the key distribution center (KDC) and therefore allows a subscriber to know which key is being used for either encryption or MAC creation. The MAC is required in R-GOOSE and therefore contents of the packet are protected from tamper.

Figure 8.80 shows a partial capture of a R-GOOSE message.

Note that since the usage of GOOSE is typically automation-related and GOOSE is designed to deliver the value of the current process variables, typically the timestamp of the process value generation is not an issue to be included in the DataSet being transmitted by GOOSE. However, without the inclusion of the

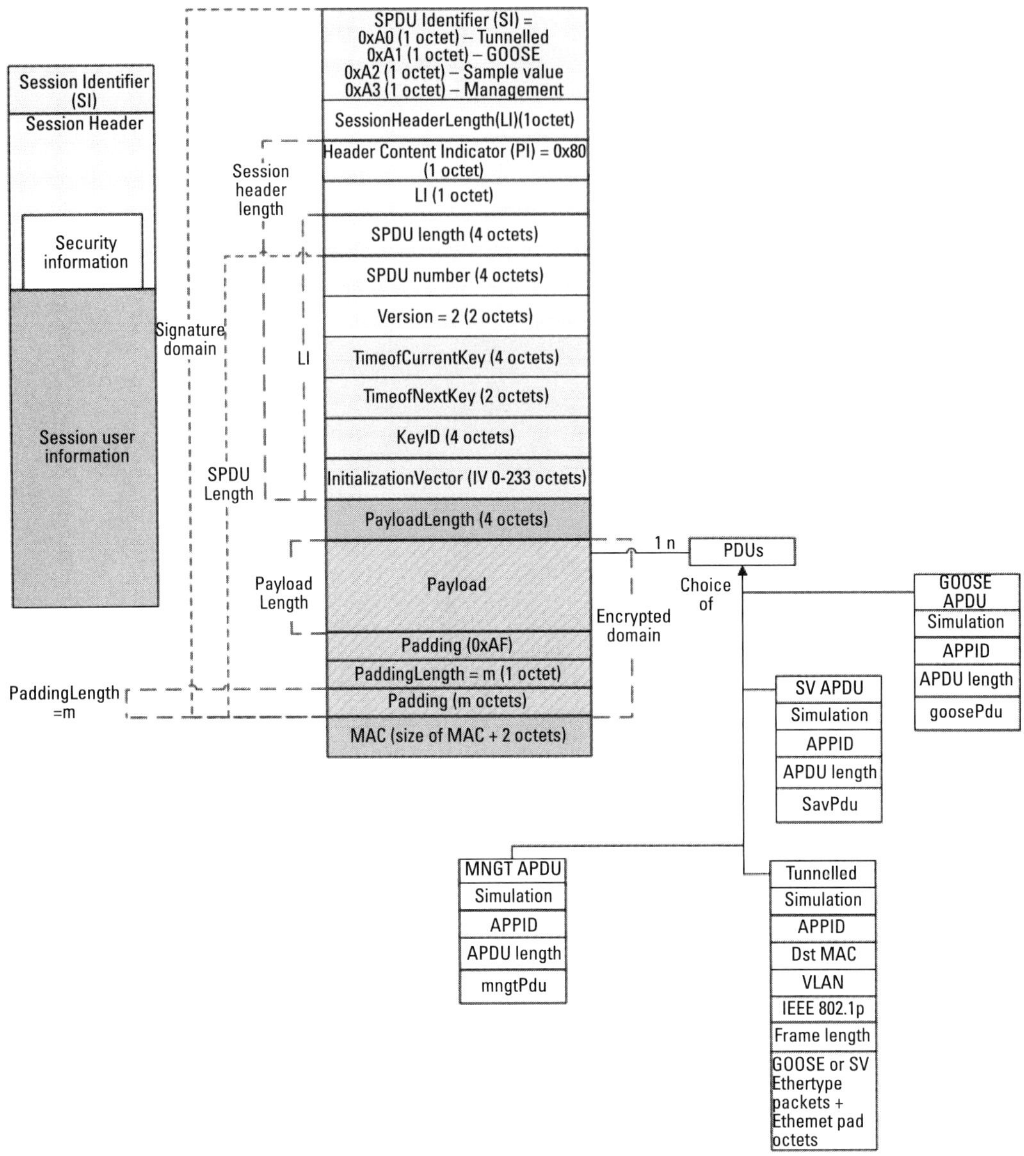

Figure 8.79 Structure of R-GOOSE message.

quality of the process value in the dataset, there is no mechanism for a subscriber to know if the value is good quality (e.g., can be used for automation decisions) or is bad. Therefore, it is strongly advised to include quality values for the process values in the dataset.

Sampled Values

Users typically utilize Sampled Values for unsolicited multicast message delivery of CT/PT or synchrophasor information (see Section 8.1.1.3). Since SV is stream-based with time-based sampling, new information is being continuously sent. SV control blocks control the configuration and control of SV multicast messages.

```
⊞ Ethernet II, Src: Ge_00:01:23 (00:a0:f4:00:01:23), Dst: IPv4mcast_01:04 (01:00:5e:00:01:04)
⊞ Internet Protocol Version 4, Src: 192.168.10.1 (192.168.10.1), Dst: 224.0.1.4 (224.0.1.4)
⊞ User Datagram Protocol, Src Port: 51765 (51765), Dst Port: 102 (102)
⊞ ISO 8602/X.234 CLTP ConnectionLess Transport Protocol
⊟ iec61850_90_5
    SPDU type: 0xa1
  ⊟ Header
      SPDU Length: 139
      PDU Number: 68
      Protocol Version: 2
      Timestamp of current key start of usage: Time not set
      Countdown time until key rotation in minutes: 0
      Current Key ID-2: 00 00 00 00
      IV Length: 0
      User Data Length: 118
  ⊟ UserData
    ⊟ Payload
        User Data Type: GOOSE
        Simulation Bit:: FALSE
        Application ID:: 0
        Payload Length:: 112
      ⊞ goose
```

Figure 8.80 Capture of R-GOOSE.

Conceptually an SVCB, shown in Figure 8.81, is used to specify which process data is to be sent via the control of a specific control block. This filtering is performed through the control block containing a reference to a DataSet (see Section 8.2.2.3). If the reference to the DataSet is Null (i.e., not defined), the control block should not begin transmitting a SV message. The members of the dataset represent the process information of interest to be sent via SV. The process information that passes the dataset filter is sampled based on time and a sample rate. The Sampled Values are buffered and the SV interface (the software that is responsible for sending SV messages) detects value changes in the buffer. It is the changes in the buffered information that triggers the creation of an SV message. It is worthwhile to note that SV messages are not required to be sent at the actual sampling rate as they can send multiple buffer entries in a single message. Thus, the sampling/buffering process is typically separate from the message creation and sending process. Enabling a control block starts a state machine (see Figure 8.82) that controls the sampling, buffering, message transmission.

Unlike GOOSE, SV messages do not contain a TAL attribute to detect undelivered messages. Therefore, the amount of time to wait before declaring failure of reception is a local configuration or programming issue. A suggested state machine is shown in Figure 8.83.

Table 8.14 shows where the various elements of the SV message are derived from.

The actual syntax of an SV message is expressed in ASN.1 syntax. However, only part of the message is encoded using BER, most of the message is encoded with fixed offsets. The BER definition of an SV message is shown in Figure 8.84.

```
   SavPdu ::= SEQUENCE {
     noASDU [0] IMPLICIT INTEGER (1...65535), -- number of SV
sample DataSets included
     security [1] ANY OPTIONAL,     -- reserved for use for cyber
security
               -- use. Use specified by IEC 62351-6
     asdu [2] IMPLICIT SEQUENCE of ASDU
```

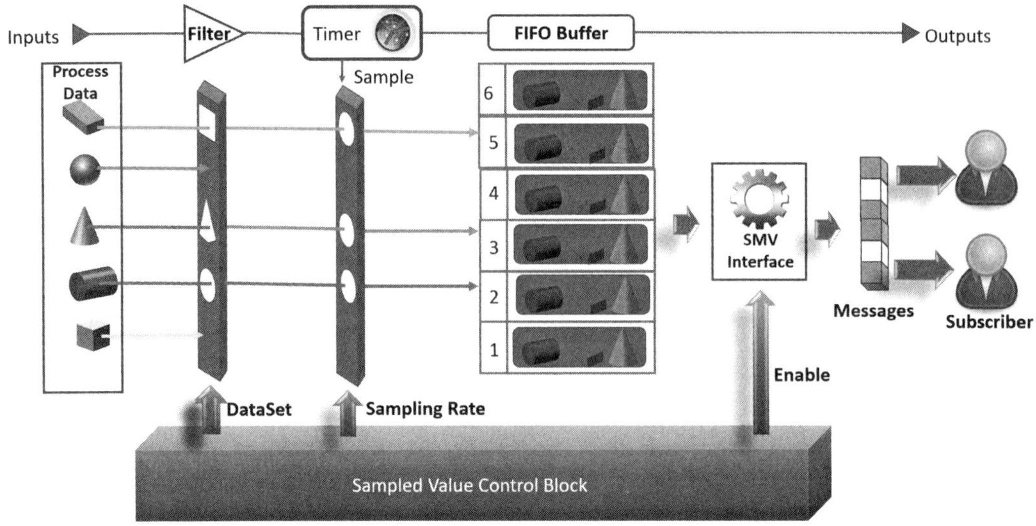

Figure 8.81 SVCB and event flow concept.

```
    };

  ASDU ::= SEQUENCE {        -- everything within this SEQUENCE
is non-BER
     MsvID [0] IMPLICIT VisibleString,
     DatSet [1] IMPLICIT VisibleString OPTIONAL,
     SmpCnt [2] IMPLICIT OCTET STRING (SIZE (2)),
     ConfRev [3] IMPLICIT OCTET STRING (SIZE (4)),
     RefrTm [4] IMPLICIT UtcTime OPTIONAL,
     SmpSynch [5] IMPLICIT OCTET STRING (SIZE (1)),
     SmpRate [6] IMPLICIT OCTET STRING (SIZE (2)) OPTIONAL,
     sample [7] IMPLICIT OCTET STRING (SIZE(n)), -- is the fixed
length encoding of the
               -- DataSet sampled values
     smpMod [8] IMPLICIT OCTET STRING (SIZE (2)) DEFAULT 0,
     gmIdentity [9] IMPLICIT OCTET STRING (SIZE (8)) OPTIONAL
      }
```

The definition of the SV message, as well as the variance of the dataset membership, initially caused interoperability issues since the use of SCL was not typical in Edition 1. Additionally, to optimize the code used in DSP (for speed purposes), a selection of information needed to be defined. This selection was a subset of the allowed SV message fields and defined the dataset values to be transmitted. Today, this definition mechanism is referred to as a profile.

The profile relevant for Edition 1 was the UCA 9-2 LE (light encoding) document. Although widely referenced and deployed, this profile has no international standard status except as a de-jour standard (through industry acceptance). This document has no standing for Edition 2 of IEC 61850-9-2 since there are profiles defined in IEC 61869-9. One of the profile options in IEC 61869-9 is backward-compatible with the profile of UCA 9-2LE. However, the SCL utilized to specify

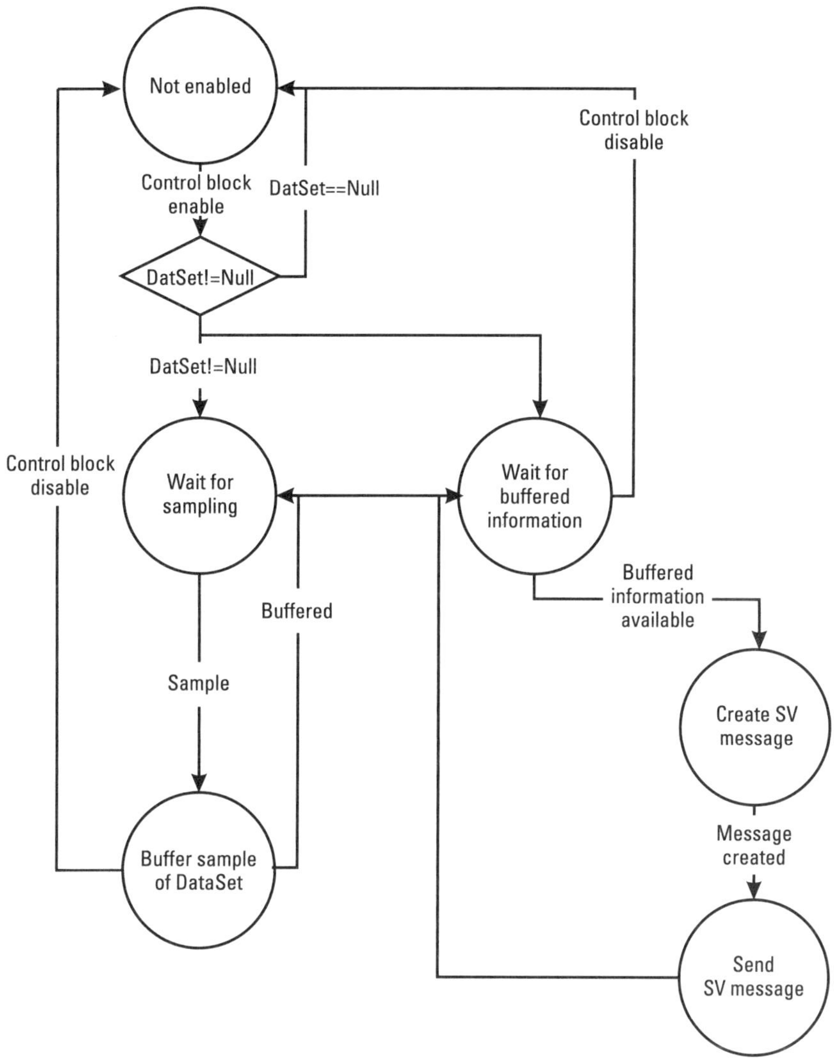

Figure 8.82 SV publisher state machine (not from IEC 61850).

the information has changed slightly. Thus, conformance tests for Edition 2 of the 9-2LE equivalent profile must be changed.

The 9-2 LE, and its equivalent Edition 2 profile, specifies the following information shall be contained in an SV message. The information contained is restricted to the information specified in the profile and no additional information can be conformed to the profile. One of the differences between the UCA and IEC document is that the UCA document predefines the control block names and their options. The IEC document reserves the UCA names but also allows additional names to be created. The 9-2LE profile defines that the dataset must contain eight dataset members representing four FCDA members for amp magnitude and four FCDA members for voltage magnitude. The dataset members also contain the quality FCDA members for each amp and volt magnitude. Each of the quality members is required to be the member directly after its associated magnitude.

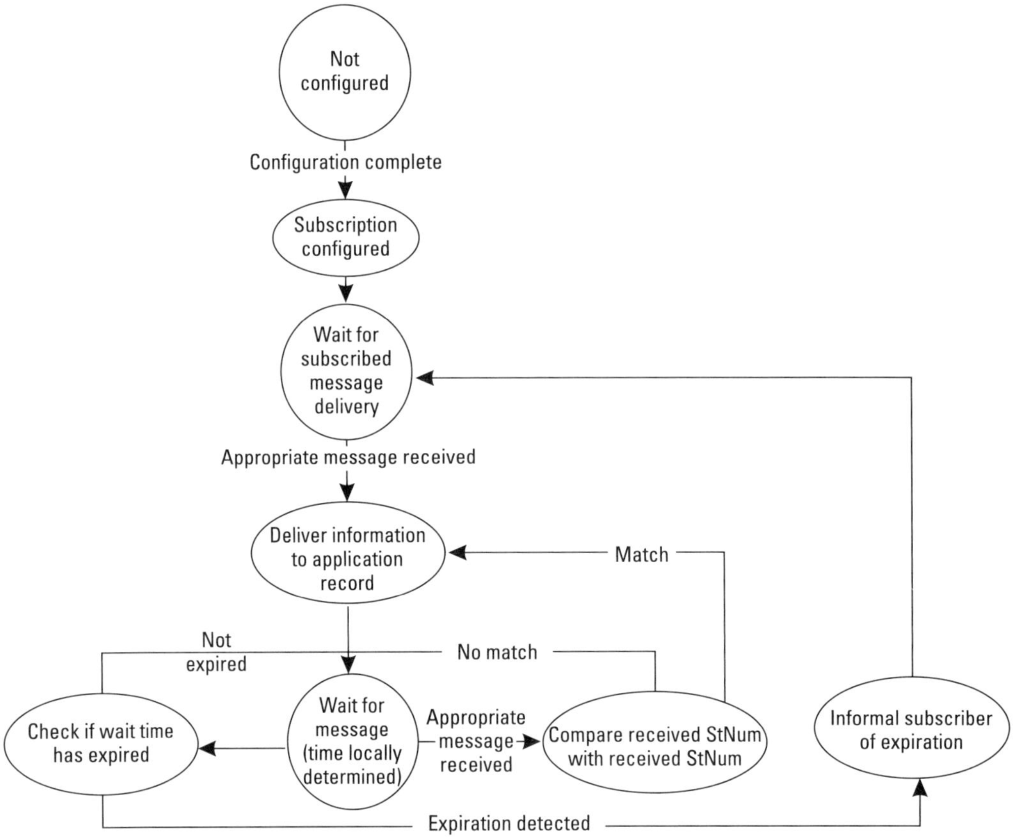

Figure 8.83 Sampled Values Subscriber State Machine (not from IEC 61850).

Figure 8.85 shows that all of the mandatory fields from the ASN.1 message production are present, the structure of the dataset, and that SmpMod is the default value. This restriction is set by the OptFld values in the control block.

The profiles also specify that the specific instances of the TCTR (current transformer LN) and TVTR (voltage transformer LN) are used to convey specific values. The instance numbers (TCTR1,2,3, and 4) are related to phases: 1 represents Phase A; 2 represents Phase B; 3 represents Phase C; and 4 represents neutral measurements.

Since there are four instances of each, it is obvious that the SV message is not occurring from a single CT or PT. It is being published by what is commonly referred to as a merging unit (a unit that combines the information from multiple CT/PTs).

The sampling rate is also specified by the profiles. 9-2LE compatibility is achieved when the control block named MSVCB01 has a rate of 80 samples per cycle or MSVCB02 has a sampling rate of 256 samples per cycle. Additionally, MSVCB01 can have only one (1) ASDU whereas MSVCB02 must have eight (8) ASDUs in an SV message.[16]

16. There is an additional profile currently under development by IEEE. This profile/dataset member definition, is intended to map the information normally conveyed by IEEE C37.118-2 into an R-SV dataset.

Table 8.14 SV Payload Contents

IEC 61850-7-2 Attribute	*Values Typically Come from*	*ED2 Only*	*Description*
MsvID	SCL		A user-configured string value that allows subscriber and diagnostic tools to easily filter SV messages. If there is not one configured, the value of gocbRef is the default value.
OprFlds	SCL		Determines which optional message fields are to be included in the SV message.
DatSet	SCL		The value is an object reference that refers to a dataset whose member values are to be transmitted.
MsvCBRef	SCL		A visible string reference to the actual control block that is contained in LN0 within a logical device.
SV Samples	DataSet Process		The transmission of one or more sets of SV data samples. The actual values of the members are also provided by the process and sampling/SV interface.
SmpCnt	Interface		A counter that increments for each transmission of SV messages. This counter is used to provide a mechanism to reorder out-of-order delivered messages. The counter resets to a value of zero if the process loses time synchronization.
RefrTm	Interface		A timestamp at which the message transmission buffer has been refreshed.
ConfRev	SCL		A number that specifies the configuration revision of the dataset. As members are added, deleted, or changed this number must be changed in order that the subscriber can be guaranteed that it is receiving the appropriate information.
SmpSynch	Interface, Process, Sampling		A value that specifies the type of clock synchronization that is being utilized. 0: not synchronized to an external clock 1: synchronized by a local area clock 2: synchronized by global area clock 5-254: ID of IEC 61850-9-3 or IEEE C37.238 clock this may not be sufficient and SyncSrcID may also be utilized.
SmpRate	SCL		This value represents the rate of acquiring samples. Its units and interpretation are defined by the value of SmpMod.
SmpMod	SCL		If this value is not present, the default value shall be assumed. 0: Samples per period—default value 1: Samples per second 2: Seconds per sample
Simulation* **	Local	ED2	A Boolean value that indicates if the GOOSE message is being issued with simulated data (e.g., not real process data). IEC 61850 reserves the right to set the value to true for test equipment although other publishers may also be able to set this value. This is conveyed outside of the actual SV message.
SyncSrcID	Interface and SCL	ED2	Is an extended ID value field for the identification of the external clock providing time synchronization.

*See Section 8.2.2.7 for a description of using Simulation.

**Edition 1 did not have the ability to interpret test/simulate. Therefore, great care needs to be taken when injecting simulation tagged Sampled Value messages into a system utilizing Edition 1 SV since it may result in mis-operation (e.g. it may look like a cyberthreat of spoofing to an Edition 1 device).

The control block also exposes configuration information related to addressing and quality of service parameters that are used typically in the header or communication profiles. There are two such profiles that result in two different SV control blocks. The first SV (present in Edition 1) is typically referred to as Layer 2 SV.

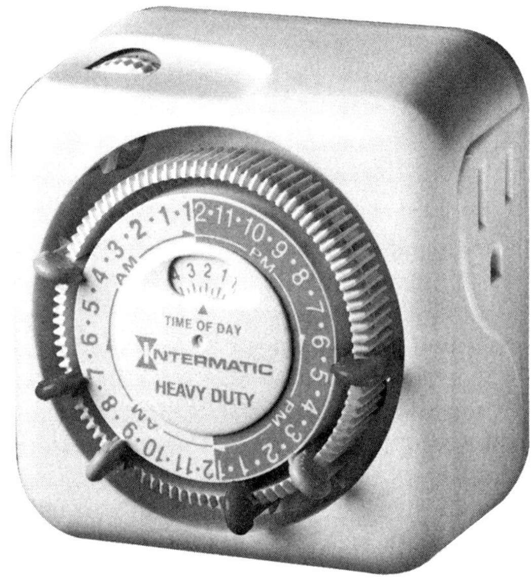

Figure 8.84 ASN.1 of SV message.

```
SMV 9-2
{
  Number of ASDUs:  1
  Start of ASDUs
  {
    ASDU
    {
      ID*:        ABBPTHSMU01
      Sample Count:  3232
      Config Rev*:        1
      Sample Synched: unknown value(255)
        ImATCTRL1.Amp.instMag.i: 1662528
        ImATCTRL1.Amp.q: validity = good Process OpB Measured
        ImBTCTRL2.Amp.instMag.i: -2812930
        ImBTCTRL2.Amp.q: validity = good Process OpB Measured
        ImCTCTRL3.Amp.instMag.i: 1150402
        ImCTCTRL3.Amp.q: validity = good Process OpB Measured
        ImNmTCTRL4.Amp.instMag.i: 0
        ImNmTCTRL4.Amp.q: validity = good Process OpB Measured
        UmnATVTR1.Vol.instMag.i: 3158804
        UmnATVTR1.Vol.q: validity = good Process OpB Measured
        UmnBTVTR2.Vol.instMag.i: -5344567
        UmnBTVTR2.Vol.q: validity = good Process OpB Measured
        UmnCTVTR3.Vol.instMag.i: 2185763
        UmnCTVTR3.Vol.q: validity = good Process OpB Measured
        UmnATVTR4.Vol.instMag.i: 0
        UmnATVTR4.Vol.q: validity = good Process OpB Measured
    }
  }
}
```

Figure 8.85 SV 9-2LE compatible trace capture.

This profile takes the SV message and transmits it directly embedding the GOOSE within Ethertype 0x88b9 (the information is not routable and does not utilize IP, TCP, or UDP). The newest version of SV is transmitted using UDP/IP multicast and is referred to as R-SV. The communication profiles are depicted in Figure 8.86.

The SV methodologies are the same as GOOSE (L2 and Routable). Descriptions of the usage of VLANs and Priority can be found in Figure 8.71 within the GOOSE section. As with R-GOOSE, R-SV also makes use of IGMP and DSCP.

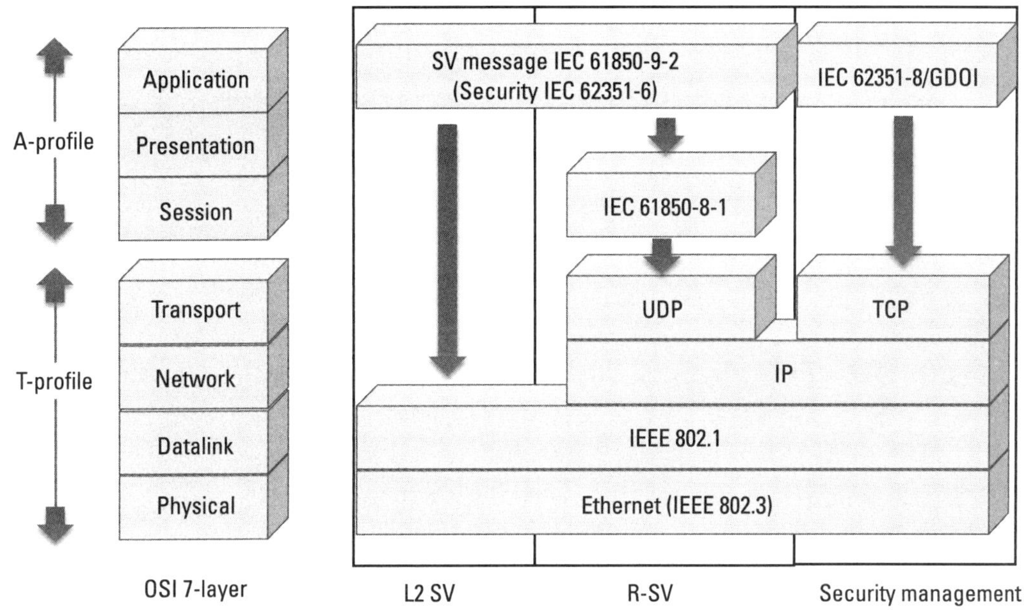

Figure 8.86 High level representation of SV communication profiles.

Figure 8.87 shows the abstract attributes that are exposed in all currently defined SV control blocks. There are currently two different GOOSE control blocks: one for Routable SV and the other for Layer 2 (L2) SV. Each of these control blocks has different physical address/destination address structures. The instantiation of a GOOSE control block is a substructure of the LLN0 named variable and is an MMS named component of the name is assigned by configuration (SCL) and will be contained by a named component of 'MS'[17] or 'RS' for L2 SV or R-SV respectively. Because of mapping rules in IEC 61850-8-1, it can also be its own named variable having the relative name of LLN0MS<name>. It is typical for most implementation to name L2 SV control blocks as gocb1 through MSVCBxxx, per the 9-2LE profile.

The mappings of the operations to an instantiated service is as follows:

Abstract Service	*MMS Service*	*Component Accessed*
GetMSVCBValue	Read	LLN0.MS.<name>.GoEna LLN0.RS.<name>.GoEna
SetMSVCBValue	Write	LLN0.MS.<name>.<x> LLN0.RS.<name>.<x>

The abstract services map to either an MMS read or write service. It is the object on which the MMS service is issued that completes the mapping.

- SetMSVCBValue, see Figure 8.88 is translated to an MMS write. This service might as well be named EnableMSVCB as the only writeable value in the

17. At one point there was a unicast SV defined, it is in the process of being deprecated and therefore will not be discussed further.

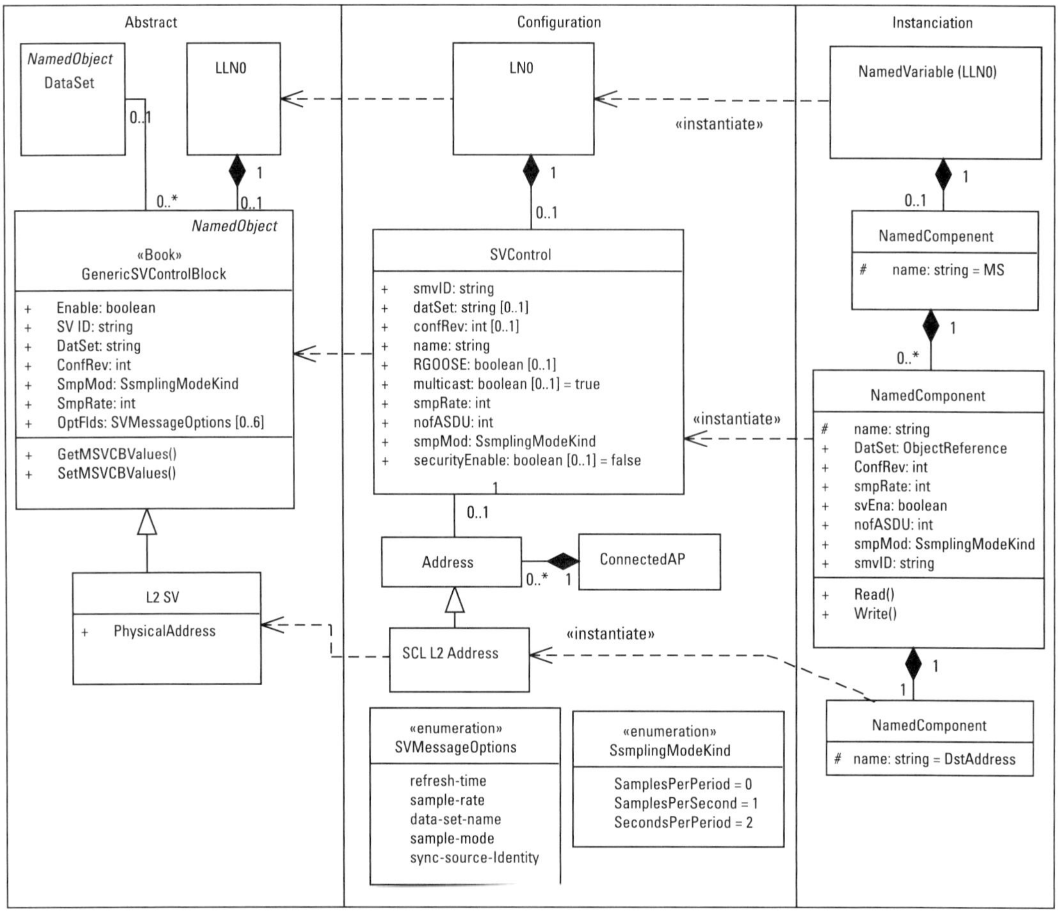

Figure 8.87 SV control block UML.

instantiated control block is the enable attributed. It is used to enable and disable the transmission of the SV message.

The sequence for SeMSVCBValue begins in the abstract with the 61850 client issuing the appropriate abstract service request (e.g., in the example SetMSVCB-Value_req). The abstract service request is mapped to an MMS write request that contains the object reference of the enable attribute to be written. The IEC 61850 server receives the request and updates the appropriate value in the internal control block. Once there is an update, the tracking structure (LTRK) is updated. A successful MMS write response is returned indicating the success of the write. This is translated to the abstract response and the select service is completed.

- GetMSVCBValue is mapped to an MMS read. This service is utilized to retrieve the current values of the control block.

The SCL configuration of both an L2 SV and R-SV control block may appear in Figure 8.89.

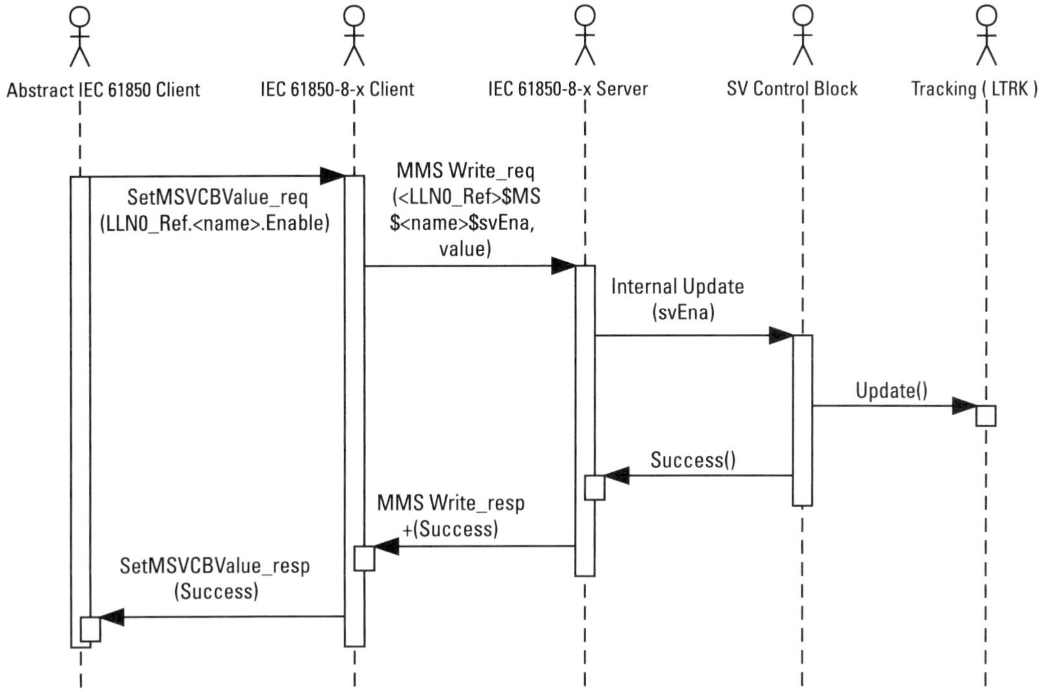

Figure 8.88 Sequence of SetMSVCBValue.

```
<Communication>....
 <ConnectedAP apName="E" iedName="IED1">|
      <Address>....
        <SMV cbName="MSVCB01" ldInst="CB2">
          <Address>
            <P type="MAC-Address">01-0C-CD-01-35-12</P>
            <P type="APPID">0001</P>
            <P type="VLAN-PRIORITY">4</P>
            <P type="VLAN-ID">2</P>
          </Address>
       </SMV>
     ....
     </ConnectedAP>
</Communication>
<IED name="IED1>
   ....
     <SampledValueControl name="MSVCB01" datSet="DS1" confRev="3"
        msvID="MSV01" smpRate="80" noASDU="1" multicast=true>
          <IEDName apRef="AP">AA1D1Q01KF2</IEDName>
     </SampledValueControl>
     <SampledValueControl name="MSVCB02" datSet="DS2" confRev="2"
        msvID="MSV02" smpRate="80" noASDU="1" multicast=true>
       <Protocol mustUnderstand="true">R-SV</Protocol>
        <IEDName apRef="AP">AA1D1Q01KF2</IEDName>
     </SampledValueControl>
   ...
</IED>
```

Figure 8.89 Example declaration of the SV control blocks.

Figure 8.89 shows the relationships of the different L2 and L3 addressing structures for SV (e.g., the abstract physical address) and the two types of SV control blocks. The example also shows the configuration of the GOOSE control block name (MSVCB01 or MSVCB02), the reference to the dataset to be transmitted (dataset1 or dataset1) and would represent the 9-2LE dataset structure, the configuration revision (confRev) and the sampled value ID (smvD). It also shows that in order to differentiate between a L2 SV and R-SV, the <Protocol> element must be specified and it is required to be understood by an SCL tooling (e.g., mustUnderstand="true").

The <IED> element within <SampledValueControl> is the mechanism that allows the SCL engineering process to configure which GOOSE messages should be subscribed to by one or more subscribers. Since SV is a multicast message, there may be multiple subscribers listed and individual <IED> elements. The apRef attribute specifies the access point through which the subscribing IED should receive the SV message. If the IED has only a single access point, this attribute is not required to be present.

Figure 8.79 shows that R-GOOSE encapsulates the IEC GOOSE PDU, as defined by L2 SV, within a different communication profile and is protected by security.

As with GOOSE, SV also shows that there is security information carried in the header as well as a security MAC that signs the entire payload. The security information provides a warning as to the next time the ID of the key is expected to change as well as the time at which the current ID took effect. The KeyID is an identification of the actual symmetric key provided by the KDC and therefore allows a subscriber to know which key is being used for either encryption or MAC creation. The MAC is required in R-GOOSE and therefore contents of the packet are protected from tamper.

Regarding the usage of SV for nonstandardized profile usage, since the usage of SV is critical for automation and synchrophasor applications, it is equally critical that the quality of the values be transmitted. Therefore, it is *strongly* advised to include quality values for the process values in the dataset. Both the UCA 9-2 Light Edition (LE) and IEC 61869-9 profiles require quality to be included.

Reporting

There are basically two ways to receive information: it is either asked for (polled) or it is delivered to you (i.e., event-driven). Imagine having to obtain ongoing information from a crowd (IEDs) and you want to know their health information. In a poll, one might ask each person about their health and continuing through all individuals in the crowd. However, just like painting the San Francisco Golden Gate Bridge, once your polling is completed, it must start again.

In a similar manner, utilities have potentially thousands of devices from which they need information. Analog values are typically acquired every 2–5 seconds and digitals need to be on change or at a much higher rate. Consider the processing requirements of a central processor going to every device and obtaining the information every 1–2 seconds (both analogs and digitals). Distributed processing lessens the processing burden, but there needs to be a better mechanism, something akin to tweeting where the provider of the information sends the information as needed by the client.

Following on Twitter forces one to subscribe to a hashtag. The generic subject of a hashtag is general (e.g., #Beyonce[18]). A follow of Beyonce would result in messages being delivered about her concerts, where she is eating, and maybe occasionally about her health. Generic tweeting isn't desirable if all that is of interest is health information. What is needed is a mechanism to define the information that is delivered and the circumstances of the tweet.

The history of SCADA goes back to 1912 and has developed over the years so that such systems need to

- Understand what information may be delivered.
- Understand the reason for the delivery or a definition of a trigger for the information to be delivered. SCADA systems typically need to be updated if data values change, quality changes, and if a timestamp has been updated (i.e., no value or quality change).
- Be able to retrieve older information if the communication path is interrupted.
- Be able to ensure that it stays synchronized with the systems providing the information. This means that there needs to be an ability to reinitialize its values for all the appropriate information once the communication path/connection is reestablished.

The older style SCADA systems did not utilize guaranteed delivery communication protocols (the information is either delivered or a connection/information delivery problem is indicated). To make sure that the SCADA system stayed synchronized with its informational sources, even if there was an undetected packet loss, the concept of integrity scans was instituted. An integrity scan polled all the appropriate information periodically and that scheme needed to be maintained, or modified, for market acceptance for IEC 61850.

The Reporting Model in IEC 61850 was designed to meet all SCADA requirements and to address other issues that relate to multiclient access to a single server.

The logical process for reporting is shown in Figure 8.90. The first design choice was to separate the event and buffer management and the creation and management of the actual reporting messages. The interface between the two logical processes is the first-in-first-out (FIFO) buffer. The buffer management process is responsible for placing/queuing events into the buffer. Buffering for buffered report control blocks is required to start when the device finishes initialization and becomes operational. Buffering of events is not contingent on the control block being enabled. The event contains a group of appropriately filtered process values and the information reason for being placed in the FIFO. This grouping, in IEC 61850 vernacular, is referred to as an entry.

Since buffering can begin prior to message generation, and the buffer has limited storage capacity, it is inevitable, and natural, that some of the device generated entries will not be delivered to the client. In other protocols, there is an

18. For historical purposes, in case this book is being read 10–20 years from the initial publication, Beyonce is a famous singer.

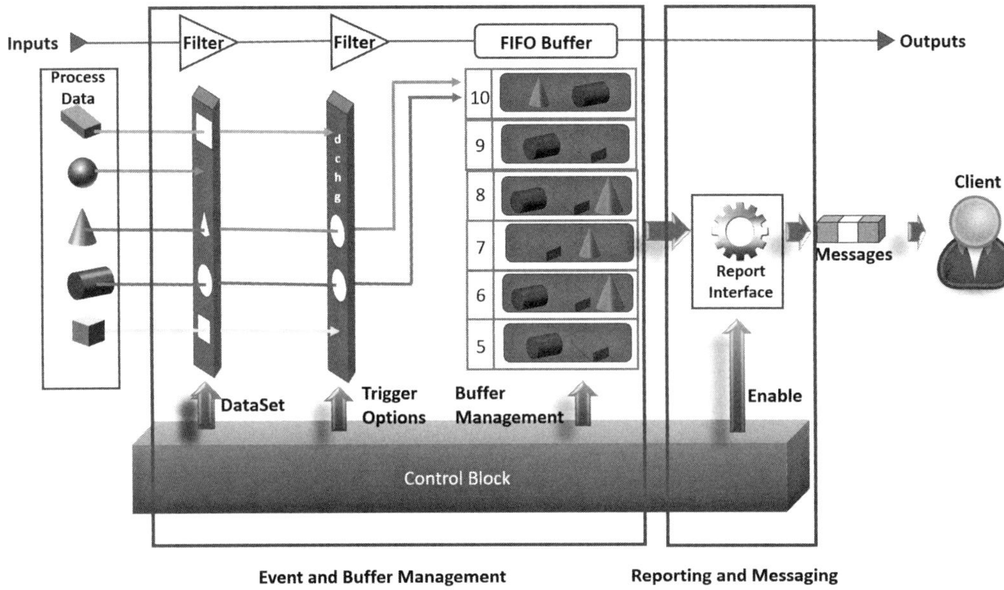

Figure 8.90 Logical process separation of reporting.

acknowledgement from the client that the report was received, and it is the reception of that acknowledgment that causes the entry to be dequeued. One of the things that makes the buffer management unique is that the removal from the FIFO is only based on the need to add new events and not the fact that the event. This allows for the abstract reporting model to be agnostic to the actual protocol implemented for a specific protocol. However, this approach places the responsibility for resynchronization on the client instead of the server. The resynchronization process is specific to buffered reporting and will be discussed in detail in the *Buffered* section on page 219.

The interface to controlling both processes is the report control block. Figure 8.91 takes liberties with the model from IEC 61850-7-2. It reflects the separation of buffer management and message generation functions and the names have been changed to provide more semantic clarity. It also provides a visual representation of the differences between the two different types of report control blocks: buffered and unbuffered.

The generic class contains the following attributes as shown in Table 8.15. Any attribute that has a "Provided By" that includes client may be set by an IEC 61850 remote client (would be considered read/write). All others may not be manipulated by a remote (would be considered read-only). No attributes may be written by a client, except enable, if the enable attribute is true;

Trigger Options To satisfy the SCADA/Client requirements, IEC 61850 developed several different trigger options for reporting and logging. The options that could cause a report to be sent are

- *Data Change (dchg):* Include the in the buffer if the data value of the data has changed.

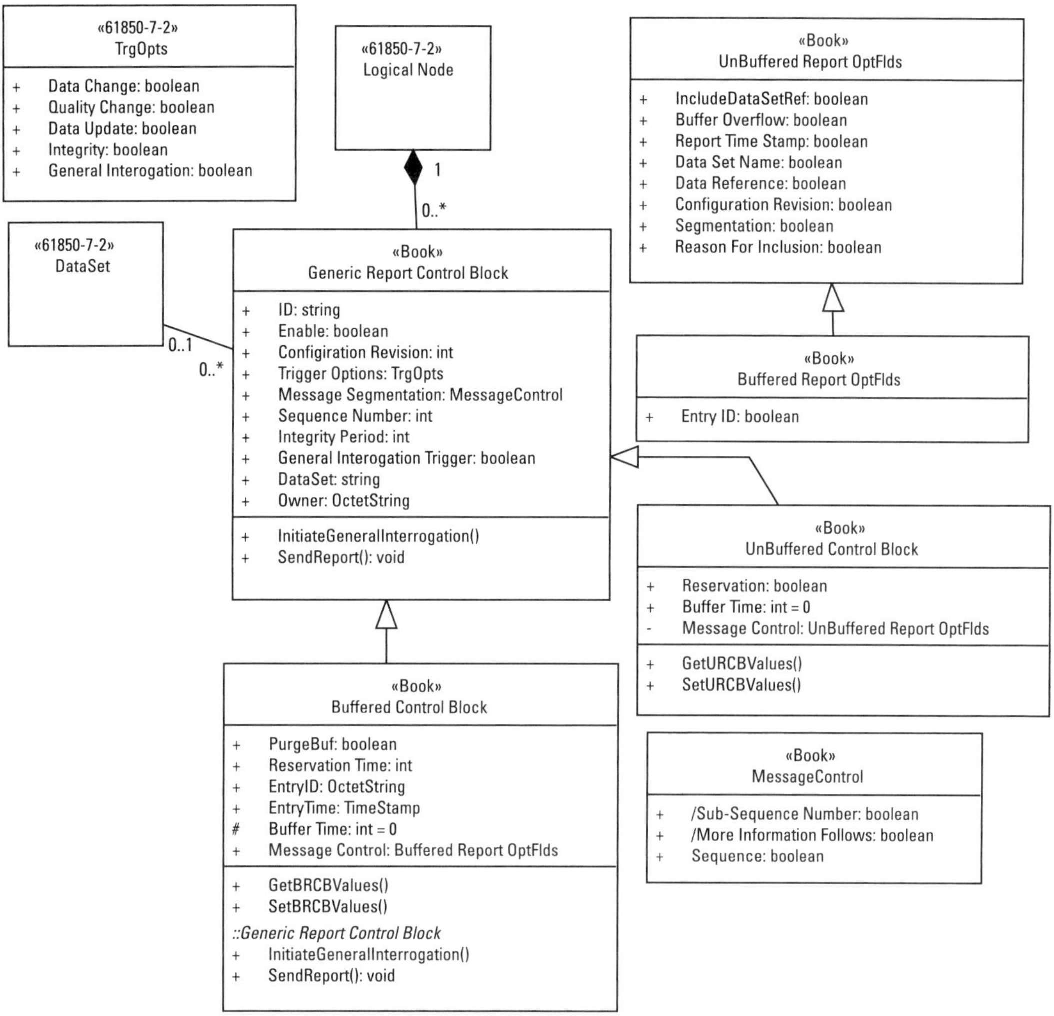

Figure 8.91 Abstract UML for report control blocks.

- *Quality Change (qchg):* Include the data in the buffer if the data quality of the data has changed (from good to bad).
- *Data Update (dupd):* Include the data in the buffer if the timestamp of the data has changed.
- *Integrity:* Causes the generation of all the appropriate information being delivered at a periodic interval.
- *General Integrity (GI):* Is a client initiate integrity request.

There are two types of reporting exchange services and therefore control blocks: buffered and unbuffered. The difference between the two exchange services are that information is lost when the communication connection is lost for the unbuffered service. Either service can lose information if the event buffer is not reported rapidly enough as it is a first-in-first-out buffer.

This book, and Figure 8.91, have taken liberties with the representation of the report control blocks so that they might be easier to explain and to protect the

Table 8.15 Generic Report Control Attribute Definitions

Attribute Name	*IEC 61850-7-2 Name*	*Provided by*	*Description*
ID	RptID	Client SCL	A user created, or SCL-assigned, name for the report. It is typically used in an implementation so that the client can easily identify the purpose and contents of a report. If it is not configured, the default value is the reference of the actual control block.
Enable	RptEna	Server	This value is written to start or stop the delivery of messages to a client.
DataSet	DatSet	SCL Client	This is an object reference to the dataset whose information is to be processed and sent by the report control block. A single dataset may be referenced by multiple control blocks.
Con-figuration Revision	ConfRev	SCL Server	This value is used by the client to ensure that the configuration of the referenced dataset is consistent with what is expected. The initial value may be set by SCL or is defaulted to 0. If the members of the dataset are rearranged or changed, the server is required to increment the value.
Trigger Options	TrgOpts	SCL Client	These values control the filtering and event creation of entries for the FIFO. Additional information can be found in the *Trigger Options* (see page 206) section.
Entry Segmentation	OptFlds	SCL Client	These values are part of the IEC 61850 OptFlds and control the formatting of the report message. It allows the selection of segmentation and the inclusion of optional message fields. Additional information can be found in the *Entry Segmentation and Report Message Generation* section on page 212.
Sequence Number	SqNum	Server	This value changes for each new report message that is sent to the client. This value is used within the message and is used by monitoring/diagnostic systems to make sure that a control block is sending reports.
Integrity Period	IntgPd	SCL Client	This value determines the periodicity of creating an event based on the current process values of the dataset members. It is only a valid value if the trigger option enabling integrity events is set True. More information follows in the *Trigger Options* (see page 206) section.
General Interrogation Trigger	GI	Client	This value is written by the client to cause the immediate execution of the equivalent of an integrity event. It is typically written so that a SCADA system can reestablish initial values of the process after being offline/disconnected. The event will only be generated if the GI trigger option is set true. More information follows in the *Trigger Options* (see page 206) section.
Owner	Owner	Server	The server sets this value based on the identification that it can establish for the client that reserves the control block. This value is nonauthoritative as it is typically an IP address that could be network address translated and thus has no real bearing on the identification of the actual client. This value will become authoritative when security (e.g., the IS version of IEC 62351-6) is implemented by both the client and server. The implementation of security will also add additional information to the owner field. Without the use of security, the value of this field has minimal, if any, value.

guilty. The figure has changed the names of some attributes, and explicitly added the InitiateGeneralInterogation since a general interrogation (GI) is an externally initiated integrity report and impacts the functional model of reporting as well as logging.

Without integrity and GI reporting, the simple model of reporting information, as shown in Figure 8.92, looks very similar to GOOSE.

Without the requirement of integrity, or GI, reporting, the trigger options of dchg, qchg, and dupd do not require the reporting infrastructure to know, or be able to obtain, the last known process values that meet the dataset filter criteria. Therefore, these trigger options can be implemented via push or the implementation of a report event creation function. However, the implementation of GI requires more complexity. The encumbrances placed on executing a GI are

- Any events detected (e.g., pass the trigger option filter criteria) that occur before a GI begins execution must be placed into the event FIFO prior to the events/data created by the execution of the GI.
- The GI must generate events/data that represent last known dataset values.
- Any additional data/event changes that occur during the execution of a GI must not cause an update to the last known values being produced by a GI.
- A set of GI events must be consecutive in the FIFO buffer with no intervening dchg, qchg, or dupd events.

Logically, these requirements lead to a natural conclusion that there needs to be a process to provide the GI service as well as managing/locking the events generated by other trigger options. Logically, a logical implementation is shown as follows.

Figure 8.93 is a logical implementation strategy that introduces three new entities that are interlocked and shows the configuration inputs to all the entities.

- *P – Process Value Change Detection:* This entity evaluates the dataset member values and updates a cache of last known values. The input to this entity is a logical queue that can accumulate changes should the entity be locked from processing the values.

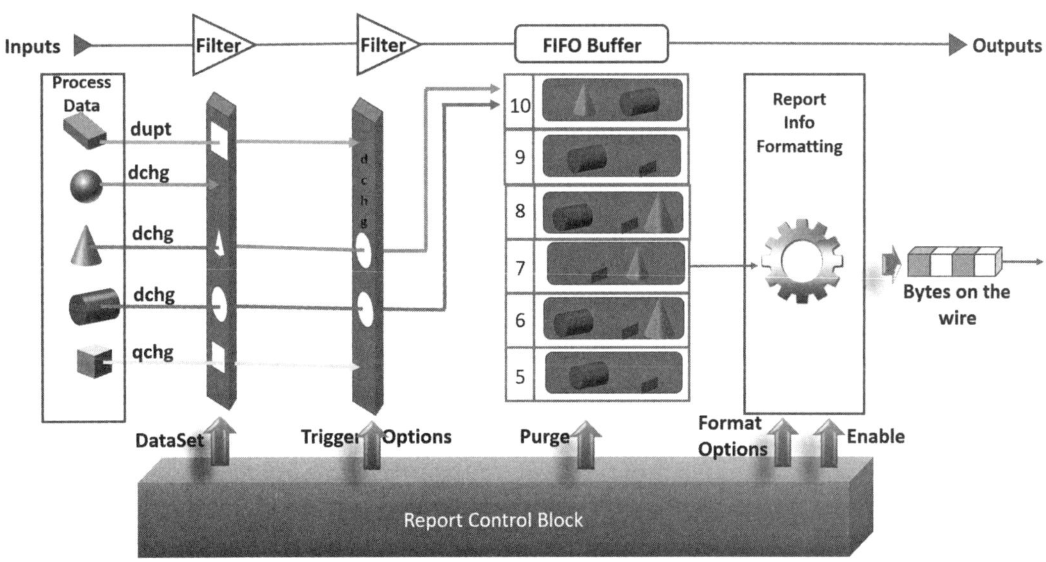

Figure 8.92 Simplified reporting information flow.

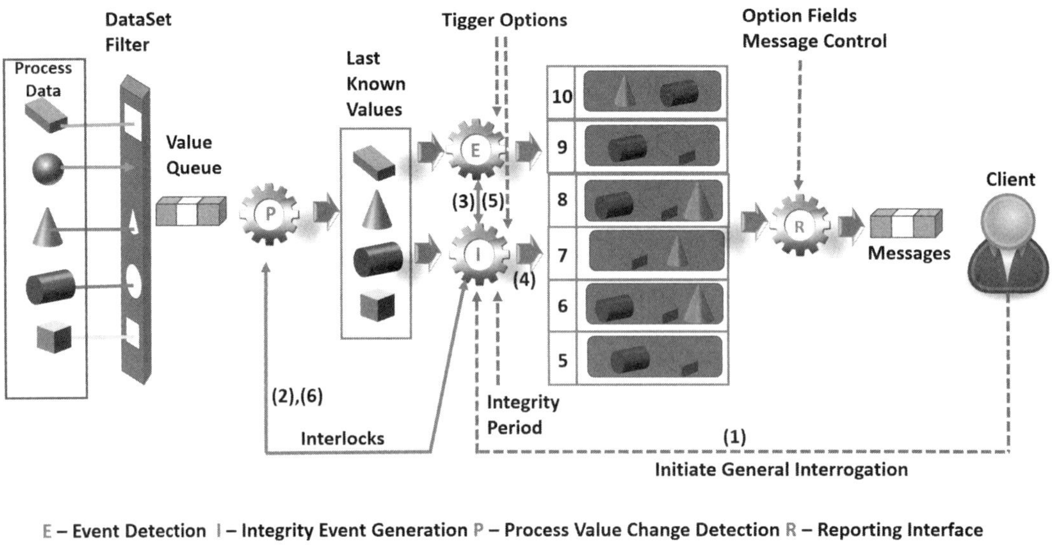

Figure 8.93 Report information flow with GI and integrity.

- *E – Event Detection:* This entity generates events based on its configured triggered options. It produces events for dupd, qchg, and dupd. If the appropriate trigger criteria is met by a value change, it is responsible for placing the events into the event FIFO buffer.
- *I – Integrity Event Generation:* This entity is responsible for executing the integrity service based on the expiration of the configured integrity period or reception of a request to initiate a GI.
- *R – Reporting Interface:* This entity is responsible for taking the information from the FIFO, formatting messages based on the optional fields including the message control options, and sending the information to the client.

Logically, the interlocking sequence for the execution of a GI might look as follows, although nobody implements in this fashion. Thus, it is being provided to discuss concepts.

Figure 8.94 depicts that the process places information into the queue, and since the process value detection entity is unlocked, it is servicing the queue. It dequeues the information and places it into the last known value cache. The diagram does not show events being placed into the FIFO and the assumption would be that the information placed into the cache did not match the configured triggered options.

The client then issues a request to initiate a GI (Step 1). This is performed by writing a control block that is enabled and has the GI trigger option set true. If either of these conditions is not met, a GI service execution will not start and an indication/error should be generated of the request failure.

The service execution continues with the integrity event generation process interlocking with the event detection process so that it can be informed when no additional events are being generated to the FIFO (Step 2). It then instructs the process value change detection process to stop servicing the process value information queue (Step 3), thereby placing the process into a locked state. Once the process

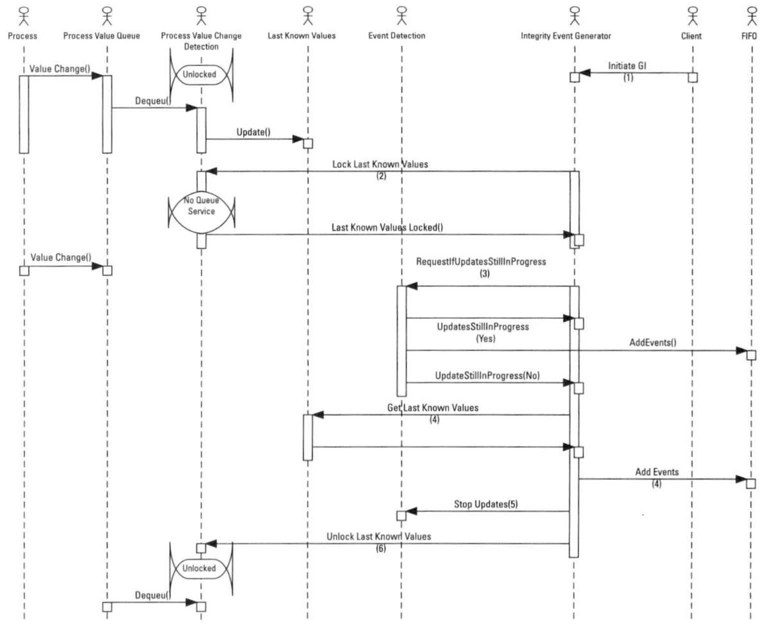

Figure 8.94 General interrogation logical sequence.

indicates it is locked (no longer updating the last known value cache), the integrity event generation process obtains the information from the cache (Step 4). Once the cache information is obtained, the process creates events and places the events into the FIFO. Since the dequeue of the process value changes is not occurring and there are no more events being generated based on dchg, qchg, or dupd, the requirement that a GI not be interspersed with other events is satisfied.

Once the information from the last known value cache has been added to the FIFO, the process and locks are unwrapped. The integrity event generation process puts a stop to receiving the updates from the event generation process (Step 5) and unlocks the process value change detection process.

A similar sequence could be envisioned for the execution of an integrity report. The trigger for such a report would be that the control block is enabled, the trigger option for Integrity is set true, and the configured integrity period and expired. Once the integrity events have been generated into the FIFO, the integrity period timer is reset to the configured interval and another trigger will occur once it expires again. Therefore, an integrity report is a periodic equivalent of a GI.

There is often a question regarding why there are two different integrity options. To the casual reader, the GI method is not required. However, since an integrity report is only generated when the integrity period expires, there are two options, which have different cause and effects:

- *Option 1—Make the integrity period small.* The effect of this option would be that client would receive the first integrity report rapidly. However, the side effect would be that the integrity event generation would occur at such an interval that the FIFO would be fully of integrity events.
- On first blush, this may not seem to be a problem. However, the design of IEC 61850 does not typically require integrity since the delivery of the messages, via the mapping to appropriate protocols, is delivered or the associa-

tion terminates. Thus, it would be suggested that IEC 61850 integrity period be 10 minutes or more. Many systems utilize 30 or 60 minutes.

- *Option 2—Implement the General Interrogation.* The effect of this option is that the client must implement the initiation of the trigger.

Entry Segmentation and Report Message Generation The need for segmenting the information in a buffered entry arises from the fact that IEC 61850 is designed for embedded devices that have limited resources. These limited resources include the size of the buffer into which a message is to be serialized (i.e., built/encoded). Without segmentation capability, and a limited serialization buffer, the information contained by a large event entry could not be delivered. This factor could be addressed through controlling the amount of information contained by an entry or by creating a streaming serializes (i.e., the entire message is not built in a buffer). Either of these options is difficult to implement and does not address protocol selections that do not allow streaming and are message-buffer-oriented, which is the protocol currently selected for IEC 61850 client/server exchanges. In the current IEC 61850 suite, the size of the messages, and therefore the serialization buffer, is negotiated. Even future mappings will probably have limited message delivery capability.

Therefore, it is always recommended that sequence is always set to true. When true, this automatically enables the inclusion of more information follows and Subsequence number. It also enables the segmentation capability for report message delivery. If the sequence option is not true, segmentation is disabled.

The state machine for report segmentation would be as shown in Figure 8.95.

The start of the state machine is tied to the control block being enabled. Upon being enabled, the sequence number and subsequence number that are to be included in the message are initialized to values of zero. SqNum with a value of zero is reserved to indicate the initial state of the control block. When there is an entry in the FIFO that needs serializations, the required serialization size is checked. If the size fits the allowed message size, the message size is serialized, including MoreFollows=False, and is sent to the client. Otherwise the amount of information from the entry that can be serialized into the allowed message is determined and as well as if more information needs to be serialized. If no other information is required, the subsequence number is incremented and MoreFollows=False will be included in the message. If more information will need to be serialized, the subsequence number is incremented and MoreFollows=True will be included in the message. When the entry serialization is completed, the SqNum is implemented and will be the number for next Entry serialization.

Many clients check for out-of-sequence sequence/subsequence numbers; the underlying protocols are based off TCP/IP. Thus, these implementations are not capable of delivering out-of-order packets or dropping messages without terminating the TCP connection. The algorithm does provide future proofing for other implementations that do not guarantee orderly delivery (e.g., UDP).

There are additional fields that will be added to a report based on the values of the option fields. The fields are defined in Table 8.16.

The recommended settings for most of the option fields, except for report-time-stamp, have already been explained. The reason to set OptFlds.report-time-stamp=True is so that a client can utilize this time-stamp to approximate the time

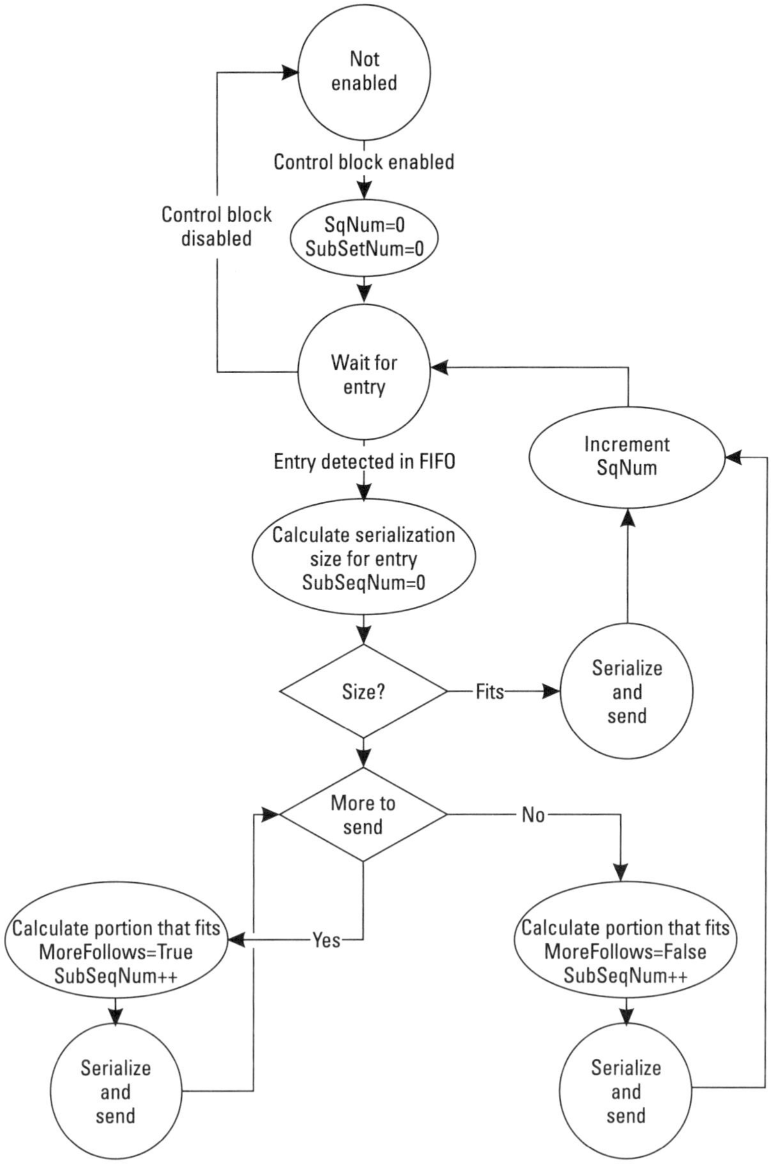

Figure 8.95 Report segmentation state machine.

of the data object values for data objects that do not contain time stamps. If it is not included, then the client would need to use the time of reception of the report as an approximation.

Generic Services Figure 8.91 also provides two generic services. The initiate general interrogation was previously discussed in length in the trigger options discussion. The send report service is used to deliver the serialized messages discussed in the previous section. The mappings of these services to MMS are found in Table 8.17.

The generic report message can seen in Figure 8.96.

Figure 8.96 shows the interrelationship of the OptFld values and the inclusion of objects in the message. It shows that if OptFld.sequence-number=True, there are three optional data objects that would be included. If the message is segmented, it

Table 8.16 OptFlds Definition

Option Field Attribute	*Description*	*Recommend Setting*
sequence-number	If true, the sequence number value will be included in the report message.	True
report-time-stamp	If true, the timestamp of the Entry being reported will be included in the report message.	True
reason-for-inclusion	If true, the trigger that caused the inclusion in the event, for a specific dataset member, will be returned in the report message.	True
data-set-name	If true, the name of the control block's configured dataset will be returned in the report message.	True
data-reference	If true, the object references for each dataset member being returned in the report message will also be returned. Although servers need to support this option, it is not recommended that clients utilize this option as it increases the bandwidth required and decreases the amount of information that can be returned in a single Report message segment.	False
buffer-overflow	If true, the flag indicating that the FIFO has lost information will be included in the report message. This is a relic from days of old and is not needed by clients as the value would typically be true. Therefore, it is suggested to keep the OptFlds.buffer-overflow false.	False
entryID	If true, the id associated with the event entry being serialized into the report message will be included. It is imperative that clients utilizing buffered reporting must set this value to True. This is due to the fact that this value is useful to re-sync with events in the FIFO if the association is dropped.	True for buffered reports
conf-revision	If true, the configuration revision information will be sent in the report message.	True
segmentation	This bit should not be in the OptFlds class/values. It is a byproduct of a modeling construct where reuse of a class supersedes* the need to appropriately represent the usage. This value is returned in a report message that is segmented and represents the ability of a server to segment messages. Since supporting segmentation is required for reports, the value of the bit in OptFlds should always be true and is set by the server. The server must override any attempts to set the value to false via client or SCL interaction.	Not to be set by Client or SCL

*Reuse of code, classes, and other common components is a trend in the industry. However, great care must be taken in the justification of sharing and especially the maintenance of a shared component. Shared components typically make maintenance and changes more difficult.

Table 8.17 Generic Reporting Service Mapping

Generic Service	*MMS Service*	*Object*
InitiateGeneralInterrogation	Write	GI Boolean of the control block
SendReport	InformationReport	The InformationReport contains the DataSet Member values, the mandatory DataObjects of the Report, and the DataObjects that are to be included due to the OptFld bits.

would contain all three fields and the segmentation bit in the report message will have the segmentation bit set.

Generic Client Interaction Pattern Many HMIs and SCADA systems hide the complexity of the interaction of clients and report control blocks. However, these high-level clients wrap IEC 61850 clients that interact with report control blocks through the following generic steps:

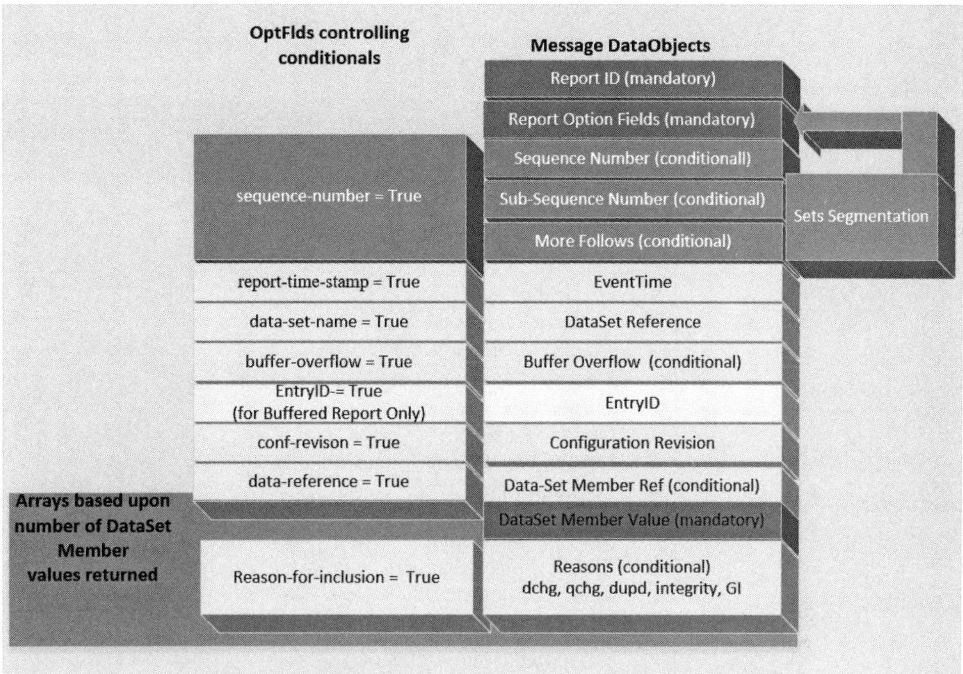

Figure 8.96 Report message contents.

- Reserve the control block so that it is locked for a client. The reservation can be performed through SCL configuration or through a specific service interaction between the client and control block. If the reservation is via a service, the server has sufficient information to restrict any other client from changing control block values.
- Once reserved, the client needs to set or verify the dataset reference being used by the control block. If the dataset reference is changed/written, the buffered events will be purged.
- The client needs to set or verify the trigger options. If the trigger options are changed, the buffered events will be purged.
- The client needs to set or verify optional field values in the control block such that the appropriate information is provided.
- The client needs to enable the control block to start sending report messages. Once enabled, the server has sufficient information to restrict any other client from changing control block values.

After the enable is successful, only the general interrogation attribute can be written until the control block is disabled.

Unbufferred An unbuffered report control block (URCB) (see Figure 8.97) inherits its attributes and services from the generic report control block. The UML for the URCB is shown in Figure 8.98. Usage of this control block type is typically reserved for clients that do not need to recover information should the connection be lost. There are typically a limited number of these types of control blocks since the buffering of information starts on device initialization for control blocks that have

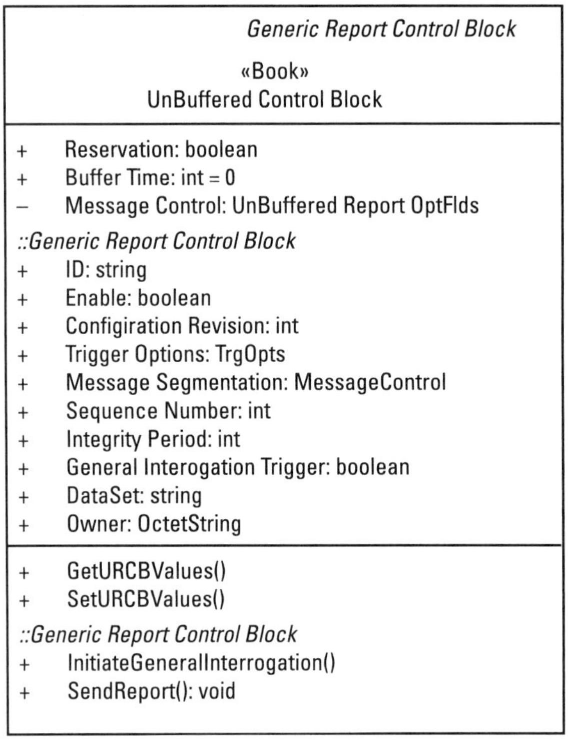

Figure 8.97 Unbuffered report control block.

a dataset value of non-null (i.e., a valid value in the control block). The additional attributes in an unbuffered control block are shown in Table 8.18.

The mappings for an unbuffered report control block are shown in Figure 8.103. The mapping of the abstract URCB is instantiated as a named component under the functional constraint of "RP" within a named variable representing a logical node. The actual name of the URCB is assigned within the SCL file and may not be changed by the SCL engineering process. The other abstract attributes are also instantiated as named components but shown as attributes in order to de-clutter the diagram. TrgOpts and OptFlds are serialized into a data type known as a BITSTRING. The actual serialization is defined in IEC 61850-8-1 and IEC 61850-8-2. The configuration of a URCB is a report control with the attribute buffered="false" and might appear as shown in Figure 8.99.

The differentiation between configuring a buffered and unbuffered report control block is the value of the buffered attribute. An unbuffered report control block has this attribute with a value of false. The example control block is a nonindexed report control block since the indexed attribute with a value of false. If the indexed attribute is missing, the control block will be an indexed control block since the default value is true. The name of a nonindexed report control block will be the value of the name attribute.

Furthermore, through the <RptEnabled> element, the "max" attribute defines the number of control blocks of this configuration that are exposed. If the attribute is not present, the SCL engineering process assigns the number. If the "max" value is greater than "1" then the report control must be an indexed type of control block.

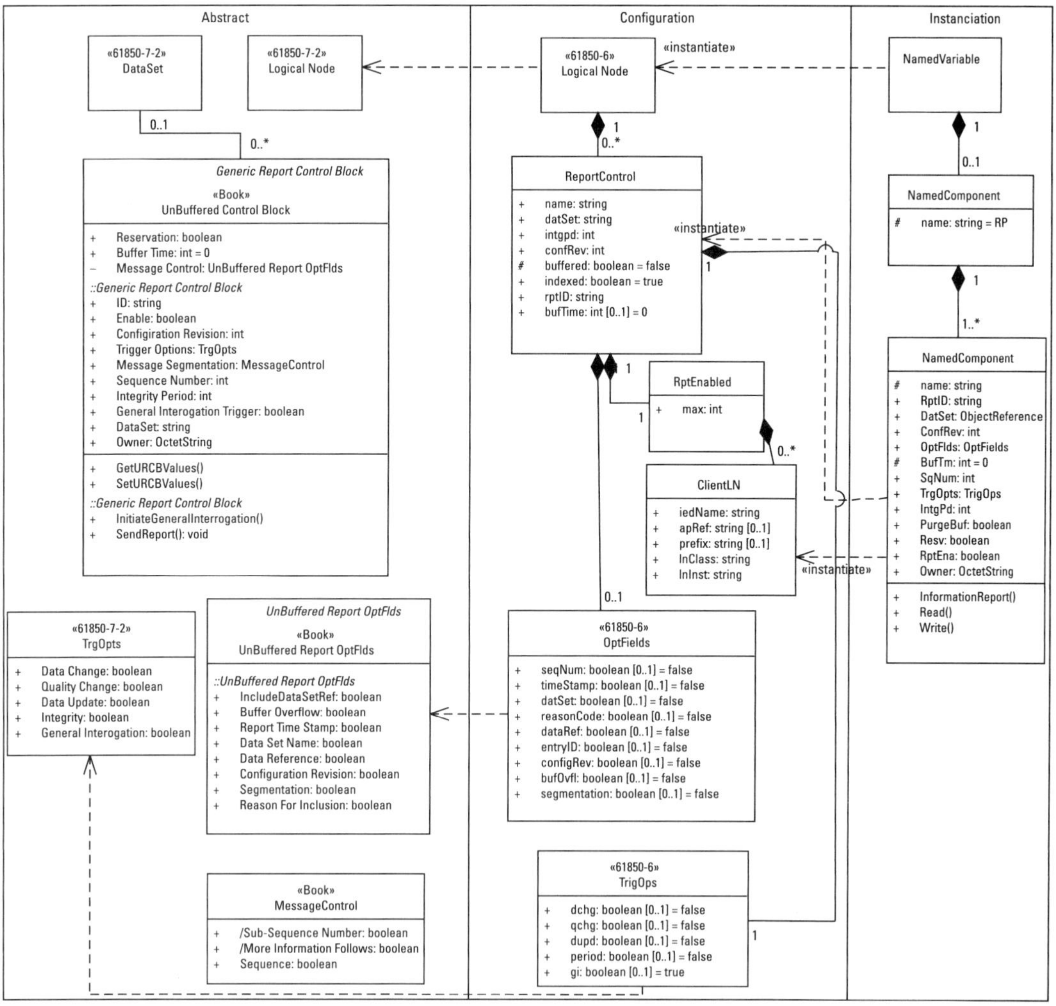

Figure 8.98 Unbuffered report control block UML.

The reservation of a control block is configured through the use of the <ClientLN> element which specifies the IED and the logical node instance for which the block is being reserved. The example shows the control block being reserved for a particular IED through the use of the <ClientLN> element. The mandatory attributes of the <ClientLN> are the iedName and LNClass attributes. It should be noted that the instantiated server will not be able to enforce the LNClass value and may be able to enforce, although it is not guaranteed, the iedName. Thus, the configuration of the <ClientLN> is providing the configuration of the client with a specification of which control blocks should be utilized instead of specifying enforcement by a server. The reservation does result in Resv having a value of true.

The indexed example, shown in Figure 8.100, takes advantage of the default value of the indexed attribute. The example shows that there will be five URCBs created (i.e., the "max" attribute value of <RptEnabled>). Use of indexing specifies that the instantiated control blocks names would be the value of the name with the number "01" through "0x" appended. In this example, the names of the

Table 8.18 UnBuffered Report Control Additional Attribute Definitions

Attribute Name	*IEC 61850-7-2 Name*	*Provided By*	*Description*
Reservation	Resv	Client SCL	A true value indicates that the control block has been reserved for a particular client. The reservation can be performed through SCL or via the SetURCBService. Once reserved, only that client may change the values in the URCB and enable the report.
Buffer Time	BufTm	Client SCL	This value indicates, in milliseconds, the period of time that the report engine will wait to aggregate additional events prior to issuing the report. A value of zero indicates that events entered into the FIFO will be sent as rapidly as possible with no additional delay.
Message Control	OptFlds	SCL Client	Although the actual IEC 61850-7-2 OptFlds applies to both buffered and unbuffered reporting, the unbuffered report control block does not contain an EntryID attribute and therefore, OptFlds.entryID has no impact on the unbuffered report message structure; thus the difference between the unbuffered and buffered abstract OptFld definition.

```
<ReportControl " datSet="ds1" name="urcb2" buffered="false"
        confRev="15" indexed="false" rptID="rpt1">
      <TrgOps dchg="true" qchg="false" dupd="false" period="true"/>
      <OptFields seqNum="true" timeStamp="true" dataset="true" reasonCode="true"
        dataRef="true" entryID="false"/>
      <RptEnabled>
       <ClientLN iedName="Client" ldInst="LD0" lnClass="IHMI" lnInst="1"/>
      </RptEnabled>
</ReportControl>
```

Figure 8.99 Nonindexed unbuffered report control block configuration.

instantiated control blocks would be myURCB01 through myURCB05. If there were reservations for specific clients, the order of the <ClientLN>'s would correlate to the index being reserved for the specified client.

Events begin to be buffered when the client writes the enable is set true. Buffering may continue after the initial enable even if it is written false. Event buffering will stop, and all events will be purged if the client association to the server is terminated.

The mappings of the operations to an instantiated service are seen in Table 8.19.

The abstract services map to either an MMS read or write service. It is the object on which the MMS service is issued that completes the mapping. The example depicts access to the instantiated Enable which is a named component RptEna. The sequence of enabling a URCB follows in Figure 8.101.

The figure depicts the translation of the abstract service parameters into a MMS write. The write is received by the Server, the URCB is located, and the RptEna value is set True. Upon setting the RptEna, the tracking logical node (e.g., LTRK) is updated. LTRK is also updated on the control block being dynamically reserved (e.g., via MMS).

```
<ReportControl " datSet="ds1" name="urcb2" buffered="false"
        confRev="15" indexed="true" rptID="rpt1">
    <TrgOps dchg="true" qchg="false" dupd="false" period="true"/>
    <OptFields seqNum="true" timeStamp="true" dataset="true" reasonCode="true"
      dataRef="true" entryID="false"/>
    <RptEnabled max="5" >
     <ClientLN iedName="Client" ldInst="LD0" lnClass="IHMI" lnInst="1"/>
    </RptEnabled>
</ReportControl>
```

Figure 8.100 Indexed unbuffered report control block configuration.

Table 8.19 Mapping of Abstract URCB Services

Abstract Service	*MMS Service*	*Component Accessed*
GetURCBValue	Read	<LN>.RP.<name>.RptEna
SetURCBValue	Write	<LN>.RP.<name>.RptEna

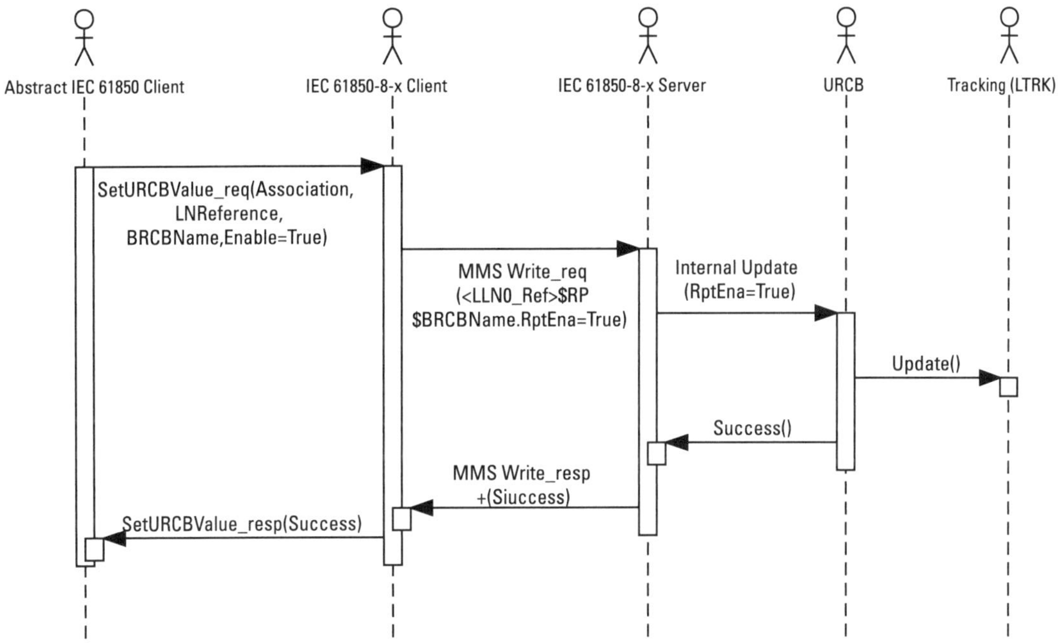

Figure 8.101 Enable sequence for URCB.

Buffered A buffered report control block (BRCB) (see Figure 8.102) inherits attributes and services from the generic report control block. Usage of this control block type is typically reserved for Clients acting as SCADA or sub-SCADA masters. There are typically a limited number of these types of control blocks since the buffering of information starts on device initialization for control blocks that have a dataset value of non-null (i.e., a valid value in the control block).

There is one attribute (Buffer Time) that is only present for easing client compatibility issues. With that said, the value should be ignored by a client using a buffered control block and by standard must have a value of zero. This value indicates

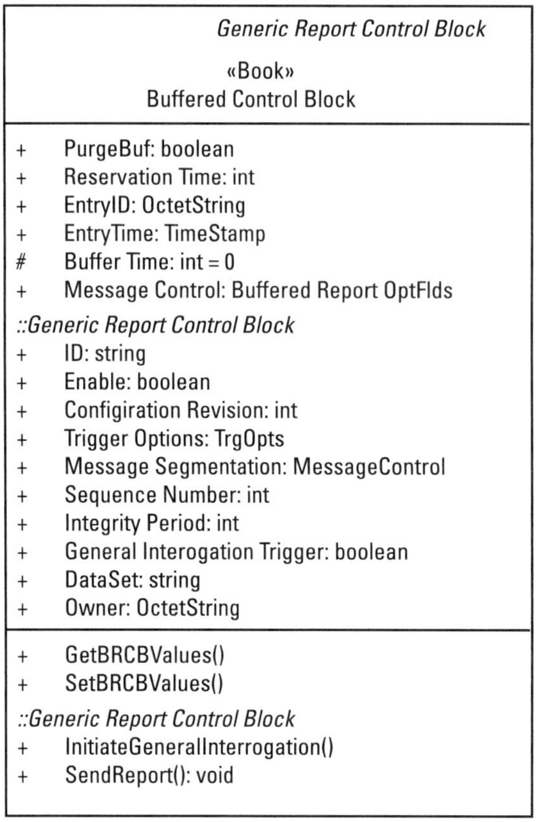

Figure 8.102 Buffered report control block.

that the report message generation will be as rapid as possible and not wait for additional events to be placed in the FIFO.

The additional attributes in a buffered control block are shown in Table 8.20.

The mappings for a buffered report control block are shown in Figure 8.103. The UML shows the abstract attributes and the configuration relationship with SCL. They are like those of the unbuffered report and thus only the differences will be covered. Besides the additional abstract attributes and therefore the instantiated corresponding data object, the major difference in instantiation is that the buffered control block is instantiated under a named component whose name is "BR" (i.e., the functional constraint is BR).

The configuration of a BRCB is like that of a URCB except that the buffered attribute must be set true. An example of SCL configuration is shown in Figure 8.104.

Since there are additional option fields (i.e., OptFields) allowed for buffered control blocks, these may also be set as part of the configuration. The example shows that the control block is reserved to a specific client. Therefore, the value of the instantiated ResvTms will be –1.

If the control block is not reserved, the value of ResvTms will be zero (0). To reserve the unreserved control block, a client must write the value of ResvTms to a value specifying the relative time of the reservation in seconds. To de-reserve the control block, the reserving client writes the value back to zero (0). However, it is

Table 8.20 Buffered Report Control Additional Attribute Definitions

Attribute Name	*IEC 61850-7-2 Name*	*Provided by*	*Description*
PurgeBuf	PurgeBuf	Client	This value is written by a client to empty the event FIFO.
EntryID	EntryID	Server Client	This value indicates the value of the last FIFO event entry. The value may also be written by the client to resynchronize with the information in the FIFO.
EntryTime	TimeOfEntry	Server	This value reflects the time at which the report interface received the indication that a FIFO entry would need to be sent. It is associated to a specific EntryID.
Message Control	OptFlds	SCL Client	These values control the inclusion of additional fields and their values in the report messages that are generated.
Reservation Time	ResvTms	SCL Client	The value represents the number of seconds that a specific client has reserved the control block.

rare that a de-reservation is performed. If the client association is lost, the reservation is still preserved until the reservation time expires and the Server is responsible for setting the value to zero in this case.

One of the major differences between buffered and unbuffered reporting is the ability for the client to resynchronize with events in the FIFO should communication be lost. This is only possible if the OptFlds.entryID is true. Therefore, clients should ensure that the following abstract Optflds are set true: Entry ID, Include-DataSetRef, Report Time Stamp, DataSet Name, Configuration Revision, Reason for Inclusion.

Event buffering, for buffered control blocks, begins once the TrgOpts and dataset reference attributes have values and typically starts on device power-up and initialization. The events in the FIFO are purged based on similar rules as unbuffered control blocks. However, the BRCB also has the PurgeBuf attribute and therefore the client can also command the events in the FIFO be purged by writing a value of true to this attribute.

It is the responsibility of the client to remember the last entry ID delivered by a buffered report message. It is this value that can be written back to the control block to have the next event in the FIFO reported. If the next entry is not able to be located, the oldest event in the buffer will be the basis of the first report sent. The client may also write a value of "00000000" to specify the oldest event in the FIFO will be the basis of the first report sent. If the client does not desire any of the buffered events, it needs to use the PurgeBuf attribute to purge all of the events in the FIFO. This resynchronization process is imperative so that the client does not receive duplicate reports and therefore be required to perform its own filtering.

The mappings of the operations to an instantiated service is as follows:

The abstract services map to either an MMS read or write service. It is the object on which the MMS service is issued that completes the mapping. The example depicts access to the instantiated Enable, which is a named component RptEna. The sequence of enabling a URCB is found in Table 8.21.

Figure 8.105 depicts the translation of the abstract service parameters into a MMS write. The write is received by the Server, the BRCB is located, and the

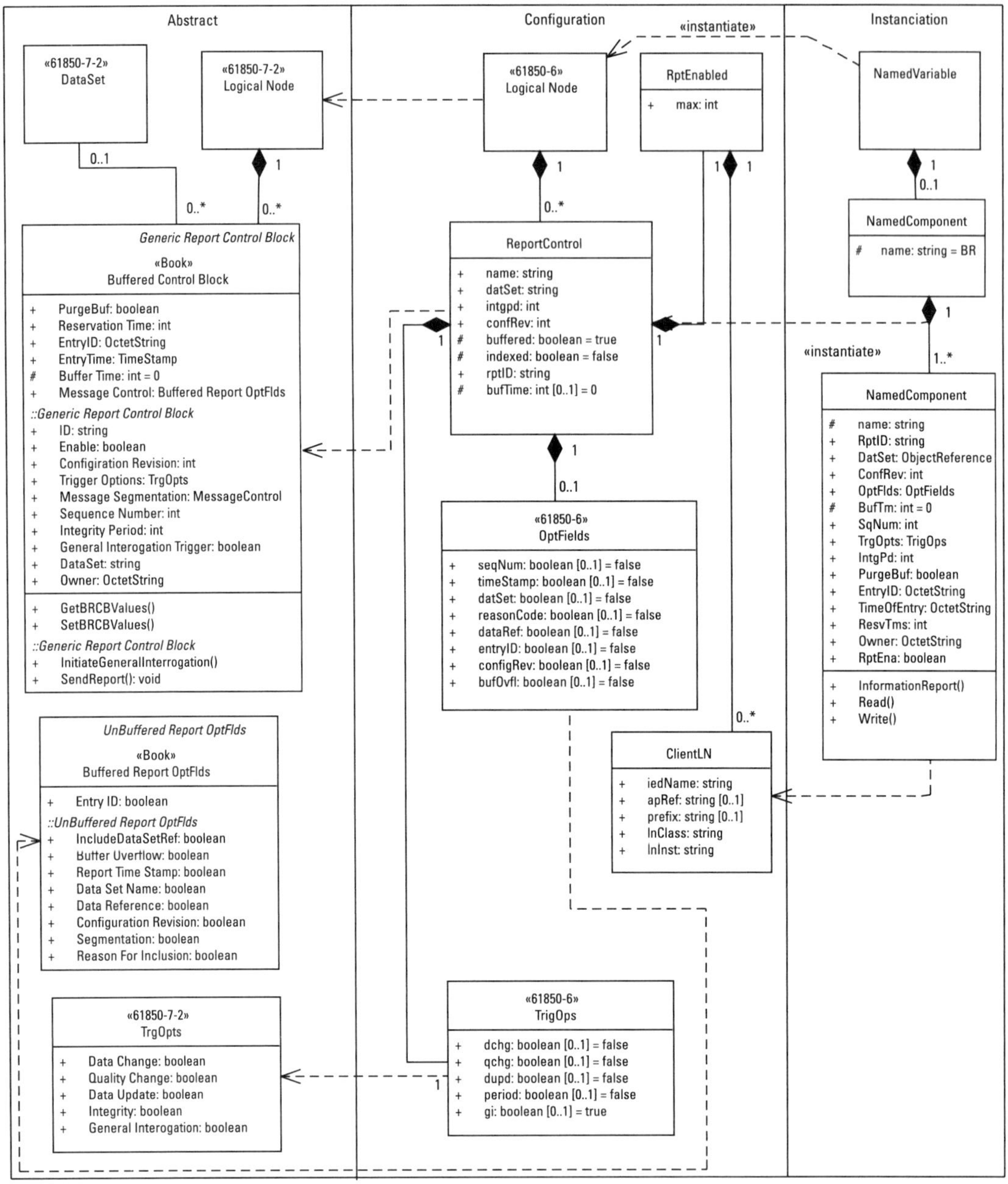

Figure 8.103 Buffered report control block UML.

RptEna value is set true. On setting the RptEna, the tracking logical node (e.g., LTRK) is updated. LTRK is also updated on the control block being dynamically reserved (e.g., via MMS).

8.2.2.5 IEC 61850 Log

An IEC 61850 log is a sequence of event (SOE) recorder. It is similar in functionality to a buffered report with the following major exceptions:

```
<ReportControl " datSet="ds1" name="brcb1" buffered="true"
        confRev="15" indexed="false" rptID="rpt1">
    <TrgOps dchg="true" qchg="false" dupd="false" period="true"/>
    <OptFields seqNum="true" timeStamp="true" dataset="true" reasonCode="true"
        dataRef="true" entryID="true" />
    <RptEnabled>
     <ClientLN iedName="Client" ldInst="LD0" lnClass="IHMI" lnInst="1"/>
    </RptEnabled>
</ReportControl>
```

Figure 8.104 Buffered control block SCL configuration.

Table 8.21 Mapping of Abstract URCB Services

Abstract Service	*MMS Service*	*Component Accessed*
GetBRCBValue	Read	<LN>.BR.<name>.RptEna
SetBRCBValue	Write	<LN>.BR.<name>.RptEna

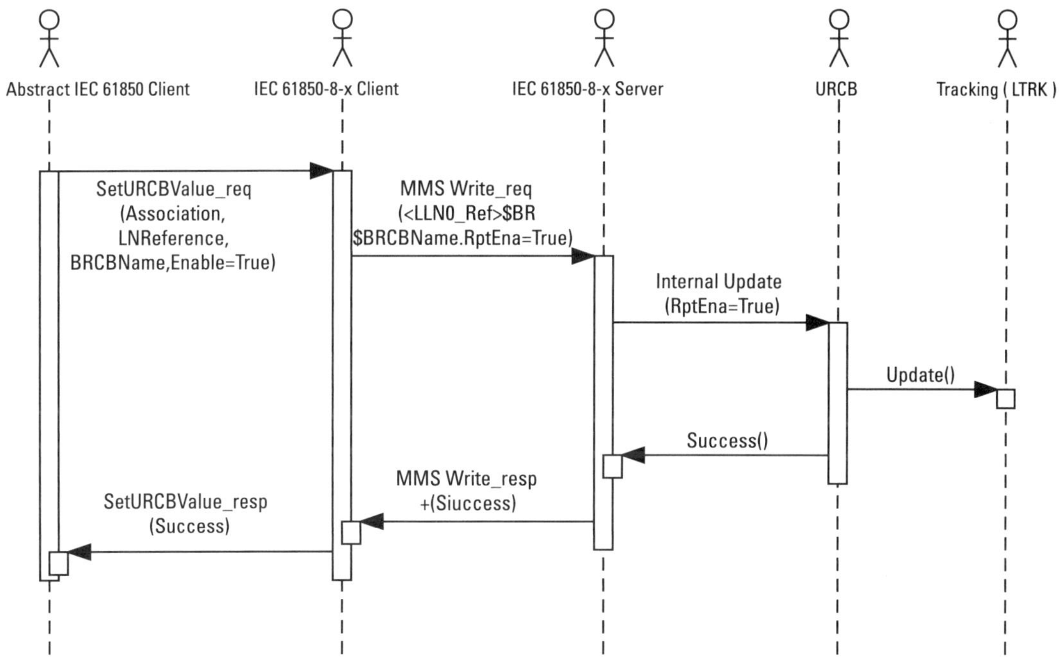

Figure 8.105 Enable Sequence for BRCB.

- Multiple log control blocks can place information into a single IEC 61850 log.
- The log consists of log entries, which could be construed as similar to the events in the reporting FIFOs.
- Multiple clients can query information from the log.
- The contents of a log can't be purged; they will naturally overflow.

It is worthwhile to note that few servers currently implement log controls or logs. This will be changing in the future as security becomes more invasive and there will be a log allocated to record security events thereby acting as a Security event log.

The relationships for log (see Figure 8.106) are more complicated than GOOSE, SV, or reporting since it also involves an additional relationship besides to a dataset. This additional reference is to the actual log. Therefore, this section will treat the log control block separately from the actual log.

Log Control Block

The log control block (LCB) has several attributes that are like those found in the report control blocks. The attributes are shown in Table 8.22.

If one does an analysis of the contents of the LCB, the configuration revision attribute is not present. This is intentional since it is not required for conveying events into the specified log. The reason this attribute is excluded is because the entire logging infrastructure is local to a single IEC 61850 server. The log processing of the server has access to the definition of the dataset even if the definition of the dataset is change. Thus, the events placed into the log can be done in an unambiguous fashion.

In a similar vein, there is no user assigned identifier in the LCB. This is because the events in the log are for multiple-client access and therefore and identifier assigned by one client would be nonsensical to other clients.

Configuration of a LCB, SCL allows the specification of an XML element that includes all the trigger options even though the general interrogation option is not allowed[19].

Figure 8.107 shows that the actual log is contained in LN0 since no other attributes specifying a different log reference are present (e.g., ldInst, prefix, lnClass, and lnInst). It also shows that the LCB is enabled due to the SCL configuration.

The mappings of the operations to an instantiated service is found in Table 8.23.

The abstract services map to either an MMS read or write service. It is the object on which the MMS service is issued that completes the mapping. The example depicts access to the instantiated Enable, which is a named component LogEna. The sequence of enabling a LCB follows.

Figure 8.108 depicts the translation of the abstract service parameters into a MMS write. The write is received by the server, the LCB is located, and the LogEna value is set true. Upon setting the RptEna, the tracking logical node (LTRK) is updated.

Log

The log is the actual object which is the FIFO of sequence of events. It is able to be queried by multiple clients. It consists of log entries that contain an array of entry data and (possibly) the reason-for-inclusion of the information in the log entry. A log is mapped to a MMS journal. The journal contains journal entries that consist of journal variables.

19. This is another by-product of re-use of common definitions and should have been avoided.

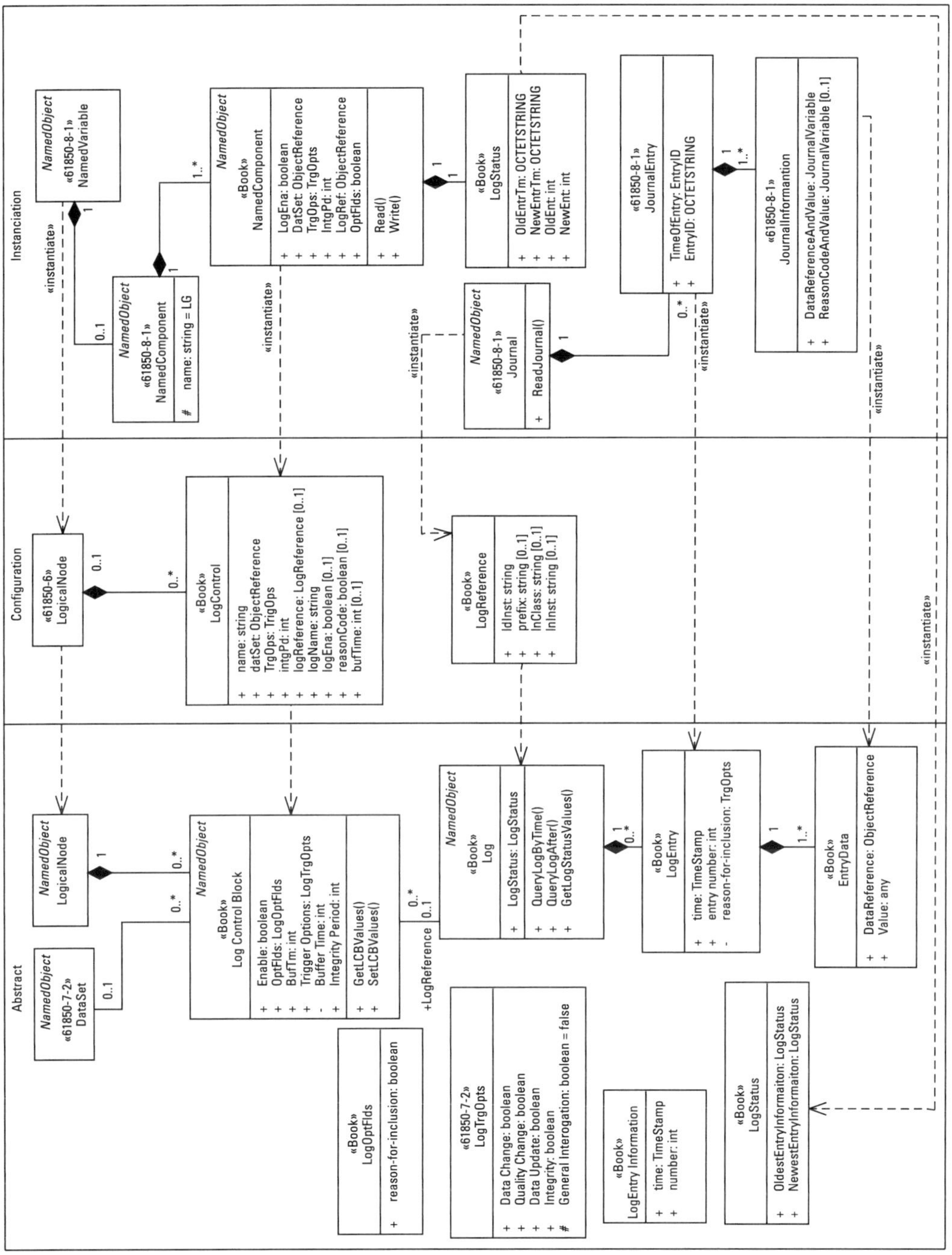

Figure 8.106 UML mappings for log control blocks and logs.

Table 8.22 Generic Report Control Attribute Definitions

Attribute Name	*IEC 61850-7-2 Name*	*Provided by*	*Description*
Enable	LogEna	Server	This value is written to start or stop the delivery of messages to a client.
DataSet	DatSet	SCL Client	This is an object reference to the dataset whose information is to be processed and sent by the report control block. A single dataset may be referenced by multiple control blocks.
Trigger Options	TrgOpts	SCL Client	These values control the filtering and event creation of entries for the FIFO. Additional information can be found in the *Trigger Options* (see page 206) section. Unlike other trigger options, the option of general interrogation is not allowed and therefore will be a value of "false" regardless of what value the client or SCL sets.
OptFlds	OptFlds	SCL Client	The only option field allowed for logs is the reason-for-inclusion option.
Integrity Period	IntgPd	SCL Client	This value determines the periodicity of creating an event based on the current process values of the dataset members. It is only a valid value if the trigger option enabling integrity events is set true. More information follows in the *Trigger Options* (see page 206) section.
LogReference	LogRef	SCL Client	This is the referenced to the actual log object into which event information is to be placed.

```
<LogControl name="lcblog" logName="PROT"
   datSet="dsLog" logEna="true" reasonCode="true" intgPd="0">
      <TrgOps dchg="true"/>
</LogControl>
```

Figure 8.107 Log control block configuration.

Table 8.23 Mapping of Abstract URCB Services

Abstract Service	*MMS Service*	*Component Accessed*
GetLCBValue	Read	<LN>.LG.<name>.LogEna
SetLCBValue	Write	<LN>.LG.<name>.LogEna

Both the log entry and journal entry include the time at which the entry was created. Unlike other timestamps, this time is instantiated as timestamp that is restricted to a resolution of milliseconds only. It is the binary time of day MMS data type. Both contain a numeric entry identifier that does not need to be sequential.

The mappings of the various log operations involve both journals and the instantiated LCB.

The mappings of the operations to an instantiated service are shown in Table 8.24.

The name of the instantiated journal must include the name of the logical node that it was configured to be part of. The mapping results in a MMS domain specific journal since all logical nodes are contained in a logical device. The name of the Journal would be <Logical Node Name>$<JournalName>. It is recommended that

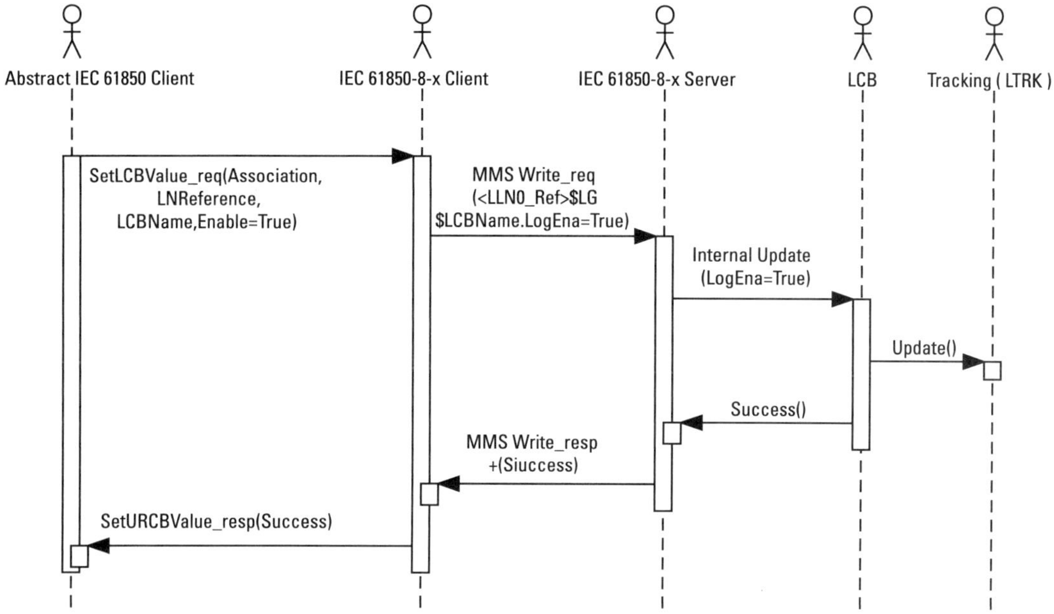

Figure 8.108 Log control block enable sequence.

Table 8.24 Mapping of Abstract Log Services

Abstract Service	*MMS Object*	*MMS Service*	*Component Accessed*
QueryLogByTime	Journal	ReadJournal	<LN>.<JournalName>
QueryAfter	Journal	ReadJournal	<LN>.<JournalName>
GetLogStatusValues	Control Block	Read	<LN>.LG.<name>.OldEntrTm <LN>.LG.<name>.NewEntrTm
			<LN>.LG.<name>.OldEnt
			<LN>.LG.<name>.NewEnt

implementation utilize LLN0 as the containing Logical Node for Journals, since its name is not modifiable by users during configuration.

The QueryLogByTime accesses the journal based on specifying the range of time that is of interest to the client. The range is specified by including a start and stop time attribute in the request. The QueryAfter service is utilized if the QueryLogByTime was not completed as indicated through a moreFollows=true being returned in the response. If the journal read response indicates that more information needs to be acquired, the client is responsible for taking the last delivered JournalEntry.EntryID and submitting a read journal request that contains the entry ID and the TimeOfEntry of the last received journal entry. The process can iterate until moreFollows=false is encountered.

The data reference is mapped to a journal tag and the values are mapped to journal variable values. The exception to this rule is the inclusion of the reason-for-inclusion. If the OptFlds.reason-for-inclusion=true, an additional journal varible will be created.

8.2.2.6 Paranoia and Control Blocks

Operating within an electric utility environment requires a good amount of paranoia and concern about acting based on the incorrect information. Action based on incorrect information can lead to automation, or humans, making the incorrect decisions and those decisions could have dire impacts to the stability of the grid. Care must be taken in the implementation and expectations of clients and subscribers for information controlled by a control block that has a dataset reference being used to specify the information to be conveyed by the service controlled by the control block. Of the control blocks discussed, the concern is valid for all control blocks with the exception of settings group.

Of the other control blocks, all contain a reference to a dataset that provides the definition of the information to be delivered by the service. If the members of the dataset are changed or rearranged, then unexpected information may be delivered and processed as something else. Somebody might say, that can't happen, but it is possible without thinking about the issues involved for the engineering and configuration process and the processing of the delivered information to the subscriber.

If configuration control and procedures are not in place, the following can cause the issue to arise. Consider a dataset whose name is Mxyzptlk,[20] the dataset contains two FCDAs of the same type. For this example, the FCDA references the ordered references of A and B. Consider if Mxyzptlk is redefined in SCL and FCDAs being ordered as B then A. Furthermore, consider the SCL engineer does not change the configuration revision number and the server is updated. In the period between the configuration update of the server and that of the client, the client has no mechanism to detect that a content change has occurred because the dataset reference and the configuration revision number conveyed in the message controlled by the control block are identical to those previously configured and used by the client. However, the client will now interpret B and being the value(s) for A and the values of A being used as B since the order was changed in the new configuration. To avoid the issue being caused by configuration, the combination of the SCL tool and SCL engineer must ensure that the combination of dataset reference and configuration revision number are unique for any change made to the dataset definition.

The unique combination of dataset reference and configuration revision number does not help unless the client/subscriber performs the appropriate check on receiving the message. IEC 61850 does not specify what a client/subscriber must check or its reaction should the checks fail. Thus, the following are suggestions based on experience. Implementations of clients and subscribers should perform the following checks:

- It is imperative that the implementations check that the received dataset reference matches the expected value (e.g., through configuration).
- It is imperative that the implementations check that the received configuration revision number matches the expected value (e.g., through configuration).

20. Mxyzptlk is a mischievous nymph from the fifth dimension and was Superman's archnemesis.

- For GOOSE and SV, it is imperative that the implementation checks that the number of dataset members received matches what is expected.
- It is recommended that the expected datatype for the dataset member being received matches with what is excepted.

If any of the checks fail, it will be a local issue of how to indicate that invalid information has been received. It is suggested that the quality of the information received be changed to indicate bad quality. The local implementation could choose to deliver the previously delivered information (e.g., prior to check failure) with the additional quality of last known value.

8.2.2.7 Common Capabilities

Besides containership, logical nodes can have their behavior controlled and to have signal inputs defined. These common capabilities will be discussed in the following sections.

8.2.2.7.1 Mode, Behavior, and Health

There are two significant aspects to controlling the behavior of logical nodes. The relationship between logical node mode and behavior data objects can be a bit confusing, as either could reflect the actual constraints placed on the operation of the logical node. In an attempt to remove the confusion, clients interact with the mode (e.g., Mod attribute). The value written to the mode causes the values in the behavior (e.g., Beh) data object to change. It is the values of the Beh that constrain the behavior of the logical node.

Figure 8.109 shows how an interaction to Mod can change the behavioral state of the logical node. Mod can be set to the following values, which in turn become Beh states. The common values are: on, blocked, test, test/blocked, and off. In IEC 61850-7-4 there is a table that shows the logical node capabilities/responses based on the behavioral state. Those with a desire for full knowledge of the table, will need to read IEC 61850-7-4.

In general, the states of Beh can be summarized as

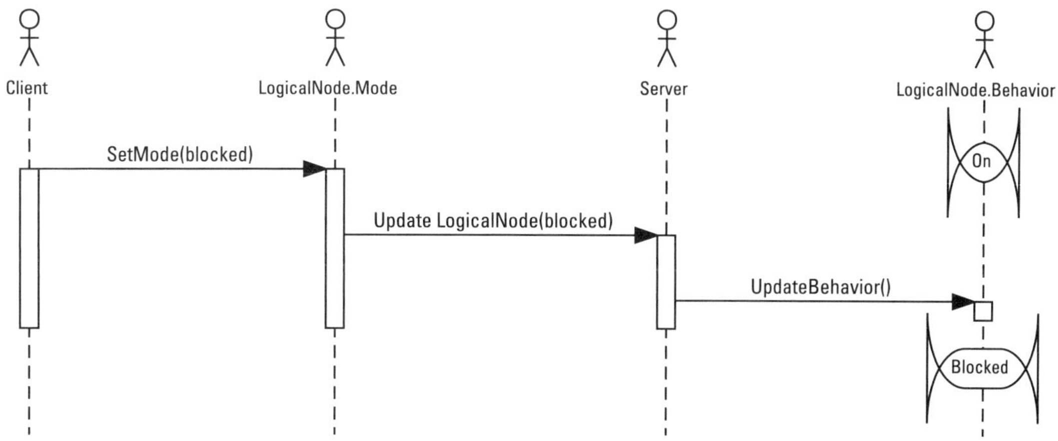

Figure 8.109 Example of Mod and Beh interaction.

- *on*: The logical node has full operational capability, will ignore incoming data marked with test quality, and will output signals to switchgear through I/O or other services. It will also process questionable quality as it would normally process such quality. It will not process incoming data marked with test quality.
- *blocked:* The logical node will provide the functionality that it proxies. It will behave in a similar fashion to on except it will not generated output signals to switchgear.
- *test:* Similar to the on state except the logical node will process incoming information marked with test quality as valid and use that information in its function. Any values produced by the logical node will have test quality.
- *test/blocked:* Is a combination of the test and blocked state.
- *off:* The logical node will not be providing its proxied function. Any information that would typically be produced by the logical node will be marked with a quality of invalid. No output to switchgear will occur.

For further information regarding IEC 61850 quality, see Section 9.1.2.

InRef, BlkRef, and ExtRef

Within the context of IEC 61850, logical nodes exchange information with each other. For example, consider a switchgear controller (CSWI) has detected that a breaker (XCBR) needs to be opened (e.g., Trips) but, needs to have a breaker open. Additionally, the CSWI may need to block the operation of the XCBR.

Figure 8.110 shows the internal logic of the XCBR where if the block coil is active, regardless of the state of input1, the breaker/XCBR will not open. It shows two virtual XCBR inputs that are used to drive the coils of input1 and block. It also shows that it is desired to connect the CSWI virtual outputs of OpOpn and Blk to the appropriate XCBR virtual inputs. It is the construct of InRef and BlkRef that allows a virtual input to connect to a virtual output within the same IED.

Both InRef and BlkRef data objects are defined as a common data class ORG. The functionality of ORG borrows a concept from Substitution. It allows two inputs with the selection of which input to utilize being determined by an input to a switch; see Figure 8.111. The two inputs represent the normal process virtual output that should be used, and the other is the virtual output that should be used for testing. The referencing of virtual outputs is based on object references. There is magic in gluing the output of ORG to the internal logic of the XCBR. This is

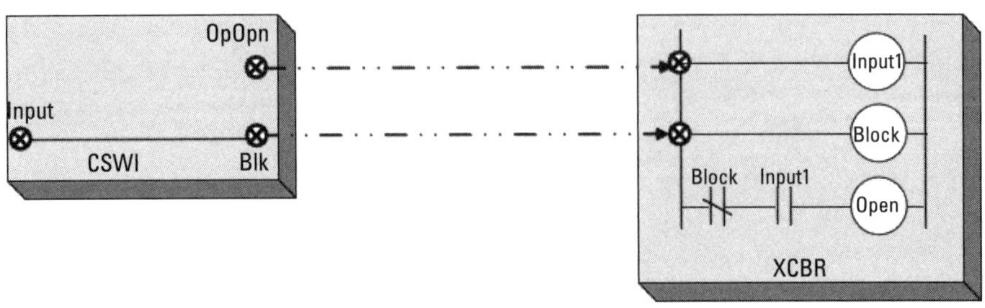

Figure 8.110 Desired CSWI control of XCBR.

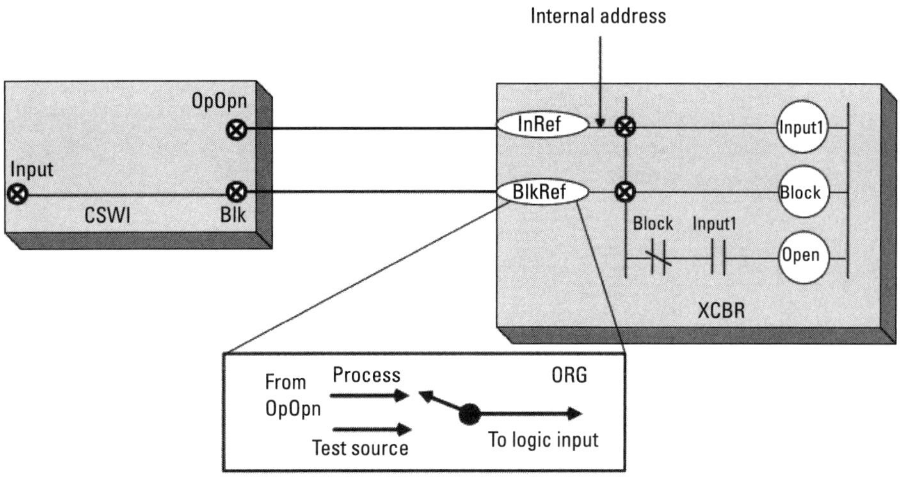

Figure 8.111 InRef and BlkRef.

an internal address that is outside the scope of the 61850 standards. The value of the internal address represents some register in the IED and that register is utilized by the internal logic of the logical node. Some of the magic is demystified due to the fact that the ORG provides a textual field that can describe the purpose of the InRef or BlkRef.

Figure 8.112 is an abstraction of the actual definition of ORG. This is because ORG can be used to reference either a data object or a control block and only one or the other should be used at any given time. The ORG definition is overloaded to address InRef/BlkRef and communication supervision (e.g., LGOS). InRef and BlkRef are local reference to data and the specification of a control block is not required and should not occur. In LGOS, the control block that is being monitored is important and not a specific data object. In this use, the control block should be present and no data reference should be present.

The figure shows two inputs being setSrc (i.e., process) and setTest (i.e., test source). The abstraction also shows the inclusion of the tstEna (i.e., the value that allows switching between the process and test sources) as well as exposing the internal address (i.e., intAddr) and the purpose value of the InRef/intaddr. The setSrc and setTst consist of the ability to have an object reference (i.e., Ref) to the virtual output that is desired as an input and an optional reference to the control block that is used to specify the service as to when to deliver a new value of the object reference to the InRef/BlkRef. In general, most applications only utilize the object reference and setSrc (i.e., only Process value).

Figure 8.113 defines the XCBR.InRef1 to reference the CSWI1.OpOpn.general attribute of the CSWI1 located in the logical device whose instance has a value of LDInst. The purpose of InRef1 is declared to be to open the breaker.[21] The initialization, because of the valImport and valKind values, specifies the value may be changed by the system configuration tool but not over the wire.

21. In many cases, the initialization of the purpose is found in the *d* (i.e., description) value, which is incorrect.

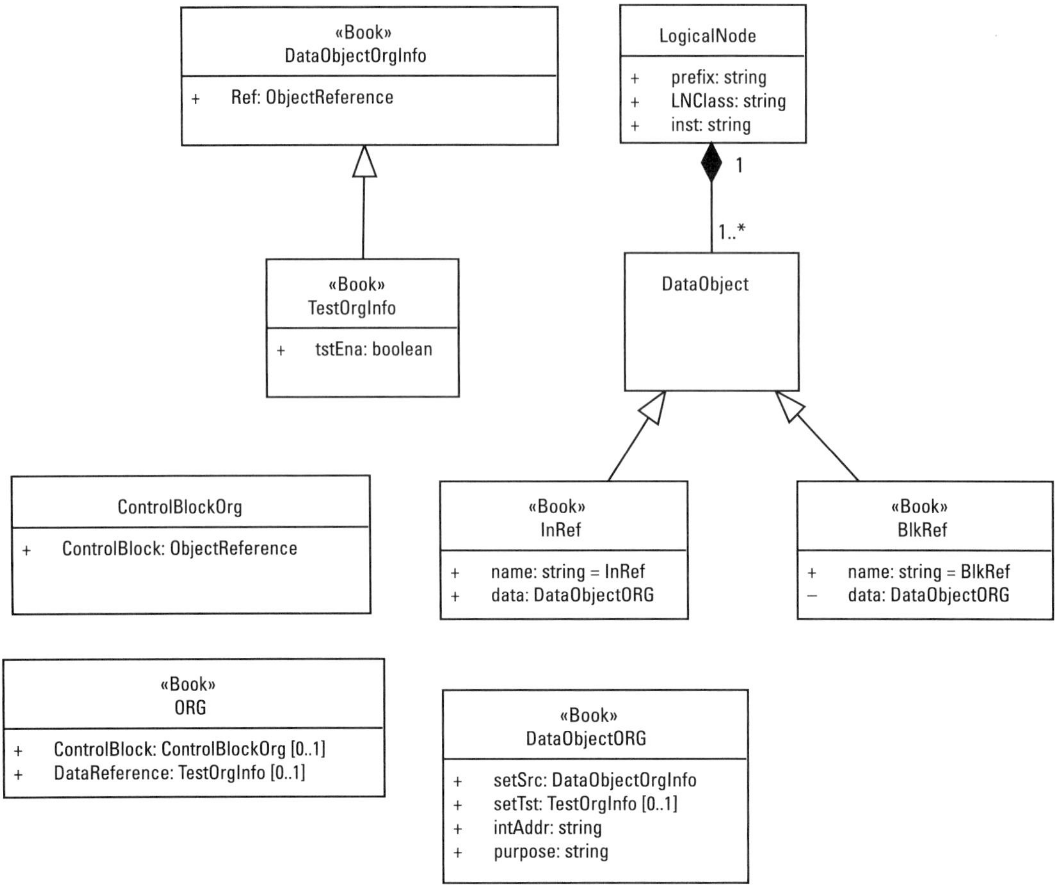

Figure 8.112 Abstracted relationship of ORG to InRef and BlkRef.

The next logical question is how to do the equivalent signal binding for virtual outputs sourced in a different IED. The answer is the use of ExtRef as shown in Figure 8.114.

In a similar manner to InRef and BlkRef, ExtRef utilizes the internal address (e.g., intAddr) to bind incoming information to an internal process variable. It also must specify that the delivery mechanism for the requested data is delivered to the entity. There are four delivery mechanisms that can be specified (see Table 8.25). There are two parts to the ExtRef: preferred bindings and the actual configured bindings. The preferred bindings can be utilized to provide a hint to the system configuration tool as to what type of information should be used to satisfy the external reference. It can include the logical node Class, DOName, DAName, and the serviceType. In SCL, these values would be serialized as pLN, pDO, pDA, and pServ1, respectively. The engineer using the system configuration tool can use this hint to assist in the design of the information delivery.

Polling and report delivery mechanisms are only valid for implementation that have client capability and thus they are seldom utilized. If polling is used for information acquisition there is no control block required since the client issues the GetDataValues request directly for the data needed.

```
<LNode iedName="demo" inst="1" lnClass="XCBR" lnInst="1"
      lnType="XCBR_0">
      <DOI name="BlkRef1">
        <DAI name="setSrcRef" valImport="true" valKind="RO">
          <Val>@LDInst/LLN0.Off.stVal</Val>
        </DAI>
      </DOI>
</LNode>

<LNodeType id="XCBR_0" lnClass="XCBR">
    <DO name="InRef1" type="ORG_0"/>
    <DO name="BlkRef1" type="ORG_0"/>
</LNodeType>

<DOType id="ORG_0" cdc="ORG">
    <DA bType="ObjRef" name="setSrcRef" fc="SP"/>
    <DA bType="ObjRef" name="setSrcCB" fc="SP"/>
    <DA bType="VisString255" name="d" fc="DC"/>
</DOType>
```

Figure 8.113 Configuration of InRef via SCL.

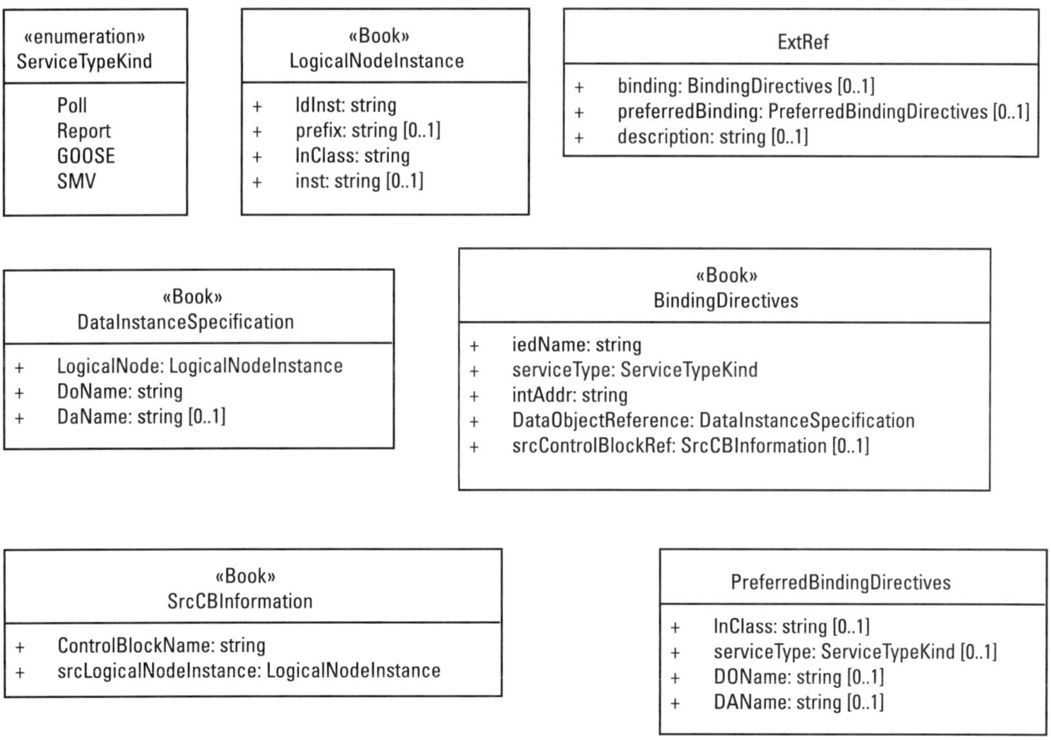

Figure 8.114 UML definition of ExtRef.

The UML show that the information being delivered may be at data object or data attribute. The SCL configuration of ExtRef, shown in Figure 8.115, lacks the specification of the Functional Constraint although this can typically be figured out algorithmically.

Table 8.25 How Extref Information is Delivered

Delivery Method	*Definition*	*Control Block Configured*	*Typically Used*
Poll	The entity issues GetDataValues request to obtain the information.	No	No
Report	Receives report messages through manipulation of the specified report control block.	Yes	No
GOOSE	Receives GOOSE information through normal configured subscription mechanism.	Yes	Yes
SMV	Receives Sampled Value information through normal configured subscription mechanism.	Yes	Yes

```
<ExtRef daName="general" doName="Op" iedName="IED1"
        ldInst="CTRL" lnClass="PTRC" lnInst="1"
        prefix="SMP" serviceType="GOOSE"
        srcCBName="gcb1" srcLDInst="LD0"
        srcLNClass="LLN0"/>
```

Figure 8.115 SCL configuration of ExtRef.

8.2.2.8 Logical Nodes and Applications of Interest

Table 8.5 shows the basic functionality of the IEC 61850 logical nodes. This chapter discusses certain specific functionality that is implemented by using specific logical nodes. The section will discuss

- How to perform testing isolation for maintenance;
- Asset information contained by IEC 61850 logical nodes;
- Logical nodes that provide communication supervision and additional communication diagnostics;
- The worst idea ever in IEC 61850, Generic I/O.

8.2.2.9 Maintenance, Testing, and Isolation: LN0, LPHD

There are a couple of major uses that require different levels of testing and isolation. The first is injecting test information to an IED and seeing how the overall system performs or recovers using the test data. The second requires the isolation of an IED from the system so that the rest of the system is not impacted by the testing.

Injection of Test Information

Typically, the testing of an IED requires a test set to be used to inject test information for a given test scenario. Hardware input typically have isolation blocks that allow the test set to be hooked directly to the device and the other hardwire I/O is isolated or removed. The use of isolation blocks for the Ethernet communication of GOOSE and SMV isn't realistic if interaction with the rest of system is still desired. The concept of simulation injection was designed to allow injection of test information via GOOSE or SMV.

This concept allows duplicate GOOSE or SMV to be published on the LAN using the exact same control block, destination address, control block, data, and so forth. It is a publication of the exact same information, except the Simulation bit is set true. This bit allows a subscriber to differentiate between test/simulation packets and a packet published by the actual process.

In order to process simulated packets (e.g., tagged with Simulation=true), the value of LPHD.Sim must be set to a value of true. Note that LPHD.Sim is an object known as a single point control (SPC). Once the status value is true, the IED can process GOOSE and SMV packets tagged as simulation. The IED will continue to process normal process published packets until an equivalent simulation packet is received. At this point, the IED will only process the detected simulated packet until LPHD.Sim is set false. This allows the IED to process a mixture of process and simulated packets.

IEC 61850 Edition 1 did not properly specify this mechanism and therefore, many Edition 1 devices do not have this capability. This ability is one of the major reasons to move from Edition 1 and to Edition 2. This also means that injecting Edition 2 simulated tagged GOOSE and SMV into an Edition 1 system may lead to unexpected and bad results. The issue is further complicated by the fact that Edition 1 SMV did not have a test/simulation bit in the Application Protocol Data Unit (APDU), thus Edition 1 devices will not be able to properly decode such a packet.

Device Isolation

The way to isolate the device under test (DUT) from the rest of the system is by setting the LN0.Mode to either test or test blocked at the highest level LN0 in the logical device hierarchy (see Figure 8.21). This will have the qualities for all of the information produced by the IED marked with either test or test/blocked. As pointed out previously, to make this viable, the datasets of GOOSE, SMV, LOGGING, and reporting need to include the quality and clients must analyze the quality.

This mode can be used with or without processing simulated tagged packets. It is this flexibility that forces utilities to write standard operating procedures (SOPs) based on their need.

Device Asset Information: LPHD, Common Information

There is a large need to get accurate asset information from the field devices. IEC 61850 provides two CDCs that are used to expose such information. The physical asset information is conveyed in the device nameplate information DPL in the LPHD or other logical nodes associated to switchgear. In this instance, the nameplate information is that of the switchgear and not the IED. There is also soft asset information (see Figure 8.116) that can be found in most logical nodes. The CDC is LPL.

The attributes of the DPL are self-explanatory except for the ePSName which is the configured name of the electrical system in which the device is operating. The soft asset related information in LPL are

- *Settings Parameter Revision (paramRev):* This value represents the number of changes of editable settings that can be in a setting group (e.g., a functional

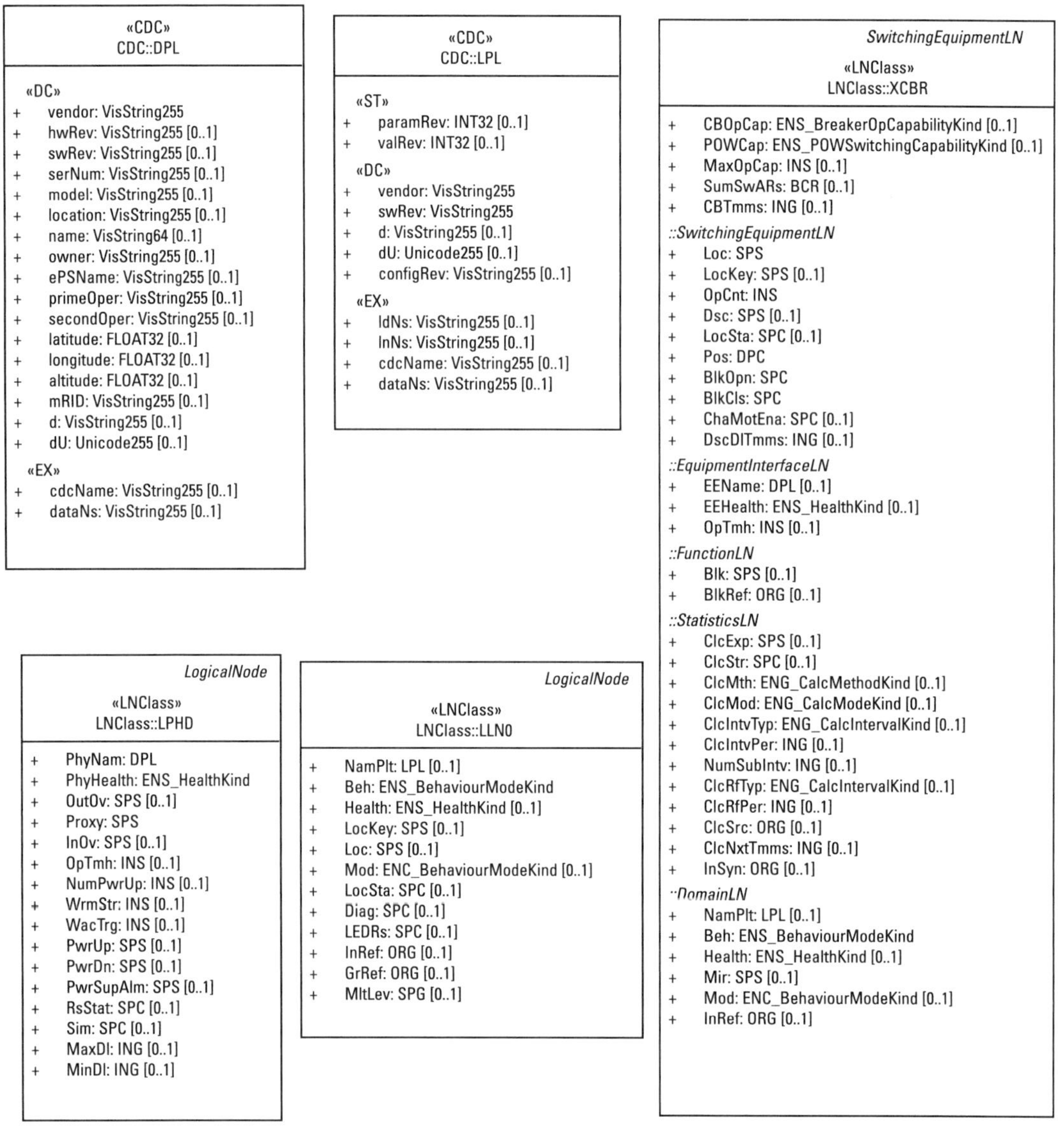

Figure 8.116 Asset related information in IEC 61850.

constraint SE) or the functional constraint SP. There is a recommendation to jump the value by 10,000 if the change is performed by SCL. It is doubtful that this is currently implemented.

- *Configuration Revision (valRev).* This value increments when a value with the functional constraint of CF is changed.

There is a good case to be made that an application should check these values to see if settings are being changed when they should be. The best place to check these values are the LN0 and LPHD at the top of the logical device hierarchy.

Communication Dianostics/Supervision: LGOS, LSVS

The transition to a digital substation changes all types of procedures including how to diagnose problems. Typical hardware diagnosis involves test tools that probe the wiring such as a volt-ohm meter as shown in Figure 8.117

In the nondigital substation, there is a schematic that can be used to beep-out a problem. In the digital substation, the beep-out needs to occur via network and exposed information. If the network is set up with a mirror-port, all network packets traversing that switch will be egressed on that port. This takes some network design, but it is always something that is recommended in IEC 61850 systems.

The control blocks for GOOSE and Sampled Values can be monitored over the network and therefore, it is possible to determine if the information is supposed to be sent to the network. This along with the mirror port will allow a maintenance engineer to determine if the information is be sourced to the network. What is lacking is an observation point on the subscriber that can be monitored. This is where LGOS and LSVS logical nodes come into play. The definition of these logical nodes can be found in Figure 8.118.

Within the IEC 61850 community, there is a debate about when the state (St) goes to a value of true which is what the standard declares as an active state. The simple interpretation is if the message packet that is being published by the ControlBlockRef (e.g., GoCBRef or SvCBRef) and is locally processed, the state should be true. If LPHD.Sim=false then the SimSt should be false since simulation tagged packets are not locally processed. Remember these logical nodes represent observation points to the local processing of the subscriber.

Complication comes if the ConfigurationRevNum does not match the expected configured value should the state be active or inactive. The typical suggestion I make is that the state should be true, but the quality should be questionable. It is something that users need to check with their device vendors about.

8.2.2.10 Synchrophasor: MMXU

As was discussed in Section 8.1.1.3, there is a need to send synchronized measurements that are calculated via IEEE C37.118.1.

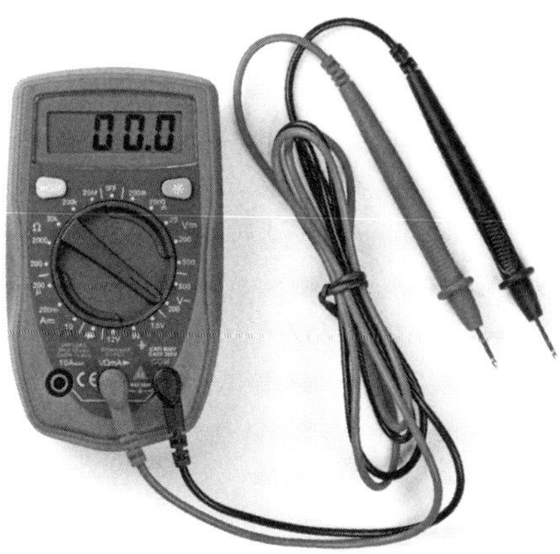

Figure 8.117 Hardware and schematic debug. (Image used under license from Shutterstock.com.)

SubscriptionSupervisionLN

«LNClass»
LNClass::LGOS

+ LastStNum: INS [0..1]
+ GoCBRef: ORG
::SubscriptionSupervisionLN
+ NdsCom: SPS [0..1]
+ St: SPS
+ SimSt: SPS [0..1]
+ ConfRevNum: INS [0..1]
+ RxConfRevNum: INS [0..1]
::DomainLN
+ NamPlt: LPL [0..1]
+ Beh: ENS_BehaviourModeKind
+ Health: ENS_HealthKind [0..1]
+ Mir: SPS [0..1]
+ Mod: ENC_BehaviourModeKind [0..1]
+ InRef: ORG [0..1]

SubscriptionSupervisionLN

«LNClass»
LNClass::LSVS

+ SvCBRef: ORG
::SubscriptionSupervisionLN
+ NdsCom: SPS [0..1]
+ St: SPS
+ SimSt: SPS [0..1]
+ ConfRevNum: INS [0..1]
+ RxConfRevNum: INS [0..1]
::DomainLN
+ NamPlt: LPL [0..1]
+ Beh: ENS_BehaviourModeKind
+ Health: ENS_HealthKind [0..1]
+ Mir: SPS [0..1]
+ Mod: ENC_BehaviourModeKind [0..1]
+ InRef: ORG [0..1]

Figure 8.118 UML definition of LGOS and LSVS.

To utilize a MMXU instance to provide synchrophasor information, two things must be specified, the calculation method and the interval of the calculations. P-CLASS and M-CLASS were enumerated values added to IEC 61850 to specify protection precision calculation or meter grade synchrophasor calculation. The interval of the calculation is typically set to MS with the ClcIntvPer being the number of msec between calculations. This calculated information can be delivered through Reporting, GOOSE (typically Routable GOOSE), or most often routable Sampled Values. The UML representing this information is found in Figure 8.119.

8.2.2.11 Avoid Using—GGIO

Sometimes the best intentions become baggage to be discarded. When IEC 61850 was semantically poor (e.g., small amount of logical nodes and data objects) and initially designed, there was comfort in register-based designs similar to DNP. With that said, GGIO was heavily utilized initially. Today, IEC 61850 has moved forward. Still, there are people that hold onto the old way. There is an implementation where an IED has 5,000 generic I/O signals. You might as well use DNP instead in this situation.

As a user, the quality of an IEC 61850 may be reflected in the number of generic I/O used. The larger the number used, the lower the potential quality of the system. If GGIOs can be avoided, they should be (at all costs) since it is the semantics that add significant value to the system.

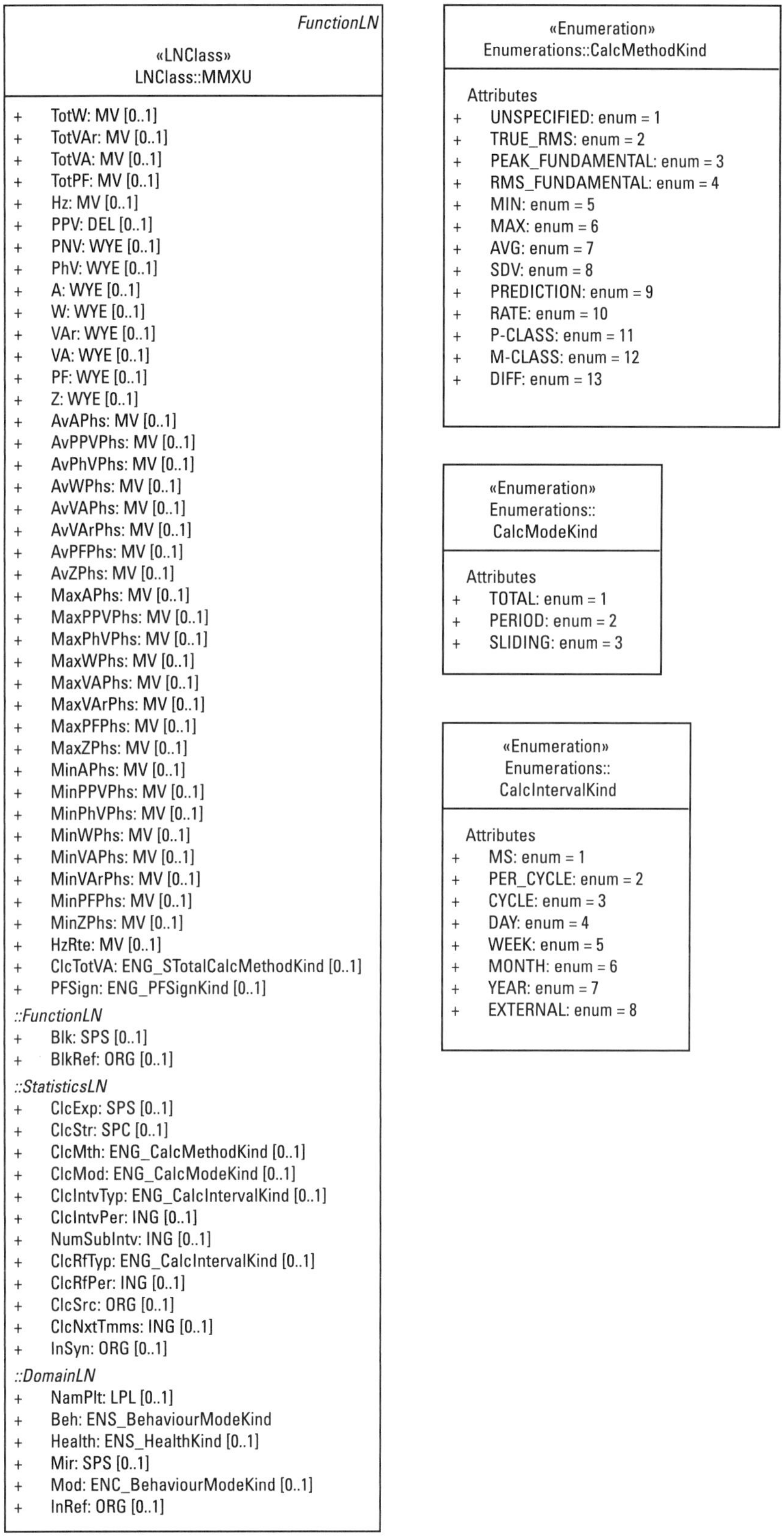

Figure 8.119 Using MMXU for synchrophasor applications.

CHAPTER 9

IEC 61850-7-2 AND IEC 61850-7-3

To understand the base information types exchanged by IEC 61850, it is necessary to discuss the base type found in IEC 61850-7-2, and quality (defined in IEC 61850-7-3).

9.1 Base Types

There are several base types defined in IEC 61850 which map to ASN.1 and MMS data types. The following table shows the mappings of these base types.

9.1.1 Timestamp and Synchronization

The timestamp represents a Coordinated Universal Time (UTC) time and is not local time. The timestamp base type consists of three parts:

- Seconds since the epoch of January 1, 1970 UTC. This is a 32-bit value. This is the same representation as NTP.
- Fractions of second. This is a 24-bit value. NTP typically specifies 32 bits but does express any unused bits shall be ignored.
- Time Quality: Is an 8-bit value and is unique to the IEC 61850 standard. This value includes information regarding leap seconds, clock failure, clock synchronization, and an estimation of the accuracy of the timestamp. Further information on this can be found within IEC 61850-7-2.
- The abstract timestamp is mapped to an ASN.1 OCTETSTRING and an IEC 61850 extension to MMS called utc-time.

One of the most controversial items in the timestamp is the leap second information. Never has such an issue caused so much time in discussion! This is critical because if a leap second occurs, it is possible for two different events to be created at exactly the same timestamp because when a lead second is applied, midnight repeats. This makes forensic analysis and some real-time analysis difficult. The leap second is tied to the speed of rotation of the Earth and by the International Astronomical Society. There have been several petitions to halt leap seconds, but to date the Astronomical Society has refused.

Table 9.1 Base Data Type versus MMS DataTypes

IEC 61850-7-2 Type	*Range of Values*	*ASN.1 Type*	*MMS DataType*
BOOLEAN	True or False	BOOLEAN	Boolean
INT8	–128 to 127	INTEGER	Integer
INT16	–32,768 to 32,767	INTEGER	Integer
INT32	–2,147,483,648 to 2,147,483,647	INTEGER	Integer
INT64	-2^{63} to $2^{63}-1$	INTEGER	Integer
INT8U	0 to 255	INTEGER	Unsigned
INT16U	0 to 65,535	INTEGER	Unsigned
INT32U	0 to 2^{32}	INTEGER	Unsigned
FLOAT32	IEEE 754 Single Precision Float	OCTETSTRING	Floating-point
Octet64	Up to 64 bytes (e.g. octets) of value.	OCTETSTRING	Octet-string
VisString64	Up to 64 characters (e.g. octets) of value.	VISIBLESTRING	Visible-string
VisString129	Up to 129 characters (e.g. octets) of value.	VISIBLESTRING	Visible-string
VisString255	Up to 255 characters (e.g. octets) of value.	VISIBLESTRING	Visible-string
Unicode255	Up to 255 Unicode UTF-8 characters.	UTF8String	MMSstring
Currency	ISO 4217 3 character values	VISIBLESTRING	Visible-string

The time basis that does not utilize leap seconds is called (TAI). The issue with TAI is that the conversion to local time becomes much more difficult and requires maintenance every time a leap second occurs until leap seconds are abolished.

To provide accurate timestamps, time synchronization is required. IEC 61850 officially supports two types of time synchronization: Network Time Protocol (NTP) and PTP. There are two unofficial synchronization techniques known as I-RIGB and GPS. This section will discuss NTP and PTP.

NTP works on the basis of a round-trip message from the entity being synced with Time1 to the time server (receives the message at T2), time server provides its current time (T3) and sends a message to the entity, which receives it at T4. The offset is calculated as shown in (9.1):

$$\text{Entity Time Offset} = [(T2\text{-}T1) + (T4\text{-}T3)]/2 \tag{9.1}$$

This calculation is intended to factor out network latency and assumes symmetry to the latency of both messages. This assumption does not typically hold true and therefore, the time offset needs to be applied is an approximation. Thus, iterative time synchronizations are needed to provide 1-msec timestamp accuracy using NTP.

PTP works via a totally different mechanism. Instead of the entity calculating the time offset, in PTP the network infrastructure calculates the offset in real time as the time message passes through the various hops of the network. The source of the original time synchronization message is known as a grand master. It is typically synchronized via GPS. It puts out a sync message that has its original time and an offset due to the delay in sending the packet to the network. When the sync message ingresses into an Ethernet switch, it must process the packet as a transparent clock. This processing requires calculation of the delay between ingress and egress and updates the offset within the sync packet. When the end device receives the sync message, it is required to estimate its ingress/processing delay and have the following available to compute a timestamp. It has the grand master original time + network offset (in the packet) + ingress delay to calculate its new time. The computation allows for time precession of better than 1 μsec typically. The additional advantage of PTP is that there can be multiple grand masters on the network and there is a predefined algorithm for selecting the best master. Thus, redundancy and failover capability is inherent in PTP.

There are multiple versions of PTP that will eventually confuse the market. There is the original IEEE 1588 standard. This standard has so many options that the power industry decided to create two power centric profiles. The first is IEC 61850-9-3, which directly references IEEE 1588. The second is IEEE c37.238, which references IEC 61850-9-3 and adds some additional fields in the PTP messages. Both IEC 61850-9-3 and IEEE c37.238 are basically compatible with each other and the industry will eventually decide which standard achieves wider acceptance.

Clock synchronization and precision is very important for IEC 61850 applications. NTP synchronization is viable for client/server and GOOSE timestamping. However, PTP, I-RIGB, or GPS must be used for CT/PT (e.g., Sampled Values) and synchrophasor applications. IEC 61850 provides the logical node LTMS to supervise the type and source of the synchronization.

LTMS, shown in Figure 9.1, provides information regarding the state and accuracy of the time synchronization that may be useful to a user. It is imperative that an instance of LTIM be provided by a server if it is possible to have different active

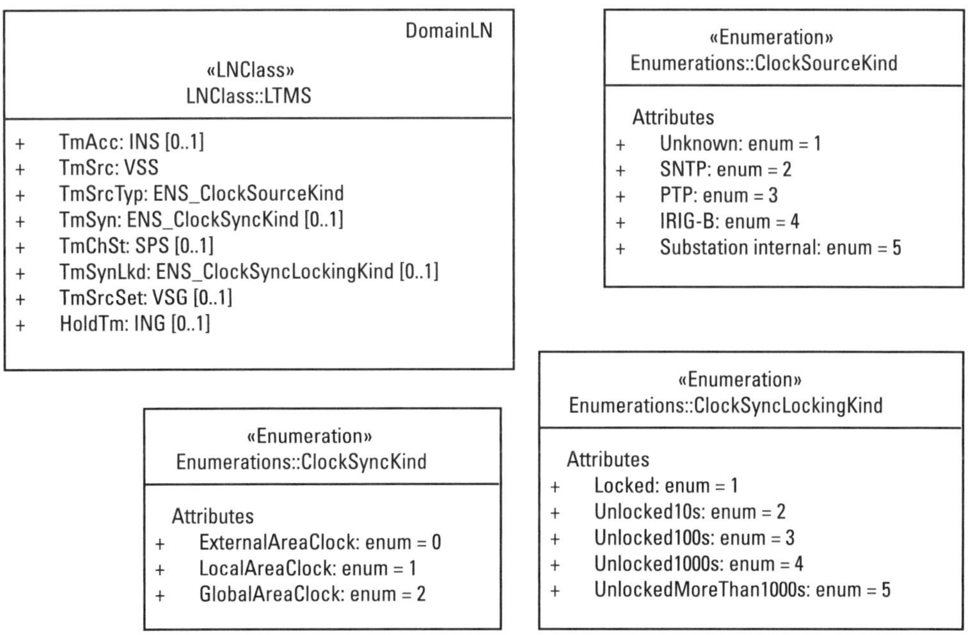

Figure 9.1 Time synchronization supervision logical node (LTMS).

Table 9.2 User Recommendations Regarding Optional Data Object for LTMS

Optional Data Object	*Number of Clock Sources Allowed by Server*	
	Single Clock	*Multiple Clock Sources*
Time accuracy	Would be nice to have, but not required as timestamp would reflect the basic accuracy.	Required.
Type of clock synchronization	Required if it can be synchronized by different types of clock sources.	Require if can be synchronized by different types of clock sources. Different clock sources offer different accuracies.
Time synchronization state	Would be nice to have, but not required as timestamp would reflect the basic state of synchronization.	Require so that clock source issues can be detected in advance for sources that are not currently being used as the current clock source.
Type of synchronization applied	Not required.	Would be nice to have.

sources of synchronization. As an example, an IED might allow PTP, IRIG-B, and SNTP inputs. It would be anticipated that other types of clocks may be added (e.g., GPS). However, for the time being direct synchronization by GPS would need to be represented by IRIG-B. Normally, only one clock source may be in use at any given time; there may be different levels of accuracy and the highest accuracy clock source available should always be utilized. The use of multiple clocks will become more common in the future as clock source tamper detection requires multiple types/locations of clock sources to be used to validate that time has not been corrupted. When LTMS is instantiated many of the data objects are optional in the standard and users should mandate that servers provide different sets of data objects based on the IED being synchronized by one or multiple clock sources.

9.1.2 Quality (IEC 61850-7-3)

The IEC 61850 quality type has different levels of detail. It is best to think of the value as an ordered list of: validity, detailed quality, source, test, and operator blocked. The validity field has four different values: good, invalid, reserved, and questionable. The detailed quality consists of several Boolean values: overflow, outOfRange, badReference, oscillatory, failure, oldData (e.g., last known value), inconsistent, inaccurate. The source value can have the value of either process or substituted. It may only have a value of substituted if the value has been substituted per IEC 61850-7-3. A true value of test indicates if the LogicalNode providing the information has a Beh value of test or test/blocked. A true value of operator blocked indicates that the logical node providing the information has a Beh value of blocked or test/blocked.

In the mappings to IEC 61850-8-1, quality values are encoded as an ASN.1 BITSTRING and a MMS Bitstring.

CHAPTER 10

Engineering

The engineering of an IEC 61850 system has a predefined workflow that starts with requirement specifications, selection of the appropriate IEDs to satisfy the requirements, and configuration of communication information. The communication information includes communication addressing, subscriptions, and signal distribution either through ExtRef, InRef, or BlkRef.

Starting with the system requirements avoids missing gaps that may lead to system integration failure. In many utilities in North America, there are IEDs on an approved vendor list that are the only devices the utility desires to use regardless of their ability to perform the IEC 61850 functions needed for the system. The requirements must be analyzed in order to have a successful integration.

Besides specification and configuration, the SCL process allows for upgrading and downgrading the various editions of IEC 61850 so that mixed systems are possible. Central to this capability are the new system configuration tools (SCTs) that implement the upgrade and downgrade rules. Some utilities are reluctant to upgrade from their Edition 1 SCT. In this case, it is not possible to engineer a mixed system.

There are multiple different SCL files that are exchanged between the tools. The files, of a given version of SCL, all utilize the same XML schema definition (XSD). This results in many optional XML elements or attributes that are needed for one function but not the others. Since the XSD is *generic* for all applications, and there are rules not defined in IEC 61850-6 and not expressible in XSD, XSD validation is not sufficient to validate an SCL file. The purpose of the files is provided in Table 10.1.

To support mixed version systems, the SCT function must support the upgrade and downgrade rules for the versions to be supported. This typically would mean that the SCT needs to support the latest version of SCL.

The tooling names have become a bit of a hybrid. There are SCTs that can produce specification files and configure devices. There are devices that can import a SCD and configure themselves (e.g., the configuration function is embedded within the device). What is consistent is the workflow as is shown in Figure 10.1.

The workflow will be broken down in the following sections. There is a question regarding the roles of the system configuration and IED configuration functions. The system configuration function is allowed to configure communication related information (addresses, subscriptions, Extref, BlkRef, etc.) it is not allowed to modify the model of the IED imported from the ICD (e.g., it can't add, delete,

Table 10.1 SCL File Types and Function

Purpose	*File Extension*	*Supplied By*	*Description*
System specification	SSD	System specification function (SSD)	Provides information regarding planned topology and required logical nodes (LNodes)
IED capability	ICD	Vendor/IED configuration function (ICT)	This file provides the generic capabilities of a specific IED.
System configuration	SCD	System configuration tool	This file contains the topology of the system, instantiated IEDs and logical nodes. The final SCD should have no remaining LNodes and have fully populated communications, subscriptions, and signal distribution definition.
Instantiated IED	IID	ICT	The resulting configuration information of a specific IED is based on the SCD and fix-ups.
System tool exchange	SED	System configuration tool	Provides the capability to take a complicated project, divide it into manageable parts, allow other engineers to work on the project, and then remerge the subprojects into the final configuration.
Additional files			
Configured IED	CID	ICT	Provides the information extracted from the SCD that is needed to configure a specific device.
IED Specification	ISD*	—	Provides a specification of capabilities required for a specific IED choice. It is intended for vendors or to be compared against vendor ICTs to determine the degree of match.
Configured IED	CID	ICT	Provides the information extracted from the SCD needed to configure a specific device.
IED specification	ISD*	—	Provides a specification of capabilities required for a specific IED choice. It is intended for vendors or to be compared against vendor ICTs to determine the degree of match.

*No official status in IEC 61850-6.

modify logical node definitions). The object model related activities are reserved for the IED configuration function. Most IED configuration functions also allow some types of communication changes. Thus, if the system configuration function contains the one truth of the system design, it must feed back its configured information in an IID. The difference between an IID and a CID is that an IID contains only configuration information related to the IED being configured and a CID contains IED information for all of the IEDs that the IED must subscribe to.

10.1 Workflow Specifics

The following sections describe the interaction of the various functions and files that are part of the SCL workflow.

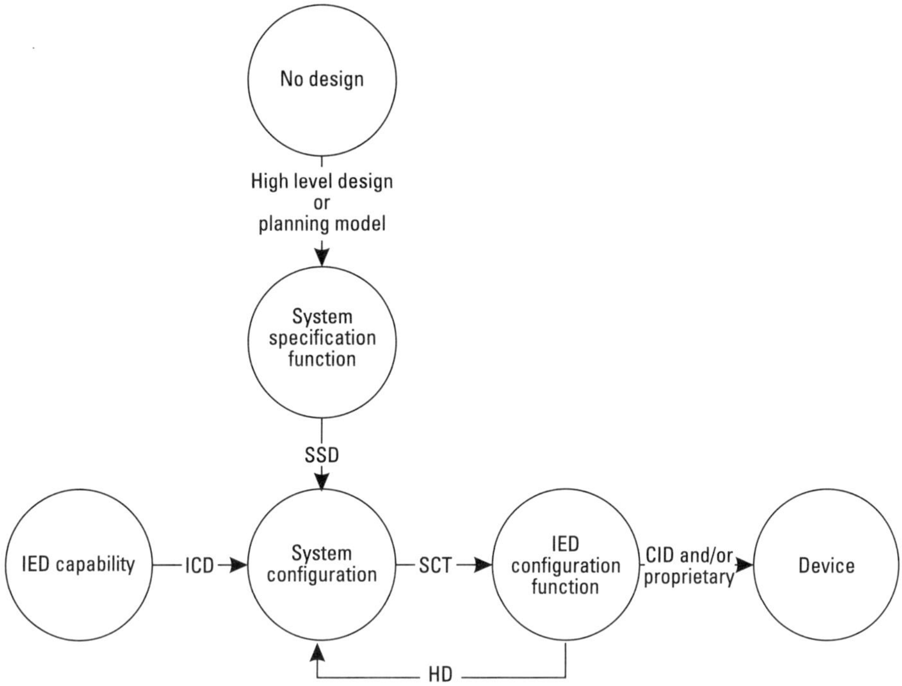

Figure 10.1 SCL workflow with no subprojects.

10.1.1 Specification Phase

A high-level design (HLD) or planning model might look similar to what is shown in Figure 10.2.

A planning model has no switches, breakers, protection devices, or anything related to the actual construction of the substation (bays). It has transformers, busses, voltage levels, and so forth. The first thing that a substation engineer/department must do is decide where the switches and circuit breakers belong.

Figure 10.3 shows the planning model where the substation engineer may decide to place circuit breakers. Based on this placement, the engineer can determine

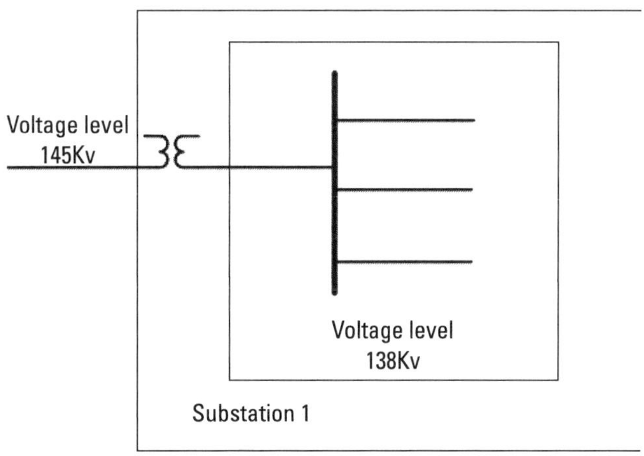

Figure 10.2 Planning model.

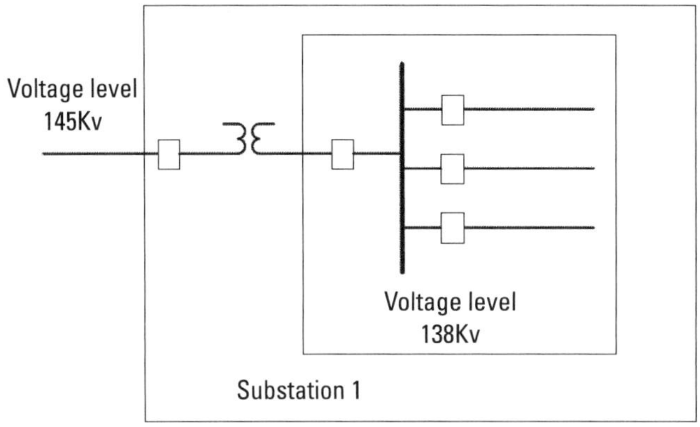

Figure 10.3 Extended planning model with circuit breakers.

the placement of CTs and PTs (this will be skipped in this example). Within the specification description, the engineer can choose which logical node definitions are to be associated to the particular primary equipment (e.g., switchgear, transformers) and to particular points in the topology (e.g., terminals). It might end up looking something like the scheme in Figure 10.4, shows that the engineer has chosen to have a CSWI/XCBR combination for each breaker in the diagram.

They have also chosen a metering function (MMTR) and measurement function (MMXU) for the ingress to the substation. There are also five additional MMXU requirements. The logical node requirements are known as LNodes and are not bound to a particular IED. They do have LNTypes that define the required data objects for the specific application. The topology, LNodes, and type definitions are all included in an SSD.

One of the reasons that the specification phase may be skipped is that a utility may have developed design/instantiations for specific types of substations and only cloning and fixups are required in those cases.

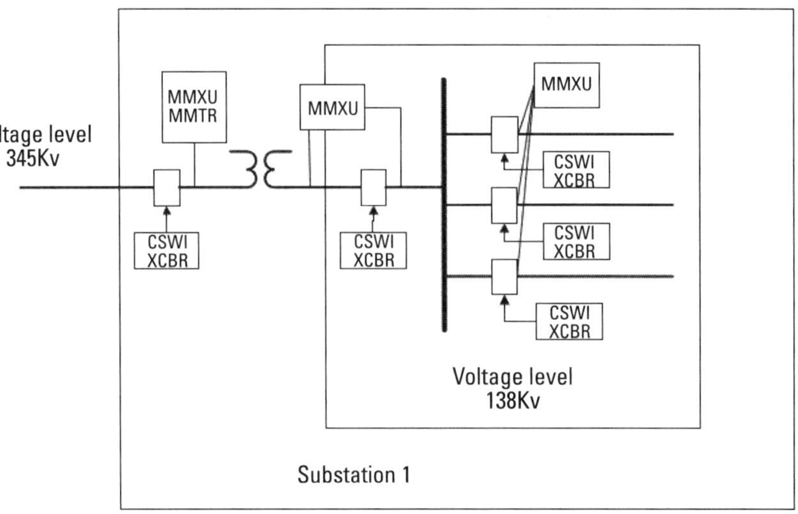

Figure 10.4 Specification with LN requirements.

10.1.2 Binding to IEDs

Once the requirements are specified, IEDs can be selected to satisfy those requirements. The selection process is through the importing a device's ICD, renaming the device, as all ICD IEDs have the name TEMPLATE. During the import, any upgrade rules are applied so that mixed mode systems can be supported. the logical nodes in the newly imported IED can be used to replace the LNodes in the specification file. This type of binding is shown in Figure 10.5. Once this occurs, the specification file is on its way to be an SCD. There may be several iterations to satisfy all the requirements, but it is helpful to think of this file as a partially completed SCD when executing that process.

Once the LNodes are bound to an IED, they become logical node instances and have all the associated naming hierarchy.

10.1.3 Information Exchange Requirements

Although this step can be done before or after actual communication configuration, the definitions of data sets, control blocks, subscriptions, and signal distribution can occur during the process.

10.1.4 Communication Configuration

The next step is typically to assign addressing information to the IEDs including the IP addresses, VLANs, multicast addresses, and more. This information is considered critical cyberinformation and is tightly controlled by IT and corporate policy.

10.1.5 Iteration and Export

As more IEDs are added, the system engineer and tool is used to manage these changes. When the system design is complete, the SCD is exported. For mixed edition systems, the SCT may need to export more than one version of the SCD (e.g.,

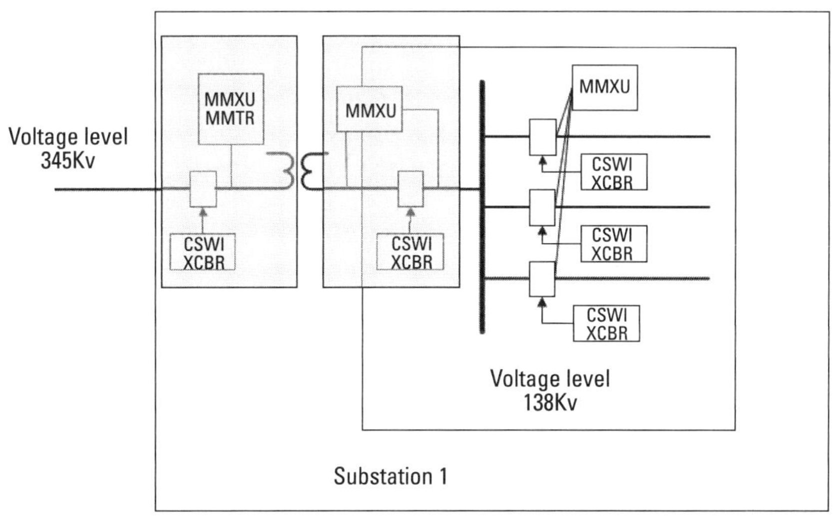

Figure 10.5 IED bindings.

one for Edition 1 and one for Edition 2) using the appropriate upgrade and downgrade rules as specified by the standard.

There is another opportunity that has arisen through the IEC effort to blend the processes of SCL and the Common Information Model (CIM – IEC 61970 and IEC 61968). Since the SCD has all the addressing information, topology, and measurement names this information could be used to populate information in energy management systems (EMS) and SCADA systems.

10.2 SCL Service Declarations

SCL is supposed to be utilized for configuration, but it also removes the need to go through the vendor documentation to find specific documentation. One issue was how to determine how many GOOSE subscriptions an IED can support. In Edition 1, this information was nowhere to be found since client capabilities (where subscription capabilities would be found) were not defined. In Edition 2, Booleans were introduced to provide an indication if GOOSE or SMV subscriptions are supported as along with the support of unbuffered reporting, buffered reporting, and logging. The Boolean is not enough information during the engineering process. The first amendment to Edition 2 (also known as Edition 2.1) started to correct this oversite. The following section gives a brief overview of what capabilities can be currently expressed.

There are two specific sections in SCL files where service capability declarations can be found.

The services, shown in Figure 10.6, may be found as part of an IED, in individual access points, or both. An IED may have different service capability depending on the utilization of an access point. As an example, there may be an access point utilized for client/server communications, a different one for GOOSE, and another for Sampled Values. In this example the access points would all have individual different service declarations. The ability to declare services at the IED level was the only mechanism allowed in Edition 1 of IEC 61850 and still may be used in Edition 2.

The actual SCL serialization of this UML does not have an XML tag for server capabilities, but does have a tag for client capabilities, as shown in Figure 10.7.

The tag for client capabilities is <ClientServices>.

Each service capability declaration is divided into client and server service declarations. Inquiring minds might want to know where GOOSE and Sampled Value service declarations occur. The answer is that they appear within client and server declaration sections.

10.2.1 Server Capabilities

The server service capabilities, shown in Figure 10.8, allow for an IED vendor within the ICD to declare the capabilities that the server aspect of the device supports. It allows declarations regarding reporting, Sampled Value, and GOOSE publishing, as well as network redundancy (e.g., RedProt) and the type of time synchronization supported.

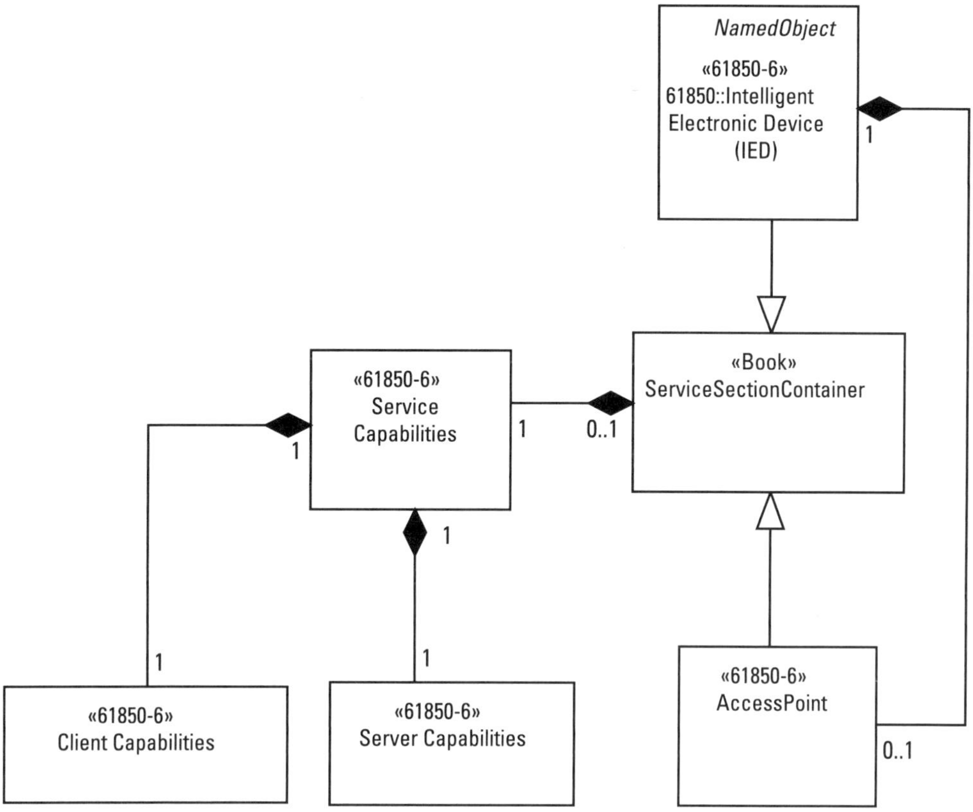

Figure 10.6 SCL service section UML.

```
<Services>
   <DynAssociation max="10" />
   <GetDirectory />
   <GetDataObjectDefinition />
   <GetDataSetValue />
   <DataSetDirectory />
   <ReadWrite />
   <GetCBValues />
   <ConfLNs fixPrefix="true" fixLnInst="true" />
   <GOOSE max="16" />
   <GSSE max="0" />
   <FileHandling />
   <DataObjectDirectory />
   <ConfDataSet max="8" maxAttributes="100" />
   <ConfReportControl max="40" />
   <GSESettings cbName="Conf" datSet="Conf" appID="Conf"
         dataLabel="Conf" />
   <SetDataSetValue />
   <ReportSettings cbName="Conf" datSet="Conf" rptID="Conf"
         optFields="Dyn" bufTime="Dyn" trgOps="Dyn" intgPd="Dyn" />
   <ClientServices sv="false" readLog="false" bufReport="false"
         goose="true" unbufReport="false"/>
</Services>
```

Figure 10.7 Example of service declaration in SCL.

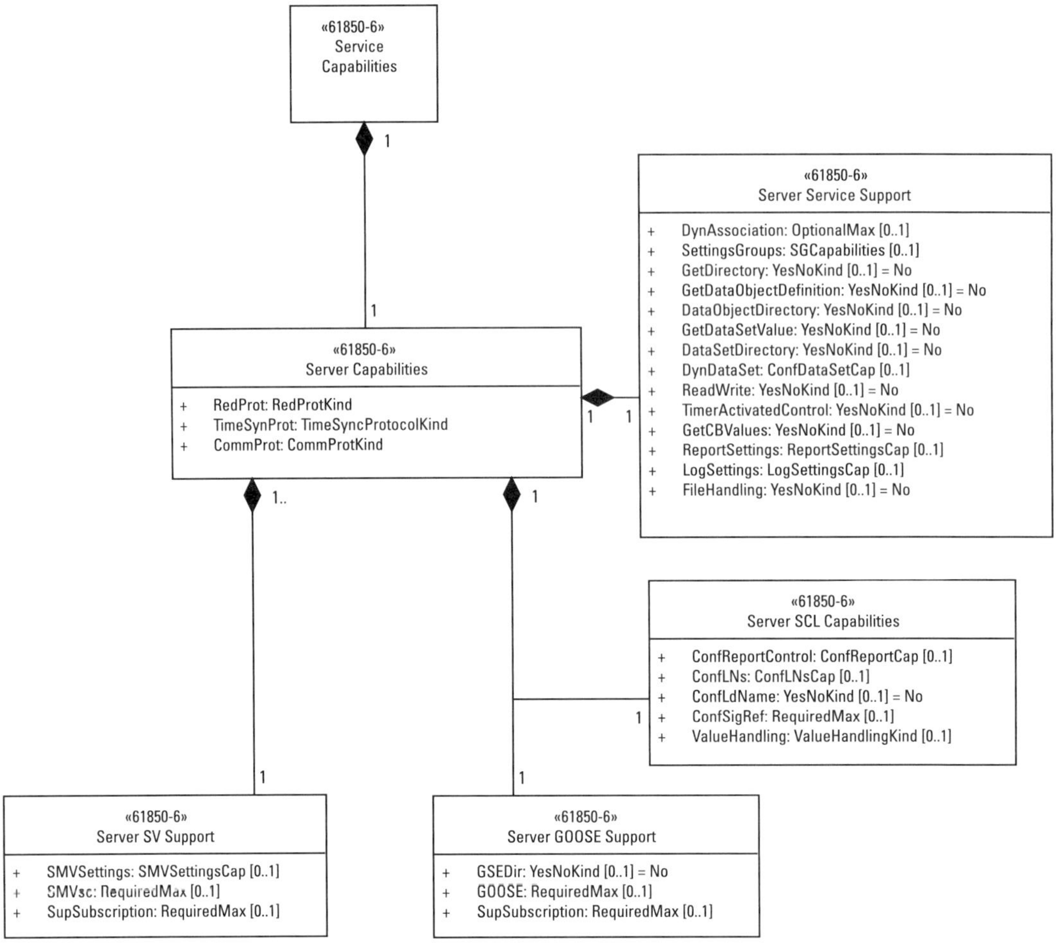

Figure 10.8 Server service capability UML.

10.2.1.1 Setting Group Capability

The server services section expresses if the edit and confirm services are supported, shown in Figure 10.9, and if there is a time reservation involved in the service interaction.

If the setting groups element, shown in Figure 10.10, is not present or has neither SGEdit or ConfSG values present, then setting groups are not supported by the

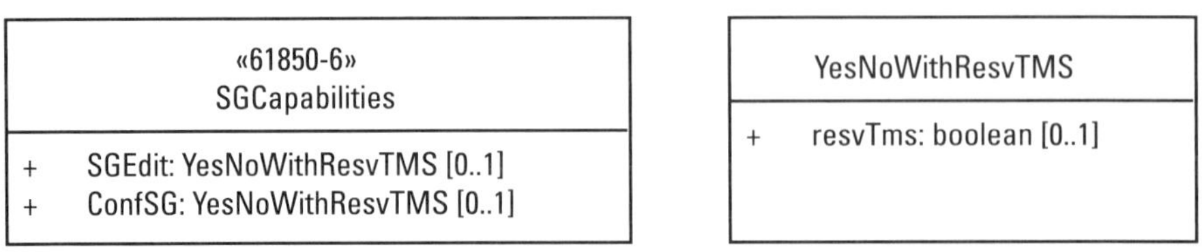

Figure 10.9 Server setting group capability UML.

```
<Services nameLength="64">
        <SettingGroups>
                <SGEdit resvTms="true" />
        </SettingGroups>
</Services>
```

Figure 10.10 Example of declaring support for setting groups.

server. If SGEdit is present, then there is the capability to select and edit a specific setting group.

If ConfSG is present, this indicates that a system configuration tool can configure the maximum number of setting groups that are to be used by the server. The rule is that the configured value must be less than or equal to the value of the numOfSGs attribute (e.g., 3 in Figure 10.11) provided by the ICD of the device.

However, the current definition requires that the SCT remember the value imported from the ICD. This requirement may impact exchanges of SCDs and/or SEDs in the future between SCL tools. Therefore, a max parameter may be added to the ConfSG declaration in the future.

When the server capabilities include setting groups, Figure 10.11 might be a potential declaration of that capability.

10.2.2 Client Capabilities

The client capability, shown in Figure 10.12, now supports both the Booleans from Edition 2 and numeric maximums. The XSD does not validate that a Boolean is set in order to have the numeric maximum. Therefore, this check needs to be performed outside of normal XSD validation.

```
<IED name="IED1>
  ....
   <Server>
     <LDevice id="FCD01">
         <LN0 lnClass="LLN0" inst="" lnType="LNN0_Type" desc="General">
   .

          <SettingControl numOfSGs="3" actSG="1" >
      </LN0>
         .....
      </LDevice>
   <Server>
</IED>
```

Figure 10.11 Example of declaring the maximum number of setting groups in ICD.

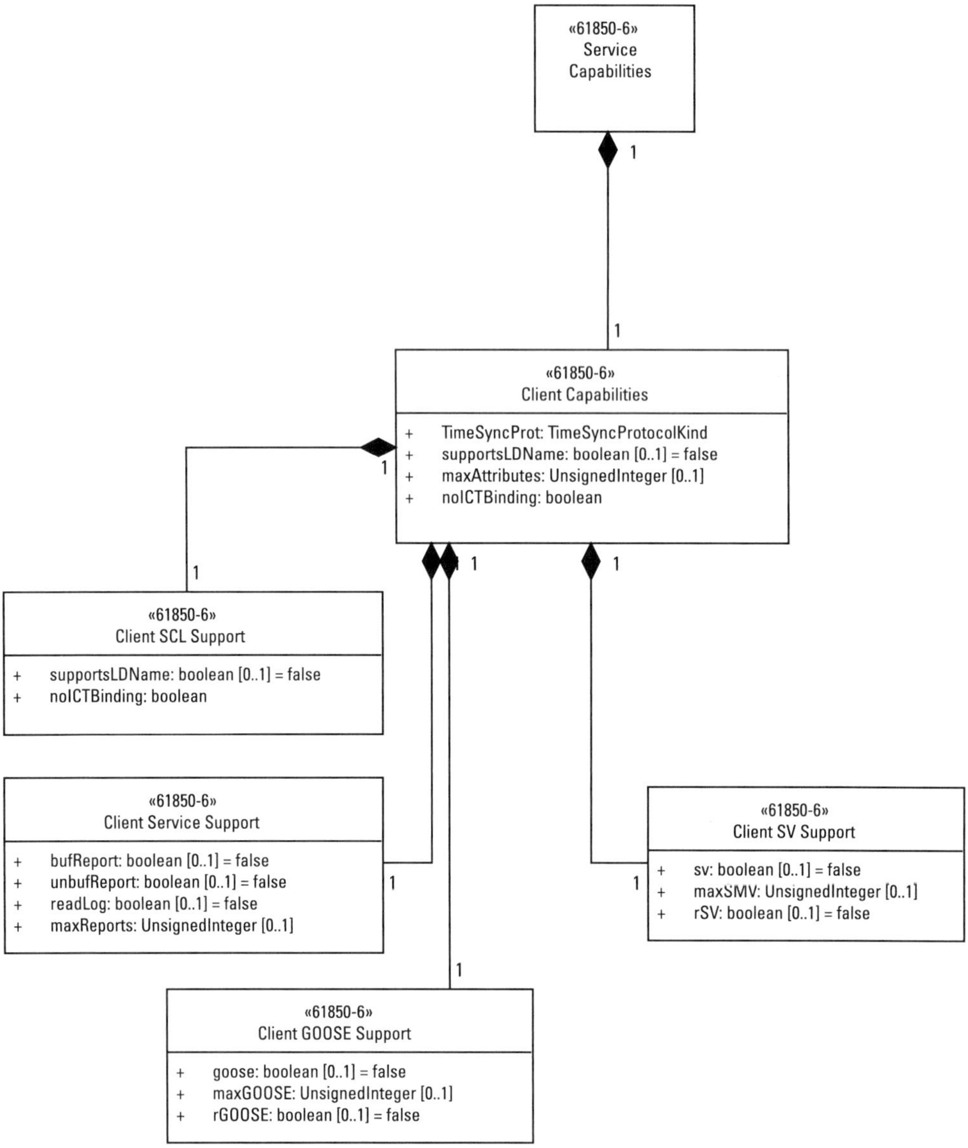

Figure 10.12 Client service capability UML.

CHAPTER 11

Client and Server Communications

Previous chapters have detailed the mappings of the abstract concepts, configuration information, and instantiation of that information in MMS. The following sections discuss how MMS is conveyed over the network.

11.1 History of IEC 61850 Client/Server

During the development of MAP/TOP, General Motors had originally developed two protocols: GMMFS and MMFS. These specifications followed the specification pattern of the normal manufacturing protocols circa 1982 through 1985. At that time, many of the prevalent protocols were specifications similar to Modbus. These protocols could best be described as a mixture of what we currently would classify as nouns and verbs. For example, Modbus defines a function code 01 whose semantic is Read Coil Status. In modern terms, Read is the verb and Coil Status is the noun. The function code utilized also caused a response. However, the concept of request and response were not explicitly discussed.

MMFS explicitly modeled and specified request and response sequences that had an enveloping request or response function code. It also attempted to define all aspects of functions for the various classes of devices that needed to be integrated within a particular manufacturing plant (e.g., PLCs, robots, CNCs). Additional function codes and messages were defined for each class of device. This provided a large dictionary of messages that allowed integration of the various devices. However, if one considered the requirement of a PLC to communicate with a robot, it meant that the PLC would need to speak the PLC and Robot set of messages complicating integration and the development of generic products. Not only did MMSF specify the messaging format, it also specified the encoding of the messages for transmission (e.g., the OSI Presentation function).

However, MAP also specified the use of FTAM. FTAM was different from the manufacturing protocols. It had a document that specified its services and a separate document which specified the protocol. The services document specified what parameters and values were needed (optional and mandatory) to be provided to accomplish work or generate a message. The document did not specify the syntax, encoding, or how to transmit the information. The protocol document specified the actual protocol in a new format (BNF). BNF was part of the emerging Abstract Syntax Notation 1 (ASN.1) standard that separated the message syntax from the encoding of the syntax. The protocol document also had an abstract mapping of the OSI Presentation Interface. This specified methods and parameters that were to be provided to and received from the layer. It was an abstract interface so that FTAM was independent of the presentation protocol used. This is the concept of OSI reference model layering.

Circa 1984, the MAP initiative decided to address the need to simplify the messaging tasks between the various devices, embrace the ASN.1 services versus protocol specifications, and concepts of layering. This started the initiative to develop MMS, which is now ISO 9506. Originally, it was thought that changing MMFS to the paradigm would only require 6 months of concerted effort. Three to four years later, the draft specification was ready for the world.

All these new concepts have provided the flexibility needed in the utility industry as part of IEC 61850.

11.2 IEC 61850 Client/Server Over the Wire

Back in the 1970s and 1980s, there were two competing Transport profiles. One was based on ISO Transport and Networking (ITU x.214/x.234[1] and x.213/x.233) the other was based on the IETF TCP (RFC 793 and RFC 791). Although TCP/IP won the transport wars, at the time this was the technical equivalent of the VHS video tape format winning the acceptance of the industry instead of the Beta format. Beta was theoretically technically superior according to video files, but the lower cost VHS won the industry. One of the innovations that allowed the OSI packet oriented upper layers (e.g., required for IEC 61850 and MMS) to adapt to the TCP streaming paradigm was the creation of RFC 1006. This RFC is integral in making TCP/IP appear as a packet oriented OSI Network Layer. Yes, within the IEC 61850 context, TCP is in the network layer from a modeling perspective.

Since 2003, the IEC 61850 Client/Server protocol MMS utilizing OSI upper layers over TCP/IP. This is the profile as specified by IEC 61850-8-1 and is shown in Figure 11.1.

The wave of Distributed Energy Resource (DER) incursion into the electrical grid has impacted IEC 61850 from a modeling and communication profile perspective. Circa 2014, work began in IEC TC57 WG17 to determine the appropriate application layer protocol and other communication specifications that could be utilized to support DER generation and systems. The intent was to use the more modern technology of Web Services. After much discussion, the direction moved away from Web Services to the utilization of Web Technology. There was a decision to align the transport with that chosen by IEC TC57 WG21. The choice was the Extensible Message and Presence Protocol (XMPP). On evaluation of application protocols, MMS was once again chosen even though politically it is not referred to as MMS. The IEC 61850-8-2 profile follows in Figure 11.2.

XMPP is a store and forward intermediary technology. It is a switchboard where both the IEC 61850 Client and IEC 61850 Server are XMPP Clients. The XMPP Clients must connect to the XMPP Server prior to communications being allowed between the IEC 61850 Client, including association establishment. The

1. The ITU specifications are being given since they can be downloaded for free.

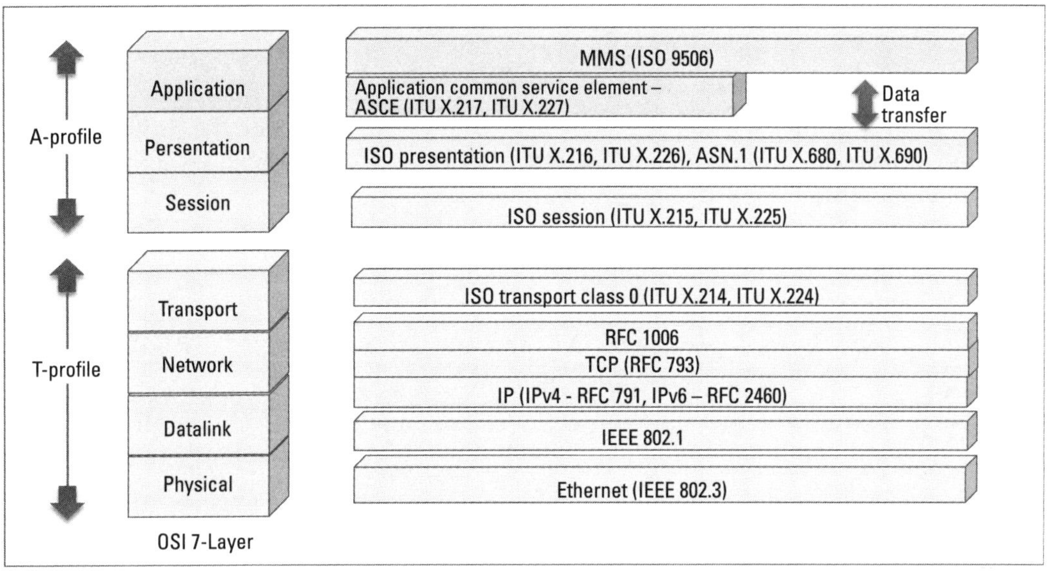

Figure 11.1 Client/Server Communication Profile for IEC 61850-8-1.

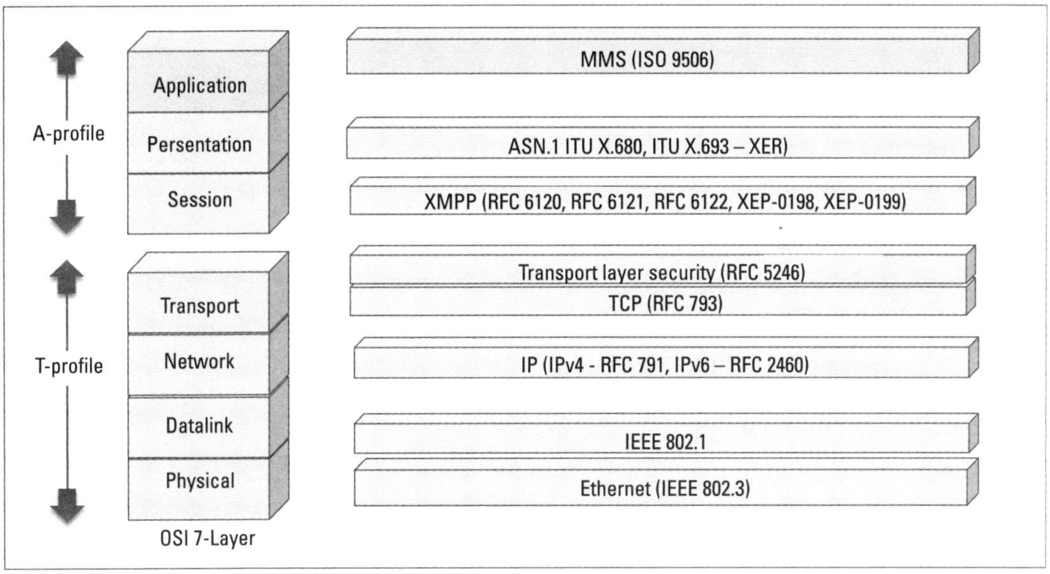

Figure 11.2 Client/server communication profile for IEC 61850-8-2.

use of this intermediate switchboard, and the termination of the TLS protection, gave rise to a legitimate concern regarding security and a man-in-the-middle threat. This caused an end-to-end security protocol to be developed and utilized in both IEC 61850-8-1 and IEC 61850-8-2 through its incorporation into IEC 62351-4.

A comparison of the application and presentation layers, between the two profiles, shows that MMS and ASN.1 are common. IEC 61850-8-1 utilizes ASN.1 BER whereas IEC 61850-8-2 utilizes XML Encoding Rules (XER). It is the fact that ASN.1 supported XML encoding that allowed MMS to be utilized directly as a Web Technology in IEC 61850-8-2.

11.3 ASN.1

Abstract Syntax Notation 1 (ASN.1) is a set of standards that dictate the representation of protocol syntax and then defines how to encode the protocol syntax for transmission between entities. Besides MMS, ASN.1 is used in other protocols besides MMS. The list includes, but is not limited to

- *X.400 Message Handling System:* This protocol is typically used for enterprise level email exchanges.
- *X.500 and Lightweight Directory Access Protocol (LDAP):* Provides addressing and directory management.
- *H.323 Voice over IP(VoIP):* Provides the standardized VoIP capability that allows us to talk, via voice, in conference calls and other conferencing technologies.
- *Kerberos:* A protocol used to provide strong cyber authentication of communicating entities.
- *BACnet:* A protocol used for building automation.
- *Simple Network Management Protocol (SNMP):* Used by most IT organizations to monitor networks, network equipment, and computing resources.

ASN.1 is a set of specifications available from ISO and the ITU. The 8824/X68x series of standards details how to express a protocol in Backus-Naur format. The 8825/X.69x series specifies how to encode the protocol. The relevant specifications, for MMS are

- *ISO/IEC 8824-1:* Information technology—ASN.1: Specification of basic notation. This standard is also available from the ITU as X.680 free of charge.[2]
- *ISO/IEC 8825-1:* Information technology—ASN.1 encoding rules: Specification of Basic Encoding Rules (BER), Canonical Encoding Rules (CER), and Distinguished Encoding Rules (DER). This standard is also available from the ITU as X.690 free of charge.[3]
- *ISO/IEC 8825-4:* Information technology—ASN.1 encoding rules: XML Encoding Rules (XER). This standard is also available from the ITU as X.693 free of charge.[4]

Besides the importance of separating protocol definition from encoding, the encoding rules found in ISO/IEC 8825-1 solve the big-endian versus little-endian problem in exchanges. The problem of big and little endian surfaced due to the two prevalent CPU architectures: Intel and Motorola. For example, the integer value of 129 would be represented by: 0xFF 0x01 for an Intel CPU and 0x01 0xFF for a Motorola CPU. If CPUs attempted to exchange their local representation of a value, Intel and Motorola CPUs would not be able to exchange information.

2. Available at: http://www.itu.int/rec/T-REC-X.680-201508-I/en.
3. Available at: https://www.itu.int/rec/T-REC-X.690/en.
4. Available at: https://www.itu.int/rec/T-REC-X.690/en.

Therefore, a neutral transfer encoding is required. It is this definition that ISO/IEC 8825 provides.

IEC 61850 utilizes the following aspects from ASN.1:

- *ISO/IEC 8824-1:* This standard is used to express the protocols found in ISO/IEC 9506, IEC 61850-8-1 GOOSE and GSE Management, and IEC 61850-9-2 Sample Values.
- *ISO/IEC 8825-1 Basic Encoding Rules (BER):* This standard is used to encode the protocols found in ISO/IEC 9506, IEC 61850-8-1 GOOSE and GSE Management, and IEC 61850-9-2 Sample Values.
- *ISO/IEC 8825-4 Extended XML Encoding Rules:* This standard is used to encode ISO/IEC 9506 in XML format for use in IEC 61850-8-2. To make an explicit XSD that solved certain syntactical issues in IEC 61850-8-2, the BNF used for MMS is slightly different between IEC 61850-8-1 and IEC 61850-8-2.

11.3.1 Protocol Definition Syntax (Bakus-Naur Form Notation)

IEC 61850 and MMS do not make use of all the syntactical options, or basic types, defined in ISO/IEC 8824-1. The following sections discuss the set of syntactical definitions used by ASN.1. The syntax is typically referred to as BNF notation. The syntax of BNF allows for protocols to be expressed without defining the actual bits/byte encoding over the wire.

The following sections delve into the various BNF syntactical definitions that are used to express MMS, GOOSE, and Sampled Value protocol definitions.

11.3.1.1 Basic ASN.1 Types

ASN.1 has a set of primitive data types. Most of these are recognizable via name. However, the precision or length of the values is not specified as part of the ASN.1 basic types. This is due in part that the lengths in ISO/IEC 8825 are expressed explicitly. For example, for BER, ASN.1 is a TLV (tag, length value). In XER, the use of XML tags defines the length.

The basic types used from ASN.1 are shown in Table 11.1.

11.3.1.2 Complex Definitions, Directives, and Rules

ASN.1 has several definitions that represent definitions of multiple basic types or complex definitions (see Table 11.2).

There are additional directives that have an impact on the encoding of the protocol (see Table 11.3).

When constructing a protocol syntax using ASN.1, there are additional rules that apply:

- Words that are all capital letters (e.g., INTEGER) are reserved for definitions found in the ASN.1 specification.

Table 11.1 Basic ASN.1 Data Types

Basic ASN.1 Name	*Description*

BIT STRING	A sequence of binary digits (true and false). In ASN.1 each bit in a BITSTRING is designated by its bit position in the value. As an example if one needed to represent information like Leap Second Known (bit 0) and Time SynchOk (bit 1) might look like this:
	`BISTRING {`
	`leapSecondKnown (0),`
	`timeSynchOK (1)`
	`}`

BOOLEAN	Used to express "true" or "false."

INTEGER	Used to express values that are whole numbers. ASN.1 does not differentiate positive or negative numbers from a syntactical perspective. In the syntax, if no explicit values are defined, then it represents a range of values whose meaning is numeric. If there are explicit values, those are noted. As an example, value 0 = Martin, value 2 = the, and value 3=Marian would be represented as
	`INTEGER {`
	`Martin (0),`
	`the (1),`
	`Martian (2)`
	`}`

NULL	Indicates that there is no value.
OBJECT IDENTIFIER	Represents a sequence of integer values that represent a distinguished identifier for an object.
OCTET STRING	The value is a sequence of octets. The name of octet is used to indicate an 8-bit value.
VISIBLE STRING	A sequence of octets that have a constrained range to be the set of values that are visible from ASCII (e.g., ISO 646). This means no values less than 0x21, or other non-visible values may not be used.

Table 11.2 BNF Directives for Complex Grouping of Parameters

Complex ASN.1	*Description*
CHOICE	Indicates that the following production contains at least one "or."
SEQUENCE	Indicates that the following production is constructed from other ASN.1 constructs.
SEQUENCE OF	Indicates that the following production is a SEQUENCE that repeats (e.g., an array).

- Words that are camel-case, starting with a capital letter, indicate that the definition is found elsewhere in the protocol definition.
- Words that are camel-case, starting with a lower-case letter, indicate that the definition follows immediately.

An example from MMS follows:

```
ReadJournal-Response ::= SEQUENCE
  {
  listOfJournalEntry    [0] IMPLICIT SEQUENCE OF JournalEntry,
  moreFollows           [1] IMPLICIT BOOLEAN DEFAULT FALSE
  }
```

Table 11.3 Additional BNF Directives

Directives	*Description*
DEFAULT	Specifies a value that is to be assumed if the production is not encoded.
IMPLICIT	Indicates that the following production or definition is context-specific.
APPLICATION	Indicates that the following production or definition is specified as used within a particular application and is similar to the IMPLICIT.
OPTIONAL	Indicates that the production need not be encoded.
TRUE	Used to specify a value of "true."
FALSE	Used to specify a value of "false."

An analysis of the example shows that all information in a production must be uniquely tagged (e.g., [0] and [1]). This example also shows that the ReadJournal-Response contains a definition for JournalEntry that is defined elsewhere. Since listOfJournalEntry is a SEQUENCE OF, this indicates that an array of JournalEntry values may be encoded. It also shows that context specific definitions (e.g., IMPLICIT) are being used. The production attribute of moreFollows indicates that if it is not encoded, the value of FALSE is assumed due to the DEFAULT directive.

11.3.2 Encoding Rules

IEC 61850 currently utilizes two of the encoding rule sets specified to be part of the ISO/ITU standard set. These are BER and XER. The following sections discuss the actual various encoding mechanisms used by IEC 61850 in a simplified manner. The following sections do not reflect the documentation of all the encoding rules of either format.

11.3.2.1 Basic Encoding Rules (BER)

BER is typically referred to as a Tag, Length, Value (TLV) encoding. Some rules for encoding the length are

- Lengths may never have the upper bit of the length set. This means for a single byte length the maximum value is 127.
- Lengths greater than 127 will be encoded with an additional byte which represents the length-of-the-length. In the example of a length of 128, the length-of-the-length would be two with the actual length following in the next two bytes (e.g., 02 00 80 hexadecimal).

The encoding of the Tag value can best be represented by the following cheat sheet (Table 11.4). It shows the encoding of a single tag that fits in a single byte.

The cheat sheet is divided into three major columns. The first two columns show the value of the various bit(s) depending on the keywords found in the BNF. The last column shows an example keyword and the resulting value which combines bits 7–5 with the actual tag to create the value of the actual encoded tag. For UNIVERSAL tags, ASN.1 assigns values shown in Table 11.5.

As with many rule sets, there are always exceptions. In BER, the encoding of BITSTRINGs is special. A BITSTRING is encoded with the following fields: tag,

Table 11.4 ASN.1 BER Encoding Cheat Sheet

Bits 7,6		*Bit 5*		*Tag*	
Value	*Keyword*	*Value*	*Keywords*	*Example*	*Tag Value (Hex)*
0 0	UNIVERSAL	0		INTEGER	02
0 0	UNIVERSAL	1	SEQUENCE, SEQUENCE OF	SEQUENCE	30
0 1	APPLICATION	0		[1] APPLCIATION	41
1 0	IMPLICIT	0		[1] IMPLICIT INTEGER	81
1 0	IMPLICIT	1	SEQUENCE, SEQUENCE OF	[1] IMPLICIT SEQUENCE	A1
1 1	PRIVATE	Not used within IEC 61850			

Table 11.5 Universal ANS.1 Values Used by IEC 61850

Keyword	*Value (Hex)*
BOOLEAN	01
INTEGER	02
BITSTRING	03
OCTETSTRING	04
NULL	05
OBJECT IDENTIFIER	06
SEQUENCE	10
IA5String	16
UTCTIME	17
GENERALIZEDTIME	18
VISIBLESTRING	1A

length, number of unused bits, and the value. This allows for variable BITSTRING encodings which when decoded, unrecognized bits (e.g., decoder is expecting 10 bits but receives 13) are able to differentiate between used and unused bits (e.g., there is an ability to send a 9 bit BITSTRING value). This would encode as 03 02 07 11 80. Bit 0 of a BITSTRING is the most significant bit (MSB) of the BITSTRING value.

11.3.2.2 XML Encoding Rules (XER)

XER encoding has much simpler rules than BER. This is largely because the intent is to develop an XSD that can then be used to serialize the BNF payload. Figure 11.3 uses an example of the identify-response to demonstrate the serialization of BNF into an XSD syntax.

As can be seen in Figure 11.3, the name of the attribute (e.g., modelName) becomes the name of an xsd:Element, The use of IMPLICIT has no bearing in XER. The type of VisibleString could have been an xsd:string, but IEC 61850 defines constraints on that type. In general, it is easier to understand the encoding of BNF to XML via XER than BER encoding.

```
Identify-Response ::= SEQUENCE {
        vendorName              [0] IMPLICIT VisibleString,
        modelName               [1] IMPLICIT VisibleString,
        revision                [2] IMPLICIT VisibleString,
        listOfAbstractSyntaxes  [3] IMPLICIT SEQUENCE OF OBJECT IDENTIFIER OPTIONAL
        }
```

BNF to XSD via XER

```
<xsd:complexType name="Identify-Response">
<xsd:sequence>
    <xsd:element name="vendorName" type="VisibleString"/>
    <xsd:element name="modelName" type="VisibleString"/>
    <xsd:element name="revision" type="VisibleString"/>
    <xsd:element name="listOfAbstractSyntaxes" minOccurs="0">
    <xsd:complexType>
        <xsd:sequence minOccurs="0" maxOccurs="unbounded">
        <xsd:element name="OBJECT_IDENTIFIER" type="OBJECT_IDENTIFIER"/>
        </xsd:sequence>
    </xsd:complexType>
    </xsd:element>
</xsd:sequence>
</xsd:complexType>
```

XSD to XML serialization

```
<Identify-Response>
  <vendorName> </vendorName>
  <modelName> </modelName>
  <revision>   </revision>
</Identify-Response>
```

Figure 11.3 Example of BNF, XER, and XML serialization.

CHAPTER 12

Impact of Cybersecurity

The world we live in has so many examples of why cybersecurity is important. From the successful cyberattack that yielded the personal information of 143 million individuals,[1] the calls that make you believe that Microsoft has detected that your computer is infected, to the power outage of the Ukrainian electric grid,[2] there are bad people (hackers) in the world that want to do bad things. It is also a truism that there is nothing absolute in cybersecurity except threat vectors, technology changes rapidly, and eventually a determined and well-funded hacker will most likely be successful. It is also true that there is a need for cybersecurity-focused people in the utility industry.

When people think of cybersecurity, they think of technology. That is the wrong place to start as technology can only provide part of a solution. Coopting Smokey the Bear's slogan,[3] "Only you can prevent successful cyberattacks." Cybersecurity starts with educationon existing threats, how to spot and report an attack, policy, and the resources required to respond to an attack. In the event of a successful attack, there needs to be a recovery plan. This is all common sense and has little to do with protocol security.

The first attempt to implement cybersecurity was in 1993 as part of a UCA initiative. This was before the work on IEC 61850 began in earnest. When the cybersecurity specification was released, only one company implemented it and none of the implementations were deployed. At that time, cybersecurity was a passing thought, and many thought it would never be needed. How times have changed.

There have been many examples of wide ranging power outages in the United States that could have been caused by successful cyberthreats but were not. However, enough concern was raised over the vulnerability of the grid to cybersecurity threats that the NERC was tasked by the federal government to provide governance and enforcement of a cybersecurity policy. The result is an initiative known as the NERC CIP.[4] Other regional reliability entities on other continents have similar programs and enforcement. The EU has similar guidances.[5]

---

1. See https://www.washingtonpost.com/business/technology/equifax-hack-hits-credit-histories-of-up-to-143-million-americans/2017/09/07/a4ae6f82-941a-11e7-b9bc-b2f7903bab0d_story.html?noredirect=on&utm_term=.8fc145a3bfb1.
2. See https://en.wikipedia.org/wiki/December_2015_Ukraine_power_grid_cyberattack.
3. See https://smokeybear.com/en/smokeys-history/story-of-smokey.
4. For the actual standards, see https://www.nerc.com/pa/Stand/Pages/CIPStandards.aspx.
5. See http://eur-lex.europa.eu/legal-content/EN/TXT/?uri=LEGISSUM:l33260 and https://iecetech.org/Technology-Focus/2017-07/Cyber-security-for-the-modern-grid for further information.

Cyberthreats continue to evolve, as does the guidance that NERC CIP provides for the Bulk Electric System (BES). The current NERC CIP guidance is version 5 and provides mostly nonprescriptive guidance regarding cybersecurity for utilities that operate transmission assets that are voltage levels of 138 kV or greater; shed loads that are 300 MW or greater; or have generation assets that are 300 MW or greater. The NERC documents are listed in Table 12.1.

Figure 12.1 shows the relationship of the IEC 62351 and IEC 61850 pertaining to security.

The following sections discuss the impact of the NERC guidance and the technology provided by IEC 61850 and IEC 62351.

12.1 SCL

Some of the SCL files contain critical system information. These files are the SSD, SCD, IID, and CID. Based on the requirements of CIP-011-5, these files need to be protected in transit and at rest. The requirements for protection involve any files that contain operational communication addresses (e.g., addresses used to

Table 12.1 NERC CIP Version 5 Documents

Document	*Name*	*Description*
CIP-002-5	BES Cyber System Categorization	Provides the definition of asset and system criticality pertaining to communicating entities in BES.
CIP-003-5	Security Management Control	"To specify consistent and sustainable security management controls that establish responsibility and accountability to protect BES Cyber Systems against compromise that could lead to misoperation or instability in the BES" (from NERC).
CIP-004-5	Personnel and Training	Provides the guidance pertaining to training and education.
CIP-005-5	Electronic Security Perimeters	Provides guidance pertaining to the requirements for communication monitoring and access control that ingress into and egress from BES assets in an ESP. The treatment of serial or LAN/WAN connectivity is the same based on the asset classification.
CIP-006-5	Physical Security of BES Cyber Systems	Provides guidance as to the requirements for monitoring and access control that ingress into a physical perimeter that contains BES cyber assets.
CIP-007-5	System Security Management	Provides guidance regarding the management of security for BES cyber assets and systems. This includes the requirements for virus scanning and malware detection.
CIP-008-5	Incident Reporting and Response Planning	Provides guidance regarding creating an infrastructure (e.g., people) that are responsible to handle reports of attacks and how to respond to those attacks.
CIP-009-5	Recovery Plans for BES Cyber Systems	Provides guidance on how to reestablish grid reliability in the event of a successful cyberattack.
CIP-010-5	Configuration Change Management and Vulnerability Assessments	"To prevent and detect unauthorized changes to BES Cyber Systems by specifying configuration change management and vulnerability assessment requirements in support of protecting BES Cyber Systems from compromise that could lead to misoperation or instability in the Bulk Electric System (BES)" (from NERC).
CIP-011-5	Information Protection	Provides guidance and classification of information that needs to be protected as part of NERC CIP. Critical information includes but is not limited to communication addresses, grid topology, and device settings.

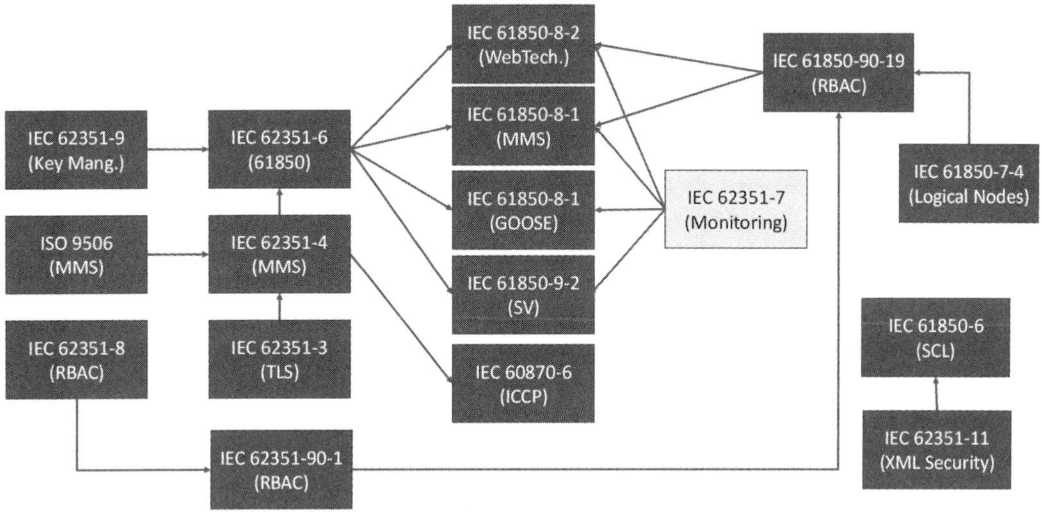

Figure 12.1 Relationship of security standards and IEC 61850.

communicate with assets that are in the field) and grid/system topology. When these files are at rest, may also be governed by IEEE *Guide to Cyber Security for Protection Related Data Files*. Looking forward, not only do the files need to be encrypted, but access control and access audit records will probably be required. Designing those additional capabilities today may prevent problems in the future.

IEC 62351-11 provides some assistance in protecting the contents when files are in transit by specifying specific mechanisms to utilize the W3C recommendations for XML Encryption Syntax and Processing and XML Signature Syntax and Processing. These W3C specifications have many capabilities that IEC 62351-7 does not utilize. Therefore, it is best to think about IEC 62351-11 as an implementation profile of the specifications. The intended use of the IEC standard is to encapsulate, sign, and potentially encrypt, XML files in their entirety such that the wrappers can be removed and normal tooling (e.g., SCL parsing) can be performed.

Figure 12.2 shows the encapsulation per IEC 62351-7. Public/private key technology is used for the creation of the Signature (mandatory) and in encrypting (optional) the noted information. The nonce is a cyber-related parameter that is a random value that protects file transfers from being spoofed. The "Data in Transition" section of the format is intended to be utilized with Common Information Model (CIM) XML files.

12.2 61850 Application Role-Based Access Control

Since cybersecurity requires defense in depth, IEC has adopted the philosophy that the IED/61850 application is the last line of defense. Since CIP-003-5 and CIP-005-5 basically require access control, IEC 61850 (at the time of publication) is working on IEC TR 61850-90-19 which specifies how role-based access control (RBAC) should be configured within a 61850 environment. To date, it has been

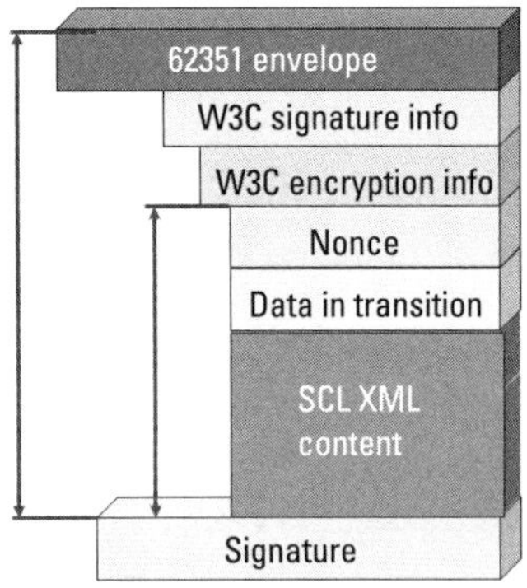

Figure 12.2 Secure XML encapsulation as specified by IEC 62351-7.

decided that the security configuration will need to be configured and controlled by a different entity than the engineer that designs the substation. This is due to the segregation of responsibilities and roles. Thus, the RBAC configuration will be in a different file created by a security configuration tool. The RBAC configuration will reference objects within the SCL file.

The concepts within IEC TR 61850-90-19 are extensions/specifications of IEC TS 62351-8 (access control), IEC 62351-9 (key and certificate management), and IEC TR 62351-90-1 (Guidance in application of 62351-8). IEC 61850 will utilize digital certificates, including attribute certificates, to provide identity and roles of connecting entities. Since X.509 certificates are being utilized, use of a PKI infrastructure is a side effect of this choice. This includes the inclusion of utilizing LDAP. One of the differences between IT RBAC and that being worked on within IEC 61850 is the fact that grid operations and reliability is more important than cybersecurity. This introduced the concept of operation constraints into the RBAC equation.

Figure 12.3 shows the standardized roles, operations, and permissions defined for IEC 61850 within IEC TS 62351-8. The configuration of RBAC is intended to allow emergency situations (e.g., fire, earthquake, etc.) to change the levels of access control. The exact mechanism to provide this capability is not complete at the time of publication.

Note: For those not familiar with areas of responsibility (AOR) this is a concept that certain individuals may only have access to assets at certain geographical locations. IEC 61850 intends to extend this concept to include an element of time (e.g., shifts).

12.3 Protocol Related

IEC 61850 protocol-related security provides counters to the threats of spoof, replay, eavesdropping, and message integrity. Although the cyber countertechnology differs between client/server, GOOSE, and SV, they all help satisfy the requirements in CIP-005-5.

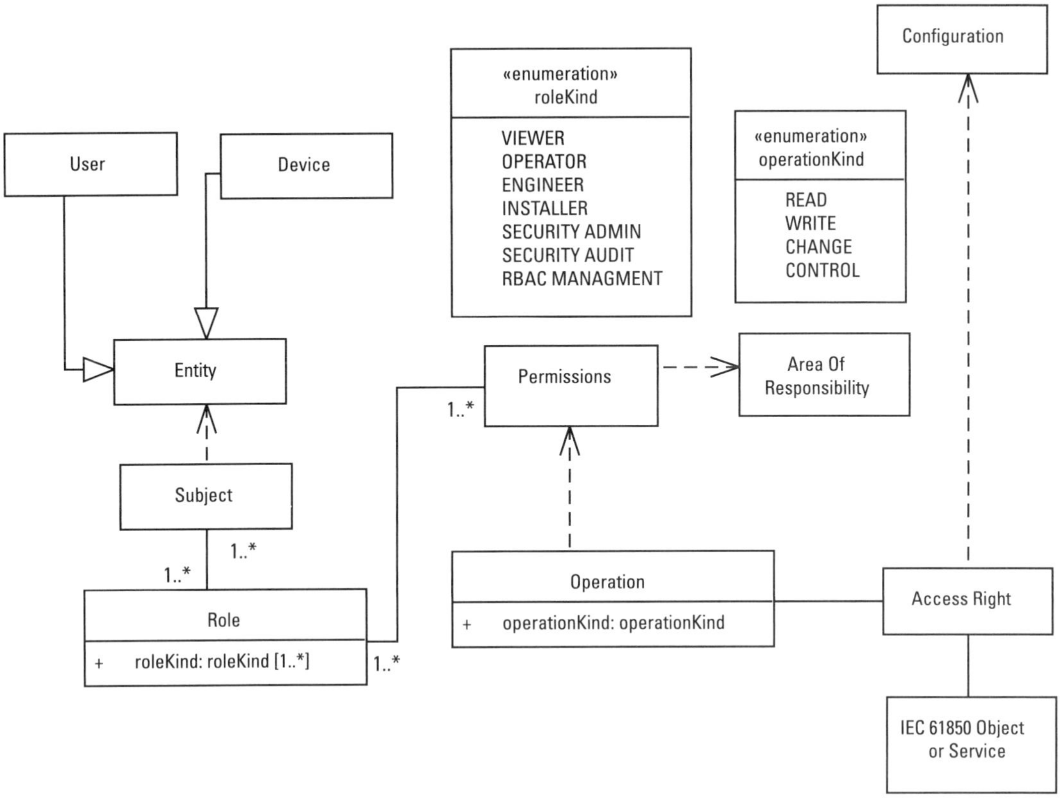

Figure 12.3 RBAC and IEC 61850.

12.3.1 Client/Server

The original security for IEC 61850-8-1 client/server services was 100% aligned with the security of the Inter-Control Center Protocol (ICCP) (IEC 60870-6 TASE.2). It used Transport Layer Security (TLS) to provide message integrity and confidentiality. Mutual authentication of the client and server is provided at the connection establishment time utilizing an exchange of X.509 certificates and digital signing.

Sometimes there are different objectives to different standards and standard bodies. As an example, IEC 61850 security needs to utilize the least vulnerable technologies. However, IEC 62351-3 mandates the use of TLS 1.0, TLS 1.1, and TLS 1.2. There are known vulnerabilities in TLS 1.0 and TLS 1.1 and the internet is in the process of deprecating their use. IEC 62351-6, security for IEC 61850, specifies the use of TLS 1.2.

The latest version of TLS is TLS 1.3 where the IETF decided privacy was more important than message integrity and tamper protection. Therefore, the IETF deprecated the use of any cipher suites that did not encrypt. The first analysis of this shift would not seem to be a detriment to the utility industry, but message integrity in general is more important than privacy. Additionally, NERC requires that packets be inspected at the Electronic Security Perimeter (ESP) boundary on packet ingress. Using TLS encryption between two nodes does not allow for this inspection to occur unless the TLS connection is terminated at the ESP boundary. This is not an advisable solution for the utility industry. How this issue will be resolved is an ongoing discussion and a new draft RFC has been submitted to the IETF that resolves this issue.

IEC 62351-4 specifies how to utilize TLS for ISO 9506, which includes IEC 61850 and the Inter-Control Center Protocol (ICCP) (IEC 60870-6 TASE.2). It specifies additional cipher suites that are to be implemented, in addition to those specified in IEC 62351-3, and specifies the symmetric key renegotiation interval. In many internet applications, it is not important to renegotiate since the application connections are transient (e.g., you connect to a website, do your business, and close the connection/browser). However, within the utility environment, most connections are long-standing and would hopefully be permanent. Grid reliability depends on being able to communicate as needed, therefore terminating and reestablishing transport level connections is not a desirable attribute. The renegotiation interval is specified to be no more that 12 hours, which is half of the normal Certificate Revocation List (CRL) update interval.

Since 2007, IEC 62351-4 also included connection establishment authentication that can be utilized to establish the identity/role for RBAC. However, the use of this authentication does not provide authentication of every IEC 61850 packet. This authentication is provided by the MAC of TLS whose use is required by IEC 62351-3. Therefore, the connection-only authentication mechanism should only be used in conjunction with the usage of TLS.

There were two events that have brought about end-to-end (E2E) security at the application layer. The first event was the choice of XMPP for IEC 61850-8-2. XMPP is a hub and spoke technology where the hub represents a security concern that needed to be addressed within the application layer. The second event, like the first, was that there were privacy concerns when TLS tunnels are terminated and reestablished. Therefore, IEC 62351-4 now specifies an E2E security protocol like TLS but enhanced to address some of the potential TLS issues. The E2E security authenticates every packet and is introduced in the 2018 version of IEC 62351-4.

Figure 12.4 shows that the E2E security can be securely utilized with or without TLS whereas the 2007 version of authentication requires the use of TLS. Security for IEC 61850-8-1 allows both forms of security to be used. In large part this is due to the desire to provide backward interoperability with previously deployed implementations. The version of security can be negotiated using the ISO presentation protocol. This allows a client to declare what it supports, and the server responds with what it has selected to utilize for the current connection. If both the 2007 authentication and E2E are supported by both the client and server, the server should specify the utilization of E2E. The E2E security encapsulates every MMS PDU and requires that the signature for each packet be provided. The utilization of encryption is negotiated during the connection establishment.

Regardless of the use of TLS or E2E, a symmetric key is negotiated in a secure key exchange channel secured through PKI and public/private key encryption. Once the symmetric key is established the secure association (SA) makes use of the symmetric key ,which can be renegotiated on a periodic basis that must be no more than 12 hours (e.g., based on half the normal 24-hour period of certificate revocation).

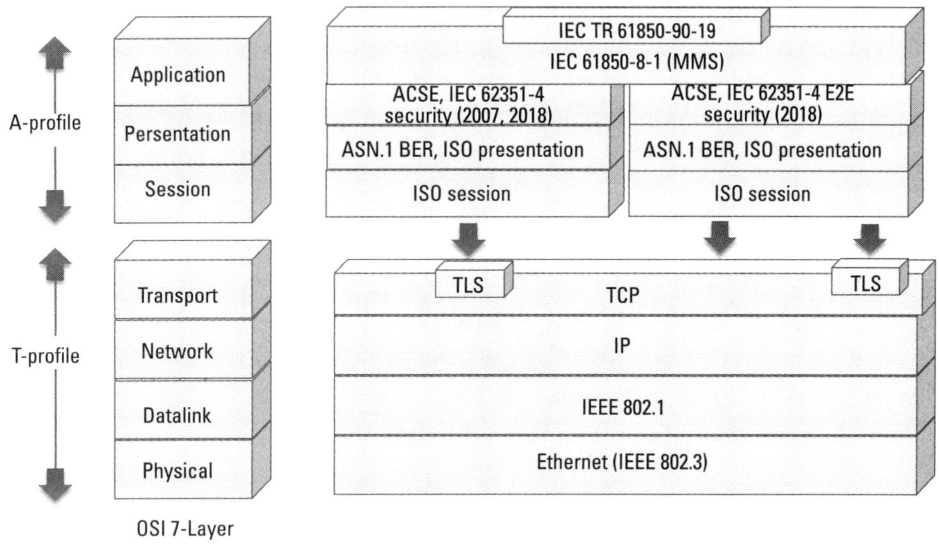

Figure 12.4 Secure communication profiles for IEC 61850-8-1 client/server.

12.3.2 GOOSE and Sampled Values

Prior to CIP Version 5, many utilities were reluctant to deploy GOOSE and SV outside of the control house due to CIP 4's requirement of a Physical Security Perimeter (PSP) being six walls. CIP Version 5 removes this restriction so that a substation fence can be considered a PSP. However, CIP-006-5 implies that control access to devices outside of the control house is needed. It is therefore suggested that the keys required to unlock the cabinets outside the control house are stored in the control house.

The 2007 version of IEC 62351-6 specified the use of public and private keys for signatures and encryption. It was found that the computation of signatures based on public/private keys was not fast enough to support high rates of sampled values. The thinking of how to fix this issue was strongly influenced by the security implementation strategy for Routable GOOSE (R-GOOSE) and Routable Sampled Values (R-SV).

The design of R-GOOSE and R-SV started with the decision to use symmetric keys that were distributed via a key distribution center (KDC) based on Group Domain of Interpretation (GDOI) (RFC 6407). However, there were several restrictions within that RFC that needed to be relaxed. RFC 8052 provides the extensions to fulfill the needs of IEC 61850 in the areas of the ID payload, the ability to use object identifiers, and extensions to the SA-TEK payload based on OID. These extensions were then used by IEC 62351-9 to provide a public/private key protected exchange that was required to exchange the shared symmetric keys based on the rights to specific streams of information. The symmetric keys are used to protect information in a secure association (SA) methodology although multicast messages are being used to deliver the information.

The current key and next key, including expiration date and time, are provided to authenticated entities as well as security policy elements. Security policy elements include specification to use encryption, cipher suite specification for the mandatory signature, encryption algorithm, and initialization vector (IV) if required. Once the symmetric keys and security policy are received, subscribers and the publisher of the stream form a group that uses the same keys and policies.

The concept of data stream access is defined as the combination of the destination address of the multicast, service (e.g., GOOSE or SV), and the data set (e.g., the content of the publication). This allows GOOSE messages published with different data sets to have different keys and security policies.

The impact of the information exchange with the KDC can be seen in the session protocol (e.g., the encapsulation layer for GOOSE and SV shown in Figure 12.5) that is used for R-GOOSE and R-SV.

The exposed security parameters in the encapsulation are when the current key became active (TimeofCurrentKey), a countdown timer that provides notification as to when the key rotation will occur (TimeofNextKey), the initialization vector

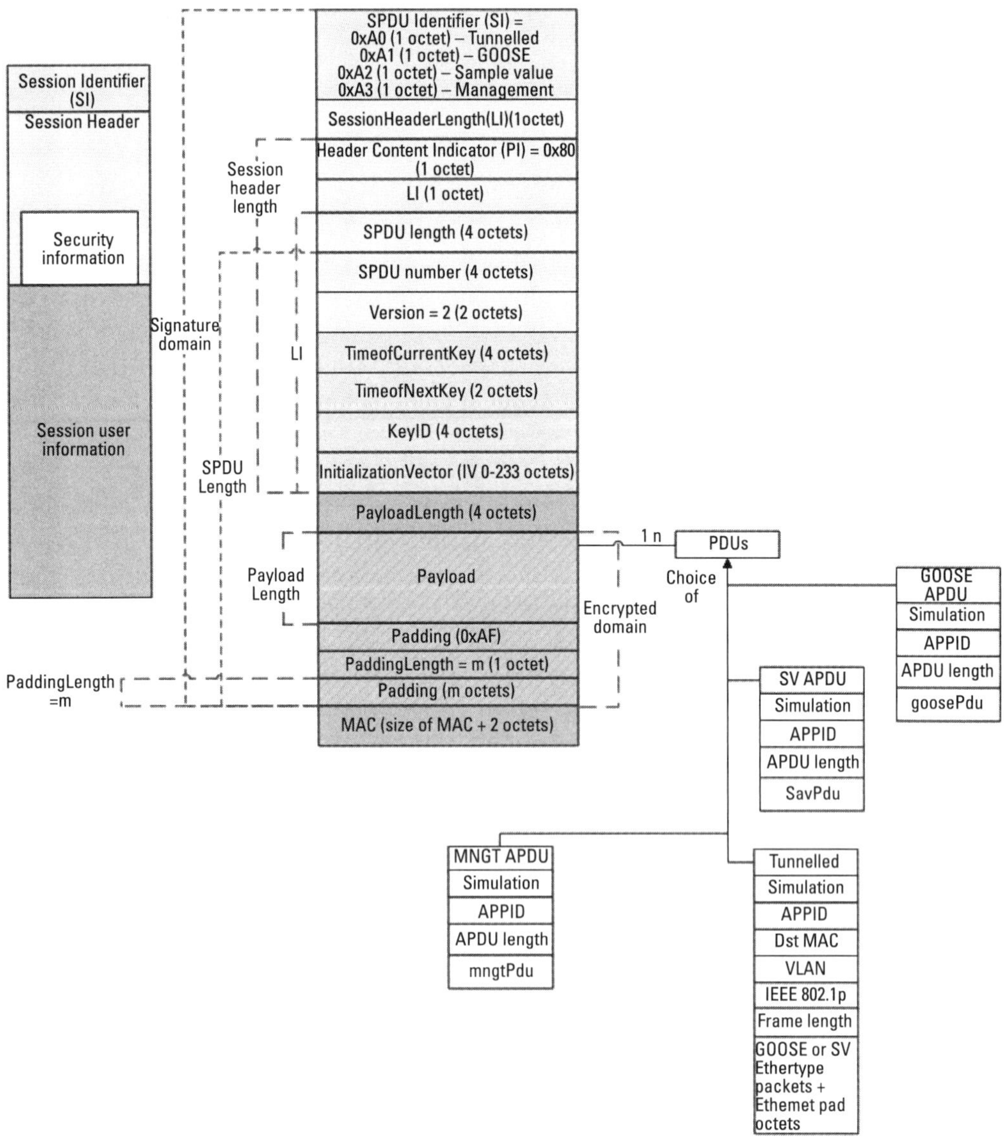

Figure 12.5 Security encapsulation and session protocol for R-GOOSE and R-SV.

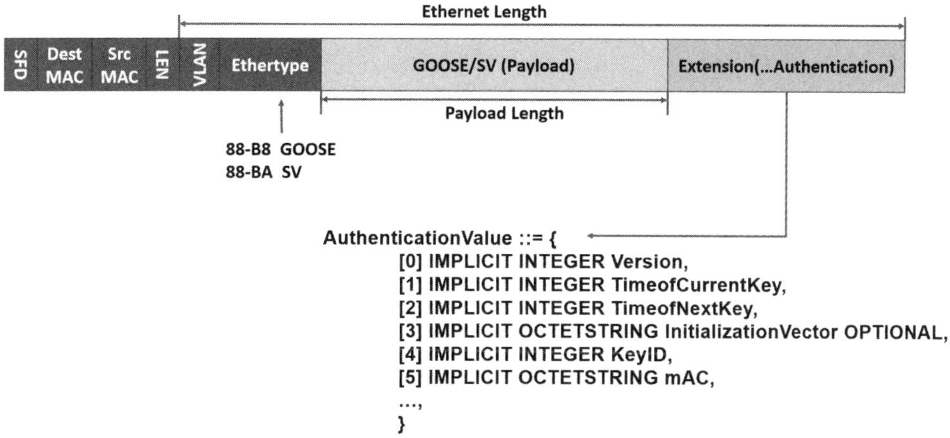

Figure 12.6 Layer 2 GOOSE and SV security frame.

(IV), and an alias for the symmetric key that is currently being used (KeyID). The KeyID is provided as part of the KDC providing the symmetric keys.

The 2018 version of IEC 62351-6 utilizes the same concepts as R-GOOSE and R-SV except for Layer 2 Ethernet multicast messages. The security parameters are appended within the Ethernet frame (e.g., after the GOOSE payload) so that non-secure subscribers see the extension as Ethernet padding and can ignore the security extension. A high-level view of the Ethernet frame is shown in Figure 12.6.

Within the GOOSE payload is a bit, that if true, indicates that security is in use.

12.4 Monitoring

To provide a utility with the capability to detect incidents for CIP-008-5, IEC 61850 management objects for SNMP have been defined in IEC 62351-7. The high-level perspective of the information defined provides information related, but not limited to, receive errors, security issues, traffic pattern analysis, and the number of communication associations. This information is provided for client/server, GOOSE, SV, and aggregated. Work is ongoing to standardize how to use Syslog as well as other IT monitoring technologies.

Appendix A
Protection Function Cheat Sheet

It is a truism that many utility engineers are more comfortable with the ANSI/IEEE Standard Device Numbers instead of the IEC 61850 protection Logical Nodes. The following table provides a cross-reference between the IEEE numbers and IEC 61850 logical nodes. The bolded definitions show the most commonly used IEEE numbers.

IEEE Standard Numbers to IEC 61850 Logical Node Cheat Sheet

Definition	*IEEE Number*	*IEC 61850 Logical Node*	*Definition*	*IEEE Number*	*IEC 61850 Logical Node*
Master Element	1	PSCH	Field Excitation	53	
Time Delay	2	*Note 1*	Power Factor	55	POPF or PUPF
Interlocking	3	CILO	Field Application	56	
Master Contactor	4		Overvoltage	59	PTOV
Stopping Device	5		Balance	60	
Starting Circuit Breaker	6		Time-Delay Stopping	62	
Rate of Change	7	PFRC	Pressure Switch	63	
Disconnecting	8		Ground Detector	64	PHIZ
Reversing	9		Governor	65	
Unit Sequence	10		Notching	66	
Multifunction	11	*Note 2*	AC Directional Overcurrent	67	PTOC+RDIR
Overspeed	12	PRTR	Out-of-Step	68	PPAM
Synchronous-speed	13		Permissive	69	
Underspeed	14	PZSU	Alarm	74	CALH
Frequency Matching	15	CSYN	Position Changing	75	
Data Communication	16		DC Overcurrent	76	PIOC
Operated Valve	20	KVLV	Phase Angle Measuring	78	MMXU
Distance	21	PDIS	AC Reclosing	79	RREC
Temperature	23	FTMP	Frequency	81	PTUF or PTOF
Volts/Hertz	24	PVPH	Transfer	83	SOPM
Synchronizing	25	RSYN	Operating	84	
Thermal	26	PTTR	Pilot Communications	85	LCCH
UnderVoltage	27	PTUV	Lockout	86	
Annunciator	30		Differential	87	PDIF
Directional	32	PDOP or PDUP	Line Switch	89	XSWI

IEEE Standard Numbers to IEC 61850 Logical Node Cheat Sheet

Definition	*IEEE Number*	*IEC 61850 Logical Node*
Polarity	36	
Undercurrent	37	PTUC
Bearing	38	HBRG
Mechanical Conduction	39	
Over/Under Excitation	40	PTHF or PRTR
Field Circuit Breaker	41	XCBR
Running Circuit Breaker	42	SSWI
Manual Transfer	43	
Reverse Phase	46	PPAM
Phase-Sequence	47	MSQI
Incomplete-Sequence	48	MSQI
Transformer Thermal	49	PTTR
Instantaneous overcurrent	50	PIOC
AC Time Overcurrent	51	PTOC
AC Circuit Breaker	52	XCBR
Regulating	90	
Voltage Directional	91	PDIR
Voltage and Power Directional	92	PDIR
Tripping	94	PTRC

Note 1: Many of the IEC 61850 protection logical nodes are available as time-based. Look for PTxx for those variants.

Note 2: IEC 61850 IEDs are multifunction.

Appendix B
CDC Cheat Sheet and Definitions

Although many users of IEC 61850 do not actually need to know the definitions of the common data classes, those with inquiring minds will eventually need to understand the use of various CDCs. CDCs provide the ability to reuse status, control, setting, and description definitions. These definitions are organized based upon the data type. The cheat sheet available in Table B.1 provides a quick lookup mechanism for those inquiring minds.

Descriptions of the various CDCs are found in Table B.2.

Table B.1 CDC Definition Cheat Sheet

Data Type	*Status*	*Control*	*Setting*	*Description*	*Other*
Analog	ACD, ACT, CMB, DEL, HDEL, HMV, HST, HWYE, MV, SAV, SEQ, WYE	APC	ASG		
Single Point	SPS	SPC	SPG		
Currency			CUG		
Curve			CSG	CSD	
Double Point	DPS	DPC			
Enumeration	ENS	ENC	ENG		
Integer	BCR, INS, SEC	BAC, BSC, INC, ISC	ISG		
Nameplate				DPL, LPL	
Object Reference	ORS		ORG		
Service Tracking					BTS, CST, GTS, LTS, MTS, NTS, STS, UTS
Time			TSG		
Visible String (V)	VSS				VSD

Table B.2 Descriptions of CDCs

Abbreviation	*Name*	*Type*	*Definition*
ACD	Directional Protection	Status	Provides status and configuration information regarding the tripping direction information.
ACT	Activation	Status	Provides status and configuration information regarding the tripping (e.g. open/closing) of various electrical phases.
APC	Analog Process	Control	Provides the ability to control an analog value. It also provides feedback regarding the status of the controlled item through a status attribute included within the CDC.
ASG	Analog	Setting	Allows the setting of an analog value that may be an individual setting, member of a setting group, or an editable setting.
BAC	Binary Analog	Control	Provides the ability to control an analog value. It also provides feedback regarding the status of the controlled item through a status attribute included within the CDC.
BCR	Binary Counter Reading	Status	Provides the ability to monitor large reading typically utilized for metering.
BSC	Binary Controlled Step Position	Control	Provides the capability to control the step position of equipment (e.g. tap changer). Typically, this would be used to increase or decrease the step position. It also provides feedback regarding the status of the controlled item through a status attribute included within the CDC.
BTS	Buffered Report	Service Tracking	Allows the tracking of buffered report control blocks.
CMV	Complex Measured Value	Analog	Provides the capability to monitor and configure analog values that contain a magnitude and angular component.
CSD	Curve Shape	Description	Provides descriptive information about the curve shape.
CSG	Curve Shape	Setting	Allows the setting the shape of a particular curve that may be an individual setting, member of a setting group, or an editable setting.
CST	Common	Service Tracking	Allows the tracking of control and other services.
CUG	Currency	Setting	Allows the setting of a currency value that may be an individual setting, member of a setting group, or an editable setting.
CURVE	Curve	Setting	Allows the setting of a CURVE value that may be an individual setting, member of a setting group, or an editable setting.
DEL	Phase to Phase values	Analog	Provides the capability to monitor and configure analog values using electrical phases based upon wiring as a DELTA connection type. Information regarding DELTA connections can be found in Section B.2
DPC	Double Point	Control	Provides the capability to control a four value object. It also provides feedback regarding the status of the controlled item through a status attribute included within the CDC.
DPL	Device Nameplate	Description	Provides descriptive information about the IED.
DPS	Double Point	Status	Allows the monitoring and configuration of a status that has four possible values. These values typically have represented the status information regarding breakers or switches (e.g. open, closed, moving, invalid).

Table B.2 (continued)

Abbreviation	*Name*	*Type*	*Definition*
ENC	Enumerated	Control	Provides the capability to control an enumerated value object. It also provides feedback regarding the status of the controlled item through a status attribute included within the CDC.
ENG	Enumerated	Setting	Allows the setting of an enumerated value that may be an individual setting, member of a setting group, or an editable setting.
ENS	Enumerated	Status	Allows monitoring and configuration of a status value whose values are defined via an enumeration.
GTS	GOOSE	Service Tracking	Allows the tracking of GOOSE control blocks.
HDEL	Harmonic Phase to Phase values	Harmonics	Provides the capability to monitor and configure harmonic values using electrical phases based upon wiring as a DELTA connection type. Information regarding harmonics can be found in Section B4.
HMV	Harmonic Measured Value	Harmonics	Provides the capability to monitor and configure harmonic information that has ONLY a magnitude component and DOES NOT contain an angular component. Information regarding harmonics can be found in Section B4.
HST	Histogram	Status	Provides statistical information regarding the distribution of values.
HWYE	Harmonic Phase to ground/neutral values	Harmonics	Provides the capability to monitor and configure harmonic values using electrical phases based upon wiring as a WYE connection type. Information regarding harmonics can be found in Section B4.
INC	Integer	Control	Provides the capability to control an integer value object. It also provides feedback regarding the status of the controlled item through a status attribute included within the CDC.
INS	Integer	Status	Allows the monitoring and configuration of a status value that is represented by an integer value.[1]
ISC	Integer Controlled Step Position	Control	Provides the capability to control the step position of equipment (e.g. tap changer). It also provides feedback regarding the status of the controlled item through a status attribute included within the CDC.
ISG	Integer	Setting	Allows the setting of an integer value that may an individual setting, member of a setting group, or an editable setting.
LPL	Logical Node Nameplate	Description	Provides descriptive information about the Logical Node.
LTS	Log	Service Tracking	Allows the tracking of logging control blocks.
MTS	Multicast Sample value	Service Tracking	Allows the tracking of Sampled Value control blocks.
MV	Measured Value	Analog	Provides the capability to monitor and configure analog values that has only a magnitude component and does not contain an angular component.
NTS	Unicast Sampled Value	Service Tracking	Allows the tracking of Unicast Sampled Value control blocks. Note that the use of unicast sampled values has been deprecated.
ORG	Object Reference	Setting	Allows the setting of an object reference value that may be an individual setting, member of a setting group, or an editable setting.

Table B.2 (continued)

Abbreviation	*Name*	*Type*	*Definition*
ORS	Object Reference	Status	Provides the ability to monitor status based upon object reference changes.
SAV	Sample Value	Analog	Provides only an instantaneous value (e.g., no dead-banding) and is utilized to convey "raw" measurements regarding CTs and PTs. It is not used for synchrophasor measurement conveyance.
SEC	Security Violation	Status	Provides an ability to track the number of security related violations.
SEQ	Electrical Sequence	Analog	Provides the capability to monitor and configure analog values based upon electrical 'sequences' such as positive, negative, sequence information. Information regarding electrical sequences can be found in Section B3.
SPC	Single Point	Control	Provides the capability to control a Boolean object. It also provides feedback regarding the status of the controlled item through a status attribute included within the CDC.
SPG	Single Point	Setting	Allows the setting of a Boolean value that may be an individual setting, member of a setting group, or an editable setting.
SPS	Single Point	Status	Allows the monitoring and configuration of a Boolean (e.g., true or false) value.
STS	Setting Group	Service Tracking	Allows the tracking of the Setting Group control block. Note that the use of unicast sampled values has been deprecated.
TSG	Time	Setting	Allows the setting of a time stamp value that may be an individual setting, member of a setting group, or an editable setting.
UTS	Unbuffered Report	Service Tracking	Allows the tracking of unbuffered report control blocks.
VSD	Visible String	Description	Provides descriptive information about the string.
VSS	Visible String	Status	Provides the ability to monitor status values that are not able to be defined as enumerations.
WYE	Phase to ground/ neutral values	Analog	Provides the capability to monitor and configure analog values using electrical phases based upon wiring as a WYE connection type. Information regarding WYE connections can be found in Section B.1.

During the transition from Edition 1 to Edition 2 many DataObjects defined as INS were re-defined as ENS in order to be more semantically clear.

B.1 WYE

A WYE electrical connection is typically established by the wiring a power transformer.

The various windings of the power transformer all have an end connected to the neutral or ground. The measurements regarding a WYE connection are named based upon the phase connection and referenced against neutral or ground. Therefore, IEC 61850 names the attributes phsA, phsB, phsC, and so forth. The phase attributes can be used to measure voltage, current, and resistance. The reference is against the neutral. Thus, positive values for current flow from the phase to neutral.

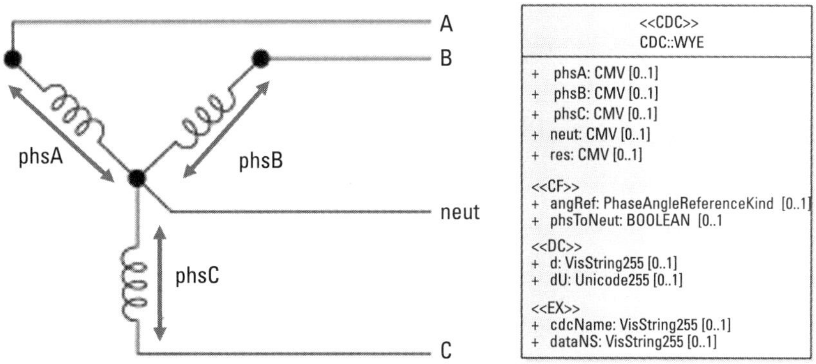

Figure B.1 WYE CDC.

B.2 Delta

A DELTA electrical connection is typically established by wiring a power transformer.

The various windings of the power transformer have connections to the other phases of the transformer. The measurements regarding a DELTA connection are named based upon the phase-to-phase connection. Therefore, IEC 61850 names the attributes phsAB, phsBC, and phsCA. The phase attributes can be used to measure voltage, current, and resistance. The reference is to the phase listed last in the attribute name. As an example, positive values for current flow from the phase A to phase B (e.g., for phsAB).

B.3 Electrical Sequences

The construct of electrical sequence is the ability to simplify electrical phasor measurements so that fault detection and protection is easier. Typically, unless quadratic calculations are utilized the attributes represent: positive-sequence (c1), negative-sequence (c2), and zero-sequence (c3). (More information regarding the calculation of the sequence components can be found at https://en.wikipedia.org/wiki/Symmetrical_components.)

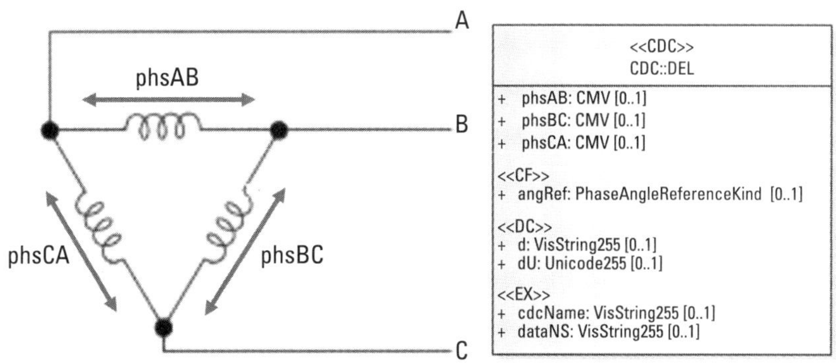

Figure B.2 DELTA CDC.

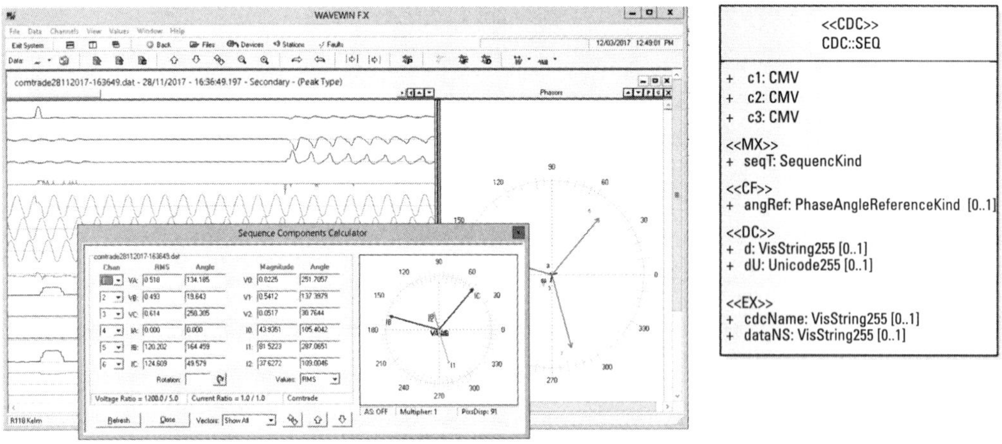

Figure B.3 Sequence (SEQ) CDC (Adapted image courtesy of PACWorld.)

B.4 Harmonics

Harmonics are calculated through applying an FFT calculation on a window of a specific value.

The FFT calculation produces 'bin' (s) of values that represent a specific frequency component range. As the magnitudes of the frequency components increase or decrease, it represents instability in the electrical system. For most AC power systems, the magnitude of the bin representing DC (0.0 frequency) is important. Abnormal values of DC may cause a power system failure or unintended protections to occur. The frequency bins are represented as an array of values within IEC 61850 ('har'). (More information regarding harmonics and their impact on the power system can be found at: https://web.ecs.baylor.edu/faculty/grady/Understanding_Power_System_Harmonics_Grady_April_2012.pdf.)

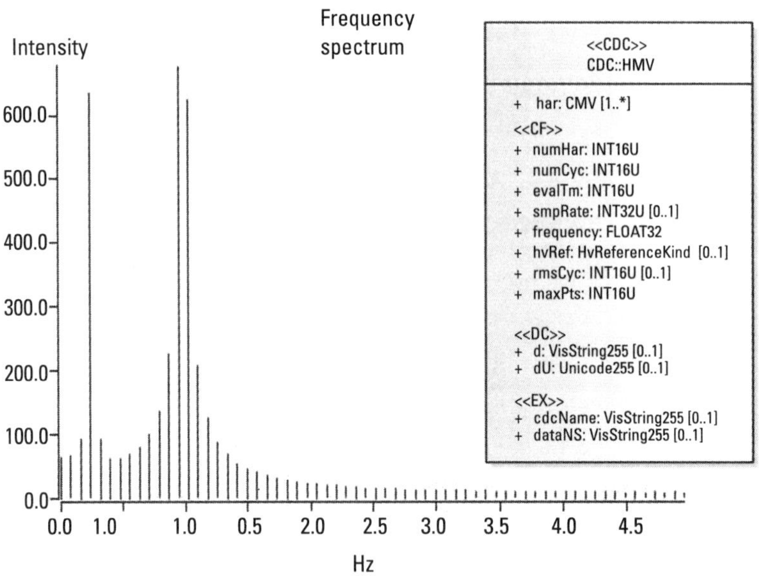

Figure B.4 Harmonic FFT CDC (HMV). (Adapted image courtesy of PACWorld.)

Acronyms and Abbreviations

A/D	Analog digital converter
ADMS	Advanced Distribution Management System
AOW	Asia-Oceania OSI Workshop
appID	Application ID
AS	Applicability Statement
ASM	Any Source Membership
ASN	Abstract Syntax Notation
BCP	Best Current Practice
BER	Basic encoding rules
BNF	Backus-Naur format
BRCB	Buffered report control block
CAD	Computer aided design
CASM	Common Application Service Models
CD	Committee Draft
CDV	Committee Draft for Vote
CIP	Critical Infrastructure Program
CLNP	Connectionless-mode Network Protocol
CNC	Computer numerical control
COS	Corporation for Open Systems
CPU	Central processing unit or computer processing unit
CRL	Certificate Revocation List
CSMA/CD	Carrier sense multiple access/collision detection
CT	Current transformer
DA	Data attribute
DARPA	Defense Advanced Research Projects Agency
DCE	Data communication equipment
dchg	Data change
DER	Distinguished Encoding Rules
DIN	Deutsches Institut für Normung also known as German Institute of Standardization
DIS	Draft International Standard
DMS	Distribution Management System
DNP	Distributed Network Protocol
DO	Data object
DOS	Denial of service and Microsoft operating system

DP	Distributed Peripheral
DSCP	Differentiated services code point
DSP	Digital signal processing
DTE	Data terminal equipment
dupd	Data update
EAI	Electronic Industries Association
ECN	Explicit Congestion Notification
EIPP	Eastern Interconnect Phasor Project
EMS	Energy management system
ENE	Enterprise Networking Event
EPRI	Electric Power Research Institute
EWOS	European Workshop for Open Systems
FC	Functional constraint
FCD	Functionally constrained data
FCDA	Functionally constrained data attribute
FDIS	Functional Draft International Standard
FDL	Field Bus Data Link
FIFO	First-in first-out
FMS	Field Bus Message Specification
FTAM	File Transfer Access and Management
FTP	File Transfer Protocol
GDOI	Group Domain of Interpretation
GM	General Motors
GMMFS	General Motors Message Format Specification
GOMSFE	Generic Object Models for Substation and Feeder Equipment
GOOSE	Generic Object Oriented Substation (or System) Event
GOSIP	Government OSI Profile
GPS	Global Positioning System
GSSE	Generic Substation Status Event
HLD	High level design
HMI	Human machine interface
HTTP	HyperText Transfer Protocol
HV	High voltage
IAB	Internet Architecture Board
IANA	Internet Assigned Number Authority
ICCP	Inter-Control Center Protocol
ICT	IED Configuration Tool
IEC	International Electrotechnical Commission
IED	Intelligent electronic device
IEEE	Institute of Electrical and Electronic Engineers
IESG	Internet Engineering Steering Group
IETF	Internet Engineering Task Force
IGMP	Internet Group Management Protocol
IP	Internet Protocol
IRIG-B	Inter-range Instrumentation Group time codes B
IS	International Standard
ISO	International Organization for Standardization
ISP	International Standardized Profile

ITU	International Telecommunication Union
IV	Initialization Vector
JID	Jabber Identification
JPL	Jet Propulsion Laboratory
KDC	Key Delivery Center
L2	Layer 2
LAN	Local area network
LDAP	Lightweight Directory Access Protocol
LLC	Logical Link Control
LNClass	Logical Node Class
MAC	Message Authentication Code
MAaC	Media Access Control
MAP	Manufacturing Automation Protocol
MHS	Message Handling System
MMFS	Manufacturing Message Format Specification
MMS	Manufacturing Message Specification
MMSF	Manufacturing Message Format Specification
MQP	Message Queue Persistence
MsgP	Message Persistence
NASPI	North American Synchrophasor Project Initiative
NC	National Committee
NCC	National Computer Conference
NERC	National Electric Reliability Corporation
NIST	National Institute of Standards and Technology
NOAA	National Oceanographic and Atmospheric
NTP	Network Time Protocol
NWIP	New Work Item Proposal
OIW	OSI/OSE Implementers' Workshop
OSE	Open Systems Environment
OSI	Open System Interconnection
PAP	Priority Action Plane
PAR	Project Authorization Request
PAS	Publicly Available Standards
PC	Personal computer
PCI	Personal computer interface
PCMCIA	Personal Computer Memory Card International Association Interface
PDU	Protocol Data Unit
PDV	Packet Delay Variation
PICS	Protocol Implementation Conformance Statements
PIXIT	Protocol Implementation Conformance Extra Information for Testing
PKI	Public Key Infrastructure
PLC	Programmable logic controller
PPS	Pulse per second
PSRC	Power System Relaying Committee
PT	Potential transformer (also known as a voltage transformer)
PTP	Precision Time Protocol

qchg	Quality change
QOS	Quality of service
RFC	Request for Comment
R-GOOSE	Routable GOOSE
RS	Recommended Standard
R-SV	Routable Sampled Value
RTU	Remote terminal unit
SA	Secure association or security association
SASB	Standards Association Standards Board
SBO	Select-before operate
SCADA	Supervisory Control and Data Acquisition
SCL	System Configuration Language
SCSM	Specific Communication Service Mapping
SCT	System configuration tool
SDO	Standards Development Organization
SEP	Stable Election Protocol
SGIP	Smart Grid Interoperability Panel
SLD	Single line diagram
SMTP	Simple Mail Transfer Protocol
SNMP	Simple Network Management Protocol
SNTP	Simple Network Time Protocol
SSM	Source Specific Membership
SUT	System under test
SV	Sampled Values
TAL	Time allowed to live
TC	Technical Committee
TCP	Transmission Control Protocol
TISSUE	Technical issue
TLS	Transport Layer Security
TOP	Technical Office Protocol
TPAA	Two-Party Application Association
TR	Technical Report
TS	Technical Specification
UART	Universal Asynchronous Receiver/Transmitter
UCA	Utility Communication Architecture
UCAIug	Utility Communication Architecture International Users Group
UML	Unified Modeling Language
URCB	Unbuffered Report Control Block
URI	Universal Resource Identifier
URL	Uniform Resource Locator
US	United States
USB	Universal Serial Bus
VLAN	Virtual local area network
VOIP	Voice over IP
VT	Voltage transformer (also known as potential transformer)
WAMPAC	Wide area monitoring protection and control
WAMS	Wide area monitoring systems
WAN	Wide area network

WD	Working Draft
WG	Working Group
XER	XML Encoding Rules
XML	eXtensible Markup Language
XMPP	eXtensible Message and Presence Protocol
XSD	XML Schema Definition

Glossary

The following are some of the definitions utilized in this book.

Electric Power Research Institute From: www.epri.com: "The Electric Power Research Institute, Inc. conducts research and development relating to the generation, delivery and use of electricity for the benefit of the public. An independent, nonprofit organization, we bring together scientists. and engineers as well as experts from academia and the industry to help address challenges in electricity."

Institute of Electrical and Electronic Engineers IEEE is an organization that produces standards. From www.ieee.org: "IEEE's core purpose is to foster technological innovation and excellence for the benefit of humanity."

International Electrotechnical Commission From www.iec.ch: "The International Electrotechnical Commission (IEC) is the world's leading organization that prepares and publishes International Standards for all electrical, electronic and related technologies.
Close to 20,000 experts from industry, commerce, government, test and research labs, academia and consumer groups participate in IEC Standardization work.
The IEC is one of three global sister organizations (IEC, ISO, ITU) that develop International Standards for the world."

Telecontrol From the IEC Online dictionary (http://www.electropedia.org): "the control of operational equipment at a distance using the transmission of information by telecommunication techniques Note: Telecontrol may comprise any combination of command, alarm, indication, metering, protection and tripping facilities, without any use of speech messages."

About the Author

Herbert Falk graduated from Northwestern University with a B.S. in electrical engineering and an M.S. in electrical engineering. He then developed manufacturing automation products for Westinghouse Numa-Logic. In 1982, while at Numa-Logic, Mr. Falk was assigned to the General Motor's initiative that eventually became the Manufacturing Automation Protocol and the Manufacturing Message Specification (ISO 9506), which is a foundational technology of IEC 61850 today. After Numa-Logic, Mr. Falk became a partner in Systems Integration Specialists Company (SISCO) which specialized in applying standard-based technology to manufacturing automation. After being involved with several manufacturing automation projects, Mr. Falk assisted in the creation of the Electric Power Research Institute's (EPRI) Utility Communication Architecture (UCA) for both versions of the architecture: UCA 1.0 and UCA 2.0. Mr. Falk became the senior solutions architect for utility solutions at SISCO and was involved with protection and automation projects globally. He has been involved in numerous projects involving the application of information systems technology and real-time communications technology to automated manufacturing, electrical distribution and automation, and power quality monitoring. Mr. Falk is a recognized expert on information integration technology, distributed object technology, and communication having served on and chaired numerous industry technical committees.

Mr. Falk was involved in the creation of IEC 61850 and is one of its editors. Additionally, he is the United States lead delegate for cyber security to the International Electrotechnical Commission (IEC) Technical Committee (57) Working Group (WG)15.

Today, Mr. Falk is an independent consultant in the utility industry and continues his involvement with related standards.

Index

C

T

Recent Artech House Titles in Power Engineering

Andres Carvallo, Series Editor

Advanced Technology for Smart Buildings, James Sinopoli

The Advanced Smart Grid: Edge Power Driving Sustainability, Second Edition, Andres Carvallo and John Cooper

Battery Management Systems, Volume I: Battery Modelings, Gregory L. Plett

Battery Management Systems for Large Lithium Ion Battery Packs, Davide Andrea

Battery Power Management for Portable Devices, Yevgen Barsukov and Jinrong Qian

Big Data Analytics for Connected Vehicles and Smart Cities, Bob McQueen

IEC 61850 Demystified, Herbert Falk

Design and Analysis of Large Lithium-Ion Battery Systems, Shriram Santhanagopalan, Kandler Smith, Jeremy Neubauer, Gi-Heon Kim, Matthew Keyser, and Ahmad Pesaran

Designing Control Loops for Linear and Switching Power Supplies: A Tutorial Guide, Christophe Basso

Electric Power System Fundamentals, Salvador Acha Daza

Electric Systems Operations: Evolving to the Modern Grid, Mani Vadari

Energy Harvesting for Autonomous Systems, Stephen Beeby and Neil White

GIS for Enhanced Electric Utility Performance, Bill Meehan

Introduction to Power Electronics, Paul H. Chappell

Introduction to Power Utility Communications, Dr. Harvey Lehpamer

IoT Technical Challenges and Solutions, Arpan Pal and Balamuralidhar Purushothaman

Microgrid Design and Operation: Toward Smart Energy in Cities, Federico Delfino, Renato Procopio, Mansueto Rossi, Stefano Bracco, Massimo Brignone, and Michela Robba

Plug-in Electric Vehicle Grid Integration, Islam Safak Bayram and Ali Tajer

Power Grid Resiliency for Adverse Conditions, Nicholas Abi-Samra

Power Line Communications in Practice, Xavier Carcelle

Power System State Estimation, Mukhtar Ahmad

A Systems Approach to Lithium-Ion Battery Management, Phil Weicker

Signal Processing for RF Circuit Impairment Mitigation in Wireless Communications, Xinping Huang, Zhiwen Zhu, and Henry Leung

The Smart Grid as An Application Development Platform, George Koutitas and Stan McClellan

Smart Grid Redefined: Transformation of the Electric Utility, Mani Vadari

Synergies for Sustainable Energy, Elvin Yüzügüllü

Telecommunication Networks for the Smart Grid, Alberto Sendin, Miguel A. Sanchez-Fornie, Iñigo Berganza, Javier Simon, and Iker Urrutia